T0291564

Smart Sensors Networks

Communication Technologies and Intelligent Applications

Smart Sensors Networks
Communication Technologies and Intelligent Applications

Edited by

Fatos Xhafa
Universitat Politècnica de Catalunya, Spain

Fang-Yie Leu
Tunghai University, Taiwan

Li-Ling Hung
Aletheia University, Taiwan

Series Editor Fatos Xhafa
Universitat Politècnica de Catalunya, Spain

ACADEMIC PRESS
An imprint of Elsevier

Academic Press is an imprint of Elsevier
125 London Wall, London EC2Y 5AS, United Kingdom
525 B Street, Suite 1800, San Diego, CA 92101-4495, United States
50 Hampshire Street, 5th Floor, Cambridge, MA 02139, United States
The Boulevard, Langford Lane, Kidlington, Oxford OX5 1GB, United Kingdom

Notices

Knowledge and best practice in this field are constantly changing. As new research and experience broaden our
understanding, changes in research methods, professional practices, or medical treatment may become necessary.

Practitioners and researchers must always rely on their own experience and knowledge in evaluating and using
any information, methods, compounds, or experiments described herein. In using such information or methods
they should be mindful of their own safety and the safety of others, including parties for whom they have a
professional responsibility.

To the fullest extent of the law, neither the Publisher nor the authors, contributors, or editors, assume any liability
for any injury and/or damage to persons or property as a matter of products liability, negligence or otherwise, or
from any use or operation of any methods, products, instructions, or ideas contained in the material herein.

Library of Congress Cataloging-in-Publication Data
A catalog record for this book is available from the Library of Congress

British Library Cataloguing-in-Publication Data
A catalogue record for this book is available from the British Library

ISBN: 978-0-12-809859-2

For information on all Academic Press publications
visit our website at https://www.elsevier.com/books-and-journals

Working together
to grow libraries in
developing countries

www.elsevier.com • www.bookaid.org

Publisher: Mara Conner
Acquisition Editor: Sonnini Yura
Editorial Project Manager: Ana Claudia Garcia
Production Project Manager: Anusha Sambamoorthy
Designer: Victoria Pearson

Typeset by VTeX

CONTENTS

PART 2   DATA STREAMING, PROCESSING, AND ANALYSIS

PART 3   HEALTHCARE APPLICATIONS

PART 4   LIVING LAB – EVERYDAY ACTIVITIES

CONTRIBUTORS

Sayaka Akiyama
Tokyo Woman's Christian University, Japan;
Toshiba Corporation, Japan

Mitra Baratchi
University of Twente, Netherlands

Mario Barbareschi
DIETI – Department of Electrical Engineering and Information Technologies,
University of Naples Federico II, Italy;
CeRICT scrl – Centro Regionale Information Communication Technology, Italy

Tomáš Bartoň
Department of Electrical and Computer Engineering, University of Alberta,
Edmonton, AB, Canada

Ray-I Chang
National Taiwan University, Taiwan

Wei-ting Chen
Aletheia University, Taiwan

Cristian Chilipirea
University Politehnica of Bucharest, Romania

Polon Chuang
Institute for Information Industry, Taiwan

Sergio Di Martino
University of Naples "Federico II", Italy

Ciprian Dobre
University Politehnica of Bucharest, Romania

Akihiro Fujihara
Department of Management Information Science, Fukui University of
Technology, Fukui, Japan

Li-Ling Hung
Aletheia University, Taiwan

Takahiro Inui
Seikei University, Japan

Yoshimasa Ishi
Cybermedia Center, Osaka University, Japan

Masayoshi Kai
NEC Solution Innovators, Ltd., Japan

Masaru Kamada
Ibaraki University, Japan

Yuka Kato
Tokyo Woman's Christian University, Japan

Tomoya Kawakami
Graduate School of Information Science, Nara Institute of Science and Technology, Japan

Shinya Kitada
Iwate Prefectural University, Japan

Chia-Yin Ko
Tunghai University, Taiwan

Masaki Kohana
Seikei University, Japan

Pavel Krömer
Faculty of Computer Science, VŠB Technical University of Ostrava, Ostrava, Czech Republic

Simon Kwoczek
Group Research, Volkswagen AG, Germany

Fang-Yie Leu
Tunghai University, Taiwan

Meng-Han Li
National Taiwan University, Taiwan

Jeng-Wei Lin
Tunghai University, Taiwan

Yi-Chen Lin
Tunghai University, Taiwan

Antonino Mazzeo

DIETI – Department of Electrical Engineering and Information Technologies, University of Naples Federico II, Italy;
CeRICT scrl – Centro Regionale Information Communication Technology, Italy

Tohru Miyazaki

NEC Solution Innovators, Ltd., Japan

Petr Musilek

Department of Electrical and Computer Engineering, University of Alberta, Edmonton, AB, Canada;
Faculty of Computer Science, VŠB Technical University of Ostrava, Ostrava, Czech Republic

Wolfgang Nejdl

L3S Research Center, University of Hanover, Germany

Shusuke Okamoto

Seikei University, Japan

Andreea-Cristina Petre

University Politehnica of Bucharest, Romania

Michal Prauzek

Faculty of Computer Science, VŠB Technical University of Ostrava, Ostrava, Czech Republic

James Rodway

Department of Electrical and Computer Engineering, University of Alberta, Edmonton, AB, Canada

Sara Romano

DIETI – Department of Electrical Engineering and Information Technologies, University of Naples Federico II, Italy;
CeRICT scrl – Centro Regionale Information Communication Technology, Italy

Hisashi Sakamoto

NEC Solution Innovators, Ltd., Japan

Goshi Sato

Iwate Prefectural University, Japan

Yoshitaka Shibata

Iwate Prefectural University, Japan

Hideyuki Shimizu

NEC Solution Innovators, Ltd., Japan

Heru Susanto
The Indonesian Institute of Sciences, Indonesia;
Department of Information Management, Tunghai University, Taiwan

Yuuichi Teranishi
National Institute of Information and Communications Technology, Japan;
Cybermedia Center, Osaka University, Japan

Maarten van Steen
University of Twente, Netherlands

Sheng-Han Wu
Aletheia University, Taiwan

Jian-hua Yeh
Aletheia University, Taiwan

Tomoki Yoshihisa
Cybermedia Center, Osaka University, Japan

Hsin-Chun Yu
Tunghai University, Taiwan

ABOUT THE EDITORS

Fatos Xhafa received his Ph.D. in Computer Science in 1998 from the Department of Computer Science of the Technical University of Catalonia (UPC), Barcelona, Spain. Currently, he holds a permanent position of *Professor Titular* at UPC, BarcelonaTech. He was a Visiting Professor at Birkbeck College, University of London (UK) during academic year 2009–2010 and Research Associate at Drexel University, Philadelphia (USA) during academic term 2004/2005. Dr. Xhafa has widely published in peer reviewed international journals, conferences/workshops, book chapters and edited books, and proceedings in the field (http://dblp.uni-trier.de/pers/hd/x/Xhafa:Fatos). He is awarded teaching and research merits by Spanish Ministry of Science and Education. Dr. Xhafa has an extensive editorial and reviewing service. He is editor-in-Chief of International Journal of Grid and Utility Computing and International Journal of Space-based and Situated Computing from Inderscience. His research interests include parallel and distributed algorithms, optimization, networking, P2P and Cloud computing, security, and trustworthy computing, among others. He can be reached at fatos@cs.upc.edu and more information can be found at http://www.cs.upc.edu/~fatos/.

Fang-Yie Leu received his bachelor, master, and Ph.D. degrees all from National Taiwan University of Science and Technology, Taiwan, in 1983, 1986, and 1991, respectively. His research interests include wireless communication, network security, Grid applications and Security. He is currently a professor of Computer Science Department, Tunghai University, Taiwan, and one of the editorial board members of at least 7 journals. Prof. Leu organizes MC-NCS and CWECS international workshops. He is an IEEE member and serves as the TPC member of at least 10 international conferences. He was also a visiting scholar of Pittsburg University, and now serves as the chairperson of Big-data Master Program, Tunghai University, Taiwan. He can be reached at leufy@thu.edu.tw and more information can be found at http://www.cs.thu.edu.tw/web/teacher/detail.php?cid=1&id=21.

 **Li-Ling Hung** received her Ph.D. degree in computer science and information engineering from the National Taiwan University of Science and Technology, Taiwan, in 2008. Her research interests include vehicular ad hoc networks, wireless sensor networks, underwater wireless sensor networks, ad hoc wireless networks, and cyber-physical systems. She is currently an Associate Professor with the Department of Computer Science and Information Engineering, Aletheia University, Taiwan. She is also a member of IEEE Computer Society. She can be reached at llhung@mail.au.edu.tw and more information can be found at http://140.131.152.31/english.html.

FOREWORD

Emerging into the current evolution of the Internet of Things (IoT) technologies, **smart sensors networks technologies and intelligent applications** are finding their way into every aspect of our daily life. As an enabler, networked wireless sensors are used to monitor and gather intelligence from our surrounding environment, allowing creation of many services for future smart homes and smart cities.

With the increasing number of users using mobile devices, wearables, and the internet, smart sensors networks is a growing trend towards providing data collection and data analysis, it has the potential to make the experience in various smart environments more contextual, effective and more user friendly. In this context, smart sensors networks and technologies are becoming more necessary to support efficient data processing and analysis from large and various sensors networks. Furthermore, smart applications development implies to deal with many disciplines, such as wireless sensors technologies, Internet and IoT architectures, and User eXperience design.

The current book *Smart Sensors Networks: Communication Technologies and Intelligent Applications* provides a valuable window on the unique characteristics of wireless sensor networks through their usage in a broad range of areas. The volume shows an impressive combination of methodologies and technological approaches to tackle both smart sensors networks and smart applications challenges by exploring the most important aspects of modern sensors technologies – such as Delay-tolerant networking, data protection, bounded error data compression and aggregation – as well as their use for different scenarios by creating intelligent applications. The authors provide excellent guidelines to handle the many issues and challenges involved in smart sensors networks management to effectively operate them or use them in a smart way to build innovative intelligent applications. I would like to stress the high quality chapters in the book, providing with views on the solutions, challenges, and research trends around both technical and implementation issues.

This book delivers on its research objectives by investigating thoroughly several topics comprising wellness and e-health, energy management, industrial automation, smart mobility systems, smart homes, and more. The multidisciplinary nature of this book is a valuable feature as it comprises key domains in a smart cities context. A number of case studies and evaluations of real-world scenarios are provided throughout the volume, which present an excellent demonstration of the successful use of the multi-disciplinary and complex approaches presented in an easy and very practical way.

I believe that the list of topics explored here, the findings, lessons learned, and challenges identified will make the readers think of the implications of such new ideas on innovative developments for user oriented smart cities applications.

I would like to commend the editors on such an achievement and wish the readers enjoy the book and will find it useful for their professional activity.

Prof. Yann Bocchi
Head of the Connected Experience Lab
University of Applied Sciences Western Switzerland, HES-SO Valais-Wallis
Sierre, Switzerland

PREFACE

In recent years, sensors and sensor networks have been increasingly and extensively employed to sense our surrounding environment and help to promote our everyday lives, without showing us their existence. Sensors and sensors networks actually exist and continuously assist us in different areas, e.g., for elderly healthcare, monitoring of environmental changes, as a tool for recognizing legal users, etc. The continuous increase of sensor networking technologies however requires the use of advanced techniques and algorithms to transform sensor collected information into knowledge and intelligence to support human activities, decision-making, businesses, sport activities, etc. Indeed, sensors do not have intelligence capabilities *per se;* sensors alone do not provide intelligence, while by integrating and analyzing data from different sensors and sensor networks platforms, it is possible to build intelligence and knowledge for various applications (monitoring systems, e-Health systems, tracking systems, surveillance systems, vehicular systems, manufacturing systems, disaster management systems, etc.). This capability of building intelligence from sensors and sensor networks is increasing due to the fact that sensors and sensor networks are nowadays the basis for the Internet of Things (IoT). With the new Internet protocol (IPv6) together with Cloud computing, there is expected to have hundreds of millions of smart devices connected to Internet.

The purpose of this book is to explore the related techniques and applications of various sensors, sensor networks, and their applications to bring to readers state of the art techniques, algorithms, and technologies that support developing sensor systems to continue improving our living convenience and quality of our environment.

The book covers various research topics in the field such as IoT, near field communication, RFID, sensor data compression, security and privacy, smart antenna, e-Health, smart home, industrial automation revolutionizing businesses, and energy consumption, among others. Chapters in the book bring recent research findings on sensor and sensor networks usage in different areas of our everyday lives. Besides being a timely publication in the field, the book aims to critically analyze the challenges in the area, thus, contributing to envisage the road ahead in the smart sensors and data-collected intelligence.

ORGANIZATION OF THE BOOK

The book is organized into four parts, comprising fifteen chapters, arranged as follows:

In the **first part "IoT and Network Communication Systems,"** chapters cover topics of IoT and Network Communication Systems, state of the art from techniques, algorithms, and software development perspectives.

Chapter 1 by Inui et al. *"IoT Technologies: State of the Art and a Software Development Framework,"* provides a comprehensive discussion on the state of the art in IoT, examples of applications, standardization trends, and basic technologies. The chapter also proposes a software framework for development of the Internet of Things applications and platforms.

Chapter 2 *"Increasing Effective Transmissions Using Smart Antenna Systems"* by Hung and Wu, introduces the characteristics and constraints of smart antennas for wireless time sensitive transmissions. Mechanisms to improve the parallelism of transmissions to faster search the routing paths are discussed in this chapter. According to the analysis and evaluation results, using smart antenna systems increases the throughput and reduces the transmission cost.

In **Chapter 3** *"A DTN-Based Multi-hop Network for Disaster Information Transmission,"* Kitada et al. study the robustness of communication network infrastructure under disaster scenarios. Network communication methods, which can be used even through in the case of disasters, are investigated and Never Die Network (NDN) based on Software Defined Network (SDN) to realize as a resilient disaster network are presented. In the chapter, the authors propose a DTN based Multi-hop network for temporal network in the case of disaster and considered transmission disruption and transmission delay. An experimental prototype by smart devices is presented as well to exemplify the findings.

Musilek et al. in **Chapter 4** *"Intelligent Energy Management for Environmental Monitoring Systems,"* present energy management in environmental monitoring system (EMS), outline the current trends in their development and evaluation. A number of power management techniques, combined into energy management strategies with the goal to minimize energy consumption while keeping the quality of service at a desirable level are discussed and extended by developing the principles for design, testing and evaluation of ensuing energy management strategies in EMS.

The **second part "Data Streaming, Processing, and Analysis"** covers topics of Data Streaming, Data Stream Delivery Technologies, Scalable Processing of Massive Traffic Datasets, Data Compression and Aggregation, and Data Analysis.

In **Chapter 5** by Kawakami et al. *"Smart Sensor Data Stream Delivery Technologies"* smart sensor data stream delivery technologies are introduced and analyzed. The authors motivate the increasing interest on sensor data stream delivery system that periodically collects and delivers sensor data. The authors propose P2P-based and cloud-based methods to distribute communication loads by relay nodes for delivering sensor data streams. According to the experimental results through simulations, the proposed method can effectively distribute the loads among the nodes in the network.

Di Martino et al. in **Chapter 6** *"Scalable Processing of Massive Traffic Datasets"* deal with the challenges of scalability and performance of processing massive spatio-temporal datasets about traffic, coming from Smart Sensor Networks. The authors present a scalable architecture aimed at exploit the computational and storage capabilities of the Cloud. Special emphasis is posed on the analysis of the underlying data models to handle massive dataset for providing vehicular traffic predictions.

Chapter 7 *"Bounded Error Data Compression and Aggregation in Wireless Sensor Networks"* by Chang et al., analyzes data compression algorithms for energy saving. The authors present an efficient and effective data compression and aggregation algorithm, called BEDCA. Under a bounded error, BEDCA determines whether the new sensed data should be compressed or not by comparing it with reference data such as the previous sensed data (for temporal correlation), the neighboring sensed data (for spatial correlation), and the codebook data (for data correlation). The algorithm is evaluated using real datasets and showed that BEDCA can reduce transmission energy and thus can make WSNs more power efficient.

Chapter 8 by Yeh and Chen *"Application of Data Analysis in Wellness and Health Sensor Network Environment"* addresses health analysis and prediction for the wellness and health sensor environment and the technologies supporting long-term care. Architecture of Wellness and Health Sensor Network Systems (WHSNS) is presented and then effective information processing steps of these systems are developed. A typical WHSNS architecture, which includes sensing subsystems and a backend server group, is introduced in this research, and a system procedure for information processing of health analysis and prediction is also proposed. According to the results of the experiments, it is shown that the health analysis and prediction functions in these systems can actually help the care recipients and caregivers.

The **third part "Healthcare Applications"** brings a variety of e-Health Applications using sensors and sensors networks. The large potentiality of using IoT technologies for healthcare applications are analyzed and discussed through four chapters exemplified with real life healthcare applications.

Susanto in **Chapter 9** *"Electronic Health System: Sensors Emerging and Intelligent Technology Approach"* provides a broad review of electronic health issues related to Sensor Emerging Technology (SET). SET as part of ICT health solutions is generally used for automated data collection, which are later processed for statistical study of data, Internet accessible shared databases, modeling and simulation, imaging and visualization of data, and investigation, Internet based communication among researchers, and electronic dissemination of research results. The author emphasizes the advantages of ICT in health system usage through sensors emerging technology for collecting data of a patients that are suffering from different diseases. Additionally, SET can be used as basis for developing sensor intelligent apps to prevent an unhealthy behavior among the patients and to provide information, which can aid to develop effective social strategies.

In **Chapter 10** by Leu et al. *"Fall Detection and Motion Classification by Using Decision Tree on Mobile Phone,"* the authors investigate the falls, which are a common cause of injury for the elderly all over the world, while it is well known that falls are hard to detect. With the ever-increasing use of smart phones, therefore, in this study, a mobile application for fall management, named the Fall Detection System using Mobile Phone (FDSMP), is presented. With such a mobile phone, an elderly fall and its type (among six classified types) can be detected almost immediately when the fall occurs. Thus, the reaction time, defined as the time period between the occurrence of the fall event and the time when caregivers arrive at the place where the elderly falls can be then shortened. We subdivided fallings into six types. The experiments showed the high accuracy of fall detection and classification.

Chapter 11 *"Approaching Hardware Solutions for Massive E-Health Sensor Data Analysis"* by Barbareschi et al. approaches hardware solutions for massive e-Health sensor data analysis emerging after innovative and not-intrusive systems to monitor in real-time the state and behavior of patients. The authors emphasize that the amount of data that has to be elaborated could represent the bottleneck of a monitoring system and it is critical in real-time applications. Therefore the authors present a layered architecture infrastructure, based on two Decision Tree predictor hardware implementations, suitable for medical data analysis in real-time and aimed at dealing with a large data volume and preserving a good hardware resources efficiency.

Akiyama and Kato in **Chapter 12** *"A Method for Estimating Stress and Relaxed States Using a Pulse Sensor for QOL Visualization"* propose a method estimating a stress state by using the instantaneous pulse rate obtaining from wearable devices in order to identify long-term stress factors in daily life. The authors define Stress and Relaxed Value (SRV) as a *QoL* (Quality of Life) indicator, considering both stress and relaxed state, and verify that SRV can be associated with lifestyle factors. The effectiveness of the proposed method is verified by conducting evaluation experiments using a wearable device.

The **last part "Living Lab – Everyday Activities"** covers topics from usage of sensor and sensor technologies from everyday life such as understanding human behavior mobility, life styles, etc.

In **Chapter 13** *"Proximity-Based Service: An Advanced Way of Extending Human Proximity Awareness,"* Fujihara deals with a recently proposed concept called proximity-based service for extending human proximity awareness using close-range wireless information–communication technologies. Proximity-based service is related to recently proposed and used beacon systems for advertising information to nearby mobile devices, such as iBeacon by Apple and Physical Web by Google. The author proposes and implements a system of multiple beacon modules for indoor route guidance. The system performance is also evaluated with a simple experiment to demonstrate the usefulness of the system to the exit in a building without using any high-tech location sensors.

Chapter 14 *"WiFi Tracking of Pedestrian Behavior"* by Petre et al. considers various techniques for tracking pedestrian behavior, and among them, the tracking through WiFi, as one of most popular one receiving increasingly more attention due to ubiquity of modern smartphones. The authors show WiFi tracking works, and explain its potentials and pitfalls. Special attention is given to the quality of data from WiFi scanning devices, and how this data can, and should be cleaned up before attempts at extracting information from sets of detected devices. As an illustration of the power of WiFi tracking, the authors bring a real case study of gathering WiFi data from a large event that attracted over 100 people geographically spread across during three days.

Finally, in **Chapter 15** *"The Life Management Platform Achieves Data Protection and Safe Sharing,"* Sakamoto et al. propose a new style of services that can respond to the diversity of citizens' needs by sharing their measured data and analysis results. To achieve such new styles, the authors propose using a IoT platform, called Life Management Platform. The platform adds access rights and browsing rights appropriate for services that maintain data and creates a good balance between data sharing and protection. As a result, the wide data collection and the analysis of service data that is only collected and stored by current services may become more active. The authors show a model scenario in which the selection, combination, and cooperation of multiple services on the platform allow for the daily detection of various subtle health changes and promote early measures for health risks for those with frail conditions.

THE BOOK READERSHIP

The book is written for a variety of readers. Students, engineers, and computer scientists who would like to learn and practice sensor-related techniques and applications will find this book useful. Professors who would like to give courses related to sensor and/or environmental monitoring can also use and assign this book as the textbook for his/her teaching reference. Furthermore, developers from industry and businesses interested on using sensors for smart homes or automating industries/business processes can find the book useful for developing such systems as well as to train employees and engineers working on and developing smart systems who would like to increase knowledge and skills on sensor based techniques and applications.

The Editors of the book

Fatos Xhafa
Universitat Politècnica de Catalunya, Spain

Fang-Yie Leu
Tunghai University, Taiwan

Li-Ling Hung
Aletheia University, Taiwan

ACKNOWLEDGMENTS

The editors would like to thank all authors for their contributions to the book, their cooperation and time in revising their chapters to meet high quality requirements. Our appreciation to reviewers for their time, constructive feedback, and valuable insights to authors for their chapters.

Fatos Xhafa's work has been partially supported by the Spanish Ministry for Economy and Competitiveness (MINECO) and the European Union (FEDER funds) under grant COMMAS (ref. TIN2013-46181-C2-1-R).

IoT AND NETWORK COMMUNICATION SYSTEMS

1

IoT TECHNOLOGIES: STATE OF THE ART AND A SOFTWARE DEVELOPMENT FRAMEWORK

Takahiro Inui*, Masaki Kohana*, Shusuke Okamoto*, Masaru Kamada[†]

*Seikei University, Japan †Ibaraki University, Japan

1.1 INTRODUCTION

Internet-connected devices such as smart phones and embedded devices other than personal computers have attracted much attention recently. Especially, small objects with a CPU, memory, sensors, and an Ethernet interface have a great potential to change our IT society. The internetworking of these devices is called Internet of Things (IoT). Data collected by IoT devices can be shared with other systems via the Internet, and they can be analyzed in real time. With their embedded technology and low energy requirements, IoT devices are expected to improve our IT experiences in the near future as they continue to make our daily life more comfortable. Recently, some studies and books are published to try to introduce early stage of personal and enterprise IoT usability (Dominique Guinard, 2016; Balani, 2016; Greengard, 2015; Familiar, 2015; Jaokar, 2015; Sula et al., 2014). However, we must bear in mind the security risks associated with them. Hijacking IoT devices or planting undesired devices can lead to invisible threats. It is important to build safe and appropriate technologies for IoT (Hu, 2016). The aim of this chapter is to give readers a proper understanding of IoT devices, the necessary technologies and its standards, as well as to present our approach in developing an IoT software.

1.2 CURRENT STATUS OF IoT

In this section, we will summarize the current status of IoT devices. They help improve lives and society, and are used in public facilities, factories, and homes. We will also describe standardization initiatives to introduce the technology aspect of IoT devices. Several companies have attempted to establish a unified standard and have also organized themselves into groups. After the discussion about standardizations, we will describe the associated technologies, such as communication technologies, power consumption, and productivity.

1.2.1 EXAMPLE OF IoT DEVICES

There are already a number of IoT devices targeted at home users. For example, iRobot Corporation is selling "Roomba 980," (iRobot Corporation) which is a home cleaning robot. It is equipped with sensors such as tracking sensors and a camera. It cleans a room autonomously without help from any person. To improve time efficiency and power consumption, Roomba 980 creates a map of the room during a cleaning task. It is designed not to go through the same place twice. This cleaning robot is equipped with a Wi-Fi module. Using a dedicated application, a user can connect to the robot and make it start cleaning from the outside of the house.

Koninklijke Philips N.V. is selling an LED lighting device named "hue," (PHILIPS) which is a type of IoT device. It is equipped with a Wi-Fi module, and, with a dedicated application, it allows users to invoke various control operations such as turning on/off the light, moving the direction of light, and setting the timer schedule. It can also turn on the light according to the command coming from outside of the house to prevent a crime. Hue links up with a GPS. It can turn on the light automatically when a user returns to his/her house. The IFTTT (IFTTT) (a web service that links a product to some application) can change the color of the lighting by sending an email to the web server.

The third example is a smart lock, which is sold by several companies. This is designed to perform locking and unlocking operations on a door; the user needs to only place his/her own authorized device such as a smart phone close to the door to perform the operations. It uses wireless protocols such as Bluetooth and Wi-Fi to communicate. It restricts access and checks the locking of the door via the Internet.

These three examples make our life convenient and comfortable by using an Internet-connected device. Next, we will introduce three use cases of IoT devices at public spaces.

The first one is a service that lets users know the congestion situation of a toilet space (Fanbright). A wireless magnet sensor is attached to a toilet door. It checks whether the toilet is vacant or occupied, and sends the information to a web server. The server shows a web page that indicates the status of the toilet space. This service is suitable for stations, office buildings, and shopping malls as it allows people to find a vacant toilet smoothly.

The second one is a bus arrival information board that was introduced in Kyoto city (Suzuki). It uses a device with Bluetooth and public Wi-Fi connection. When a bus is approaching, the device detects the bus by using a Bluetooth beacon and sends the arrival information to the information board through public Wi-Fi. This system can be installed at a low price. Moreover, the arrival information is available to the public.

The third one is a smart meter (TEPCO). It is an electrical device that measures electricity consumption and communicates it with a server. A conventional device

Table 1.1 List of Standardization Organizations

Objective	Corporate Name	Description
Standardization	OneM2M	This is a global standards initiative for M2M (machine to machine) communications and IoT, which aims to standardize global network architectures, protocols, and services. It is composed of 230 companies.
	Thread Group	This is a development group led by Nest Thread Group Labs (Google), ARM, and Silicon Lab, which aims to develop a low-power mesh network protocol for products in the home.
Platform	AllSeen Alliance	This is a cross-industry consortium dedicated to enabling the interoperability of billions of devices, services, and apps that comprise the IoT. It is composed of 185 companies such as Microsoft, PHILIPS, and SONY.
	Open Connectivity Foundation	This is the successor organization to the Open Interconnect Consortium (OIC). Members include ARRIS, CableLabs, Cisco, Electrolux, GE Digital, Intel, Microsoft, Qualcomm, and Samsung.
Ecosystem	Industrial 4.0	The fourth industrial revolution is a German initiative for improving manufacturing technologies. Together with the Industrial Internet Consortium, it aims to realize the concept of smart factories the German government.
	Industrial Internet Consortium	This organization aims to accelerate the development, adoption, and widespread use of interconnected machines. It consists of 220 companies such as AT&T, Cisco Systems, GE, Intel, and IBM.

measures the electricity consumption. However, someone has to read the cumulative value monthly. On the other hand, using a smart meter enables a company to receive the data automatically every 30 min. In addition, the company can offer a service that encourages saving electricity consumption with the report data. A smart meter can be configured to cooperate with a home energy management system (HEMS), which is a control system for optimizing the energy consumption of a house. It can be connected to a consumer electronic device and/or electrical equipment, allowing it to operate them automatically, and it presents the consumption to the user as visual data. HEMS can utilize data from smart meters and create a graph that changes every 30 min. It promotes less energy consumption.

1.2.2 STANDARDIZATION TREND

As a related topic of IoT technologies, we will introduce a few initiatives that aim to standardize the connection methods for communicating with different equipments and platforms. They are necessary for developers to build a common platform. Table 1.1 shows a list of the standardization organizations (Hachiyama). It is divided

Table 1.2 Communication Standards

Communication Standards	Distance	Speed	Energy Consumption	Cost
Zigbee	10–100 m	250 kbps	Low	Inexpensive
Bluetooth low energy	10 m	1 Mbps	Medium	Inexpensive
Wi-Fi	100 m	54 Mbps	High	Inexpensive

into three categories: standardization, platform, and ecosystem. These are mainly aimed at unifying the telecommunication technologies. OneM2M (oneM2M) group targets machine-to-machine communications. Thread Group targets wireless communications at home. Its technology achieves scalability that allows connecting 250+ devices into a wireless network with lower energy consumption. There are two platform groups: AllSeen Alliance (AllSeen) and Open Connectivity Foundation; they offer open source software for the IoT platform, which enables bidirectional communication between home electronics and electric devices. At the government level, the German government promotes the computerization of manufacturing. The terms "Industrie 4.0" and "Smart Factory" have drawn attention recently (Open-Connectivity-Foundation; Industrial-Internet-Consortium; Industrie-4.0). In Japan, a government-sponsored organization named "Industrial Value Chain Initiative" has been established, which aims to create a new style of manufacturing.

1.2.3 IoT TECHNOLOGIES

IoT technologies include various communication standards, operating systems, and data utilization. We will start by explaining the communication standards. IoT devices are used in various situations and locations. These devices use various communication methods, which are selected based on the communication distance, speed, and energy consumption. Table 1.2 shows a list of communication standards (Lee et al., 2007; kanda.com). Zigbee is the lowest energy consumption method, and it transmits data through a mesh network. On the other hand, Bluetooth low energy (BLE) and Wi-Fi constitute a star network that has a central hub. The speed of BLE is faster than that of Zigbee, but its energy consumption is higher than that of Zigbee. The speed of Wi-Fi is the fastest among them, and its distance is the longest. However, the energy consumption of Wi-Fi is the highest level among them.

The next topic is platforms and operating systems. We can find a system that has a central control for sensor management, data analysis, and display of results.

Google publicly announced an Android-based OS for IoT named "Brillo," (GoogleBrillo) which is a type of embedded operating system and is composed of core services. It has a variety of hardware functions and customizable options, and it enables the transformation from a prototype to a finished product quickly. It is possible to utilize OTA (over-the-air) update, software metrics, and crash report. A communication platform named "Weave" (GoogleWeave) enables a device to communicate with other devices such as smart locks and smart lighting.

There is a multiprotocol SoC (system on a chip) for IoT devices, which is named "Wireless Gecko" (WirelessGecko). It has an ARM processor, a memory, sensor interfaces, and a hardware encryption function. It makes IoT device development easy by combining different communication standards (such as Thread, Zigbee, and BLE) and providing an integrated development environment named "Simplicity Studio" (SILICON-LABS). Wireless Gecko's energy consumption is 63 µA when active. Wireless Gecko can be awakened in 1.4 ms from the standby state. It can use a piece of software that monitors the power efficiency and a packet of network in real time.

"OPTiM Cloud IoT OS" (OPTiM-announce; OPTiM) is an operating system that intuitively and safely controls IoT devices. It incorporates data analysis, artificial intelligence (AI) technology, and cloud services. IoT connection information for the equipment can be checked through a desktop-based application, like Windows Explorer; OPTiM Cloud IoT OS allows users to check and store the obtained information in an intuitive way. It provides visual data consisting of various numerical and video data. The obtained data from each IoT device can be analyzed using AI tools and big data analysis engines. OPTiM Cloud IoT OS also supports multiple accounts and makes it easy to manage an organizational hierarchy and its group permissions.

1.3 IoT SECURITIES

As IoT devices are used widely, IoT security becomes a matter of concern. Information security is defined by ISO/IEC 27002 as the preservation of confidentiality, integrity, and availability (CIA) (ISO). Confidentiality is designed to prevent sensitive information from reaching the wrong people, while making sure that the right people can get it actually. Integrity means maintaining and ensuring the accuracy and completeness of data over its entire life cycle. Availability refers to ensuring that authorized parties are able to access the information whenever needed. Consumer electronics has become the target of cyber-attacks because its internet connectivity increases the number of entry pathways. For this reason, the damage by intrusion also becomes severe. With the expanding use of IoT, it has become more difficult to deal with information security. We will introduce three examples related to information security incidents.

The first is an example of data confidentiality. There are a number of websites that expose video surveillance data captured by a security network camera. In Japan, a number of Panasonic cameras have been the targets of unauthorized access. Using Google, one can find network cameras that have already been intruded. They have become targets because many of these network cameras use a default or a simple password.

The second is an example of data integrity. There is a possibility that security risks may occur if industrial control systems (ICSs) adopt IoT. ICS is indispensable for running today's society because it is used for pumping water, generating power, and operating plants. There is a possibility that an unauthorized user might access an ICS and falsify its data or destroy the system if it is connected to an unsecured network. It can also be a big security risk if an industrial IoT keeps on running without any attendant after it starts. It is difficult for the administrator to find the causes when an error occurs. It can be a serious damage to stop the infrastructure equipment or leak the confidential information. For example, the "Stuxnet" malware stopped the nuclear equipment in Iran in 2010. The malware infected the control system via USB, and it made a centrifuge unavailable. The infection in this case was via USB; however, using a network is more susceptible and data integrity is threatened.

The third is an example of availability. A big earthquake struck Kumamoto in April 2016. At that time, the Japan Meteorological Agency revealed that a few seismometers stopped working owing to the strong swing. The agency could not receive the data from these meters. The main reason for this trouble was power supply. After the first swing, a power outage happened. The power supply source of the seismometer was changed to an emergency diesel generator; however, a failure occurred in the generator. The seismometer continued to work for several more minutes using its battery. Then, it finally stopped working. The swing data were lost until someone manually fixed the seismometer. The power supply and communication method must be preserved even if the incident is not due to an earthquake.

1.4 A SOFTWARE FRAMEWORK FOR IoT

Finally, we will describe a development of our planned software framework for IoT. This framework meets the standardization requirements of IoT. For example, some IoT software that controls sensors has to do a number of things such as collecting periodic data, analyzing data, and transmitting data to a server. These processes are similar among most pieces of IoT software. To make it easy and improve productivity, we will develop a software framework for the basic control. It enhances security and allows us to develop a software with a higher precision timer just by running the dedicated application on the target machine. We will provide a simple development environment using visual programming.

Our framework consists of a GUI-based editor for the state transition diagram and a translator for the diagram data to control the software for IoT. The state transition diagram represents a behavior of IoT that includes an event-driven processing. Usually, the CPU of an IoT device is different from that of a PC. Therefore, our framework is a cross development environment.

1.4.1 OVERVIEW OF THE SOFTWARE FRAMEWORK

Fig. 1.1 shows a flowchart of the development using our framework. It utilizes a software called "Islay" to edit a state transition diagram. Islay translates the state

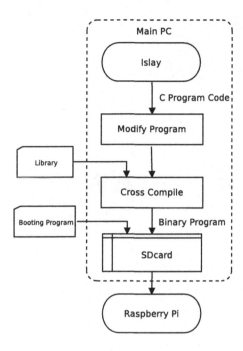

FIGURE 1.1

Flowchart of framework.

transition diagram to a C program code. Then, the whole program is compiled for the target machine, Raspberry Pi. Moreover, a program code for the boot-up process is attached. The resulting binary program is copied to an SD card, and the target machine boots with it.

1.4.2 ISLAY

Islay is an interactive animation authoring tool, which takes a classical state-transition diagram as input and runs a built-in interpreter for animation scripts. It also generates an interactive animation written in a programming language, such as C, Java, and JavaScript, as well as a Flash binary animation. Fig. 1.2 shows a screenshot of the Islay editor while editing a state-transition diagram. This diagram represents the behavior of a fire engine animation character, which goes back and forth.

Fig. 1.3 shows a C program code related to the state of the above mentioned fire engine. In this code, each state is converted to a function. This function represents the state of going left. Each function has the same structure among all functions for states. It consists of an action part and a transition part. Once a fire engine transitions to a different state, the actions specified by the state are taken. If there are multiple characters in the animation, all actions for each character are performed. After that,

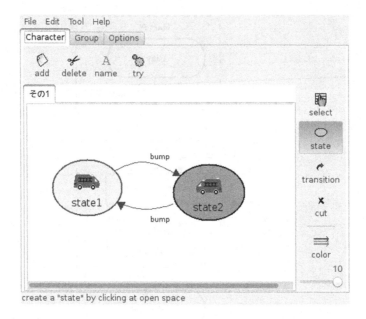

FIGURE 1.2

Screenshot of Islay.

```
static void
trans_dgm0_s0(inst_t *e, bool out, bool delayflg)
{
    /* action */
    if (out) {
        setImage(e, e->img[0]);
        move_rel(e, -8, 0);
        return;
    }

    /* transition */
    if (e->pending) { e->pending = 0; return; }
    if (e->bumped) {
        int_list_t *x;
        for (x = e->bumpedlist_av; x; x = x->next)
            target_push(e,x->x);
        e->func = trans_dgm0_s1; return;
    }
    {
        e->func = trans_dgm0_s0; return;
    }
}
```

FIGURE 1.3

C function code generated by Islay.

each state transitions to the next state. This function is called twice with a different value for the second "out" parameter. The state of the character is kept in a function pointer. Moreover, a transition can be simply presented by an assignment statement.

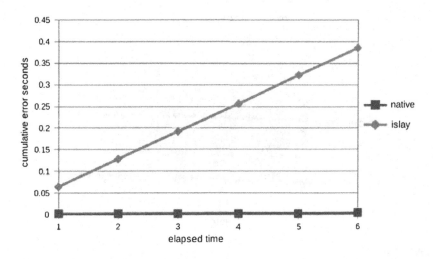

FIGURE 1.4

Accuracy of timings.

The resulting program can be used as a control program that processes something periodically. We would like to use this software to control an IoT device. However, there is a problem for this purpose and it is about the accuracy of periodic processing. For example, a generated program by the current version of Islay uses the GTK+ library. The accuracy depends on the implementation of a library function, which probably uses a user-mode functionality of the operating system. It might not be accurate enough for an application that controls some sensor devices directly.

Therefore, we measured the accuracy of the timings of a clock application made by Islay. It counts periodically for 100 ms. Then, we measured the accuracy of the timing done for each processing every minute. Fig. 1.4 shows the accuracy of the timing. The x-axis is the elapsed time and the y-axis is the cumulative error in seconds. The Islay-generated application increased the cumulative error every minute. At 6 min, the cumulative error was about 0.38 s. Thus, to use Islay for IoT software development, we must address this inaccuracy.

1.4.3 RASPBERRY Pi AND A LINUX SCHEDULER

We are using Raspberry Pi (RaspberryFoundation) as the target device for our short-term task. It is a single-board type of computer with an SoC. It was developed for educational purposes in England and has since been available for sale all over the world. Fig. 1.5 shows a whole view of the Raspberry Pi machine. It is equipped with a GPIO interface, USB device ports, and an Ethernet port, which are useful for building an Internet-connected device. The first generation of Raspberry Pi had an ARM processor, which implemented ARMv6 ISA with VFP2. It is compatible with

GPIO   RCA Video   Audio

SD Card

USB

Power

Ethernet

HDMI

FIGURE 1.5

Raspberry Pi.

the Linux/Debian armel port, which emulates floating-point number calculation, although the CPU has a hardware floating-point unit.

We will use a Linux operating system as the base software for the IoT runtime environment. Linux is an open-source operating system that can boot on Raspberry Pi. We can customize it to let it control a higher precision timer and security-related tasks. A typical user can choose Raspbian among the Linux distributions. It is Debian armhf (for the hardware floating-point unit) rebuilt for the processor. The second-generation Raspberry Pi 2 has a multicore version of ARMv7 and is compatible with Debian armhf.

Our main concern is the process schedulers on the Linux kernel. The purpose of a scheduler is to select the next process to run from a number of ready processes. The current version of the Linux scheduler is designed to allow developers to select from several different scheduling algorithms, each of which is suitable for the nature of a process. CFS (Completely Fair Scheduler) is the default scheduler. It is used for normal processes. FIFO (first-in, first-out) scheduler is used for real-time processes (Bovet, 2007).

CFS and FIFO are short-term schedulers, each of which treats processes in run queues. When a process waits for some event to occur, the process is put in a wait queue. Moreover, if the event is related to a timer, a dynamic timer is used. It is one of the timer subsystems within the Linux kernel. A timer entry for the dynamic timer is created to execute a kernel task at a specified time. It is automatically deleted after the task has finished. Thus, the dynamic timer is used to move a process from a wait queue to the run queue. Thereby, the process becomes a target of the short-term scheduler (Naoya).

Table 1.3 Experimental Environment	
Machine	**Raspberry Pi 1 (Model B)**
CPU	ARM1176JZF-S 700 MHz
RAM	512 MB (GPU shared)
OS	Raspbian Jessie (4.1.15)
Compiler	crosstool-NG (1.15.2)

1.4.4 IMPROVEMENT OF THE DYNAMIC TIMER

We plan to use Islay to generate a program that controls some sensors. Since the program is made from state transition diagrams, it functions at regular intervals. Scheduling is important for periodic processing. Historically, study of scheduling is very important field in operating system and co-operative processing. Recently, many studies also focus on this research field (Kashyap and Vidyarthi, 2011; Andrei et al., 2012; Xilong and Peng, 2015; Senobary and Naghibzadeh, 2015). The nanosleep system is the interface of Linux scheduler for controlling user task. Thus, it is important for us to know the time accuracy related to the system call behavior. The system call is implemented with both short-term and medium-term schedulers.

We surveyed the performance of CFS and FIFO, which are short-term schedulers. To know the difference between the accuracy of the CFS and FIFO scheduler, we created a C program that uses the usleep library function. Table 1.3 shows the experiment environment. We used a Raspberry Pi machine with the Raspbian operating system, ran a program at regular intervals, and then measured the accuracy of the intervals with two patterns. The first was a pattern where the program ran alone. It means an execution without any other loads. The second was a pattern where the program ran with a number of other programs.

We specified the program as an argument for the "init" program on Linux for the first pattern. This made our target program obtain a process ID of 1, and the other programs were not executed in this measurement. Furthermore, we modified the Linux kernel to measure the kernel time. Our program calls the usleep library function, which calls the nanosleep system call to enter the kernel. Then, the "printk" function in the kernel is called and the accurate time is printed. We did not use a program that measures the time interval by itself and prints it because it is not accurate owing to the system call overhead of obtaining the time and printing it.

The following four programs were run with the target program to add some load.

- A CPU-bound program that calculates matrix values
- An I/O-bound program that reads and writes to a file
- A terminal I/O program
- A program that calls the usleep library function repeatedly

Fig. 1.6 shows the result of the comparison of accuracy. The x-axis is the measurement condition. There were four combinations of two intervals (1 s and 1 ms) and

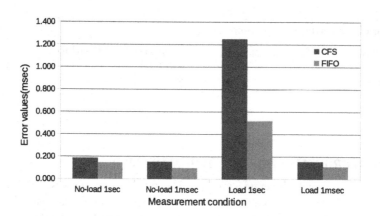

FIGURE 1.6

Comparison of the CFS and the FIFO.

two specified runtime schedulers (CFS and FIFO). These conditions were applied to the cases with and without loads. In the case with loads, it was measured under the same condition in which 20 (four types by five) programs were run. The y-axis shows the error values against the specified time interval. For the case with the 1-s interval, the value was calculated from the measured value minus 1 s. For the 1-ms interval, the value was calculated in the same way. From this result, we can see that the error value was 1.24 ms for CFS and 0.51 ms for FIFO. The error value of CFS was 2.4 times bigger than that of FIFO. As a whole, the mean error value of FIFO was about 40 μs lower than that of CFS. Therefore, the FIFO scheduler is suitable for real-time processing. However, there was still an error value of 150 μs even for the case without load. Modifying the scheduler will not solve the error because the error includes the overhead of medium-term schedulers.

From the above result, we modified the way the dynamic timer was used to implement the nanosleep system call. Fig. 1.7 shows the timing chart of this modification. The "ideal" label in this figure is the ideal time of processing when the nanosleep system call is processed. A dynamic timer is registered with the time specified in the user program to sleep the process that is related to the program, and the process is put in a wait queue. The dynamic timer wakes up the process at the specific time, and then the call of nanosleep ends. However, the current implementation of the nanosleep system call behaves as the "normal" label in this figure. It needs some extra time until the timer actually wakes up the target process. This extra time is used as an allowance for the timer, and it becomes an overhead of the nanosleep system call. The "plan" label in this figure shows the way of our modification. A dynamic timer is set to a time that is slightly shorter than the specified time in the user program. The "udelay" function, which is a kernel function, is used in kernel mode to wait for less than 1 ms. The function spends a few microseconds of time in a busy-waiting loop. It could achieve

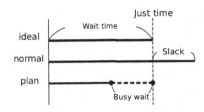

FIGURE 1.7

Time chart after modification.

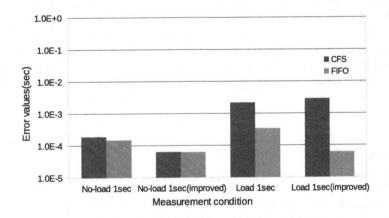

FIGURE 1.8

Error of improved nanosleep.

a higher accuracy, but all the other process would not be able to run on the CPU during the busy-waiting period.

We compared the modified version of the nanosleep function with the original one. The conditions of the experiment were the same as in the above survey for the CFS and FIFO schedulers; however, we used 40 processes as the load of the calculations and I/O. Fig. 1.8 shows the result. The x-axis is the measurement condition, which is a case with or without a load. The y-axis shows the error values against the specified time interval and is displayed on a logarithmic scale. From this result, we can see that the mean error value of no loads is 61 µs for FIFO with 1-s intervals. It achieves about 60% improvement against the original implementation and has the same performance as the original FIFO without any loads.

1.5 CONCLUSIONS

IoT is now a state of art technology. A lot of trials to confirm its usability are going on. In this chapter, we described the current state of IoT. We can now find several applications of IoT, such as a cleaning robot, a remote control lighting device, smart doors, and so on. There are also standardization organizations that include one under the government initiative. The key technologies of IoT are low power communications. We should also care about data security risk. CIA (Confidentiality, integrity, and availability) is a central of information security.

Our IoT software framework focuses on a periodic process that collects data with a fixed time interval. The accuracy depends on the implementation of nanosleep system call and the scheduler of Linux operating system. The current Linux real-time scheduler has an error value of 150 µs even for the case without load. We modified the nanosleep system call to increase the accuracy of the time for IoT development. As a result, we managed to achieve a mean error value of 61 µs for the FIFO real-time scheduler under a no-load condition. We will work on designing API on a state transition diagram for driving sensors and sending data. And then we will surveys the relationship between a security measure and a scheduler, as well as evaluating our framework.

REFERENCES

AllSeen. Framework – AllSeen alliance, https://allseenalliance.org/framework (accessed 15 October 2016).

Andrei, S., Cheng, A.M., Grigoras, G., Radulescu, V., 2012. An efficient scheduling algorithm for the non-preemptive independent multiprocessor platform. International Journal of Grid and Utility Computing 3 (4), 215–223.

Balani, N., 2016. Enterprise IoT: A Definitive Handbook. CreateSpace Independent Publishing Platform.

Bovet, D.P., 2007. Understanding the Linux Kernel, 3rd edition. O'REILLY.

Dominique Guinard, V.T., 2016. Building the Web of Things: With Examples in Node.js and Raspberry Pi. Manning Publications.

Familiar, B., 2015. Microservices, IoT and Azure: Leveraging DevOps and Microservice Architecture to Deliver SaaS Solutions. Apress.

Fanbright. Visualization of the toilet use situation, http://www.fanbright.jp/iot/manage/toilet/ (accessed 15 October 2016).

GoogleBrillo. Brillo, https://developers.google.com/brillo/ (accessed 15 October 2016).

GoogleWeave. Weave, https://developers.google.com/weave/ (accessed 15 October 2016).

Greengard, S., 2015. The Internet of Things. The MIT Press Essential Knowledge Series. The MIT Press.

Hachiyama, K., 2016. Current status of initiatives related to IoT from New York in August 1, the United States, https://www.ipa.go.jp/files/000047543.pdf (accessed 15 October 2016).

Hu, F., 2016. Security and Privacy in Internet of Things (IoTs): Models, Algorithms, and Implementations. CRC Press.

IFTTT. Learn how ifttt works, https://ifttt.com/ (accessed 15 October 2016).

Industrial-Internet-Consortium. Industrial internet consortium, http://www.iiconsortium.org (accessed 15 October 2016).

Industrie-4.0. Plattform industrie 4.0 – homepage, http://www.plattform-i40.de/I40/Navigation/EN/Home/home.html (accessed 15 October 2016).

iRobot Corporation. iRobot roomba vacuuming robot, http://www.irobot.com/For-the-Home/Vacuuming/Roomba.aspx (accessed 15 October 2016).

ISO. Iso/iec 27002:2013, http://www.iso.org/iso/catalogue_detail?csnumber=54533 (accessed 15 October 2016).

Jaokar, M.A.V., 2015. IoT and Data Science. Ajit Jaokar.

kanda.com. Zigbee wireless interfaces and modules, http://www.kanda.com/zigbee-wireless.html (accessed 15 October 2016).

Kashyap, R., Vidyarthi, D.P., 2011. Weight-balanced security-aware scheduling for real-time computational grid. International Journal of Grid and Utility Computing 2 (4), 313–325.

Lee, J.-S., Su, Y.-W., Shen, C.-C., 2007. A comparative study of wireless protocols: Bluetooth, UWB, Zigbee, and Wi-Fi. In: The 33rd Annual Conference of the IEEE Industrial Electronics Society (IECON), pp. 46–51.

Naoya. Linux of sleep processing, view the details of the timer processing, http://d.hatena.ne.jp/naoya/20080122/1200960926 (accessed 15 October 2016).

oneM2M. Onem2m home, http://www.onem2m.org/ (accessed 15 October 2016).

Open-Connectivity-Foundation. Ocf membership list, https://openconnectivity.org/about/membership-list (accessed 15 October 2016).

OPTiM. Optim cloud IoT OS a new type of OS optimized for IoT, https://en.optim.co.jp/cloud-iot-os/ (accessed 15 October 2016).

OPTiM-announce. Announcing optim cloud IoT OS, http://en.optim.co.jp/news-detail/11051 (accessed 15 October 2016).

PHILIPS. Meet hue, http://www2.meethue.com/en-XX (accessed 15 October 2016).

RaspberryFoundation. Raspberry Pi teach, learn, and make with Raspberry Pi, https://www.raspberrypi.org/ (accessed 15 October 2016).

Senobary, S., Naghibzadeh, M., 2015. Semi-partitioned scheduling for fixed-priority real-time tasks based on intelligent rate monotonic algorithm. International Journal of Grid and Utility Computing 6 (3/4), 184–191.

SILICON-LABS. Simplicity studio, http://www.silabs.com/products/mcu/Pages/simplicity-studio.aspx (accessed 15 October 2016).

Sula, A., Spaho, E., Matsuo, K., Barolli, L., Xhafa, F., Miho, R., 2014. A new system for supporting children with autism spectrum disorder based on IoT and P2P technology. International Journal of Space-Based and Situated Computing 4 (1), 55–64.

Suzuki, J., 2016. Convenient the Kyoto municipal bus with IoT – the back of the "high-tech bus arrival guidance system", http://www.itmedia.co.jp/enterprise/articles/1501/06/news093.html (accessed 15 October 2016).

TEPCO. Smartmeter, http://www.tepco.co.jp/en/announcements/2015/1257023_6902.html (accessed 15 October 2016).

WirelessGecko. Wireless Gecko IoT connectivity portfolio, https://www.silabs.com/WirelessGecko (accessed 15 October 2016).

Xilong, Q., Peng, X., 2015. An energy-efficient virtual machine scheduler based on CPU share-reclaiming policy. International Journal of Grid and Utility Computing 6 (2), 113–120.

ACRONYMS AND GLOSSARY

List of acronyms with explanation

AI Artificial Intelligence
API Application Programming Interface
ARM Acorn RISC Machine

BLE   Bluetooth Low Energy
CFS   Completely Fair Scheduler
CIA   Confidentiality, Integrity, Availability
CPU   Central Processing Unit
FIFO   Firts-In, First-Out
GPIO   General Purpose Input/Output
GUI   Graphical User Interface
HEMS   Home Energy Management System
ICS   Industrial Control System
IoT   Internet of Things
LED   Light-Emitting Diode
OTA   Over The Air
RISC   Reduce Instruction Set Computing
SoC   System on a Chip

Glossary of terms with explanation

Bluetooth a wireless technology for short distance.
GTK+ a library to build graphical user interface.
Raspberry Pi a tiny computer provided by Raspberry PI Foundation.
Wi-Fi a technology that allows devices to connect to wireless networks.

INCREASING EFFECTIVE TRANSMISSIONS USING SMART ANTENNA SYSTEMS

2

Li-Ling Hung, Sheng-Han Wu
Aletheia University, Taiwan

2.1 INTRODUCTION

Antenna technology has increased the development of network communication. Depending on the directionality, we can classify antennas to omnidirectional and directional antennas. Using an omnidirectional antenna, a host can send messages to the direct neighbors in the range of signal transmission, the direct neighbors named one-hop neighbors. When it has a common message to all its neighbors, the job can be performed in one transmission. In other words, when a host transmits a message, all its one-hop neighbors receive the message and are interfered if they are communicating with others. Due to the spatial bottleneck, the transmissions using omnidirectional antennas may be limited. The directional antennas include unidirectional and multiple-beam antennas. Using a unidirectional antenna, a host can transmit a message to its neighbors in some direction without interfering its neighbors in other directions. The unidirectional antenna (Nasipuri et al., 2000; Ko et al., 2000) takes the advantages of smaller interference area. When using unidirectional antenna, a host can send messages to the neighbor in some direction at a given time. However, using a unidirectional antenna, a host can only transmit to or receive from one neighboring host at a given time. When a message is for the neighbors in multiple directions, named multicasting, the transmission time for the message is much varied (Li and Luo, 2010). To improve the usage of unidirectional antennas, a number of research (Raman and Chebrolu, 2005; Jain et al., 2005; Bao and Garcia-Luna-Aceves, 2005) proposed protocols with multiple-beam smart antennas, which include antenna array and controllers. Smart antenna systems can support simultaneous transmission (or reception) of multiple packets on different beams using the same channel. The smart antennas can fairly reduce the interference problems and power consumptions for the transmissions (Mehrotra and Bose, 2014; Chou et al., 2014). Furthermore, the smart antennas have been proposed for using in many domains including WiMAX multihop networks, cellular networks, or mesh networks (Wendt et al., 2011; Chen et al., 2015). Therefore, we attempt to exploit smart antenna systems efficiently.

Smart Sensors Networks. DOI: 10.1016/B978-0-12-809859-2.00003-6

19

In some applications, the postponed information may lose the plot of itself; then we call the transmissions time-sensitive transmissions. Thus the delay constraints of transmitting packages should be taken into account (Chou et al., 2014; Mavromoustakis et al., 2013; Bazan and Jaseemuddin, 2011). The postponed message may be caused by the interference or the deafness problems. Thus, to exploit a smart antenna system efficiently, we should consider the deafness, interference, and postponed problems.

2.2 BACKGROUND AND LITERATURE REVIEW

To increase the spatial reuse and network throughput, Gossain et al. (2005) proposed a scheduling based on 802.11. The scheduling has to wait for the RTS/CTS communication and backoff time. Jain et al. (2005) introduced a TDMA scheduling where a host can transmit or receive packets simultaneously without RTS/CTS and backoff. However, the network performance may not be significantly improved if not taking the transmission of neighboring hosts into account because the parallel transmissions should be scheduled in advance. Bao and Garcia-Luna-Aceves (2005) and Chang et al. (2013) proposed a cluster-based scheduling, which, considering the transmissions of the neighboring hosts, can allocate the scheduling in polynomial time. However, these mechanisms do not deal with the delay constraint of transmissions. Hung et al. (2011) proposed transmission mechanisms with smart antenna based on cluster topology, which considers the delay constraint of each transmission.

As routing paths, Jawhar et al. (2010) proposed a mechanism that finds an efficient route from the source host to the destination host. However, the amount of searching packets is large. Hence, it needs to be improved. Liu and Lin (2005) built a routing path based on a cluster topology. An efficient mechanism for building route with Omni directional antenna was presented. Both Jawhar et al. (2010) and Liu and Lin (2005) do not consider the delay constraint of transmission packets. When a packet is time sensitive, the packet is useful in a period of time. If the packet arrives after the period, then it is useless. Therefore, for time-sensitive packets, a routing mechanism that takes the delay constraint of packets into account is needed.

To avoid packet collisions, Rafique (2013) claimed that hosts can arrange the receiving and transmitting of beams by means of sending ready-to-receive and ready-to-send messages. But a host cannot communicate and negotiate with its interfering neighbors and neighbors with deafness problem. The interfering and deafness problems can be eliminated when the transmissions are arranged by a proper third party. Therefore, we propose a cluster-based model where the transmissions are arranged by the cluster heads. Wendt et al. (2011) proposed an approach that builds a routing tree periodically by means of scanning and detecting the situations of neighbors, and scheduling the transmissions that satisfy the requests of hosts. In addition, the requests of hosts are transmitted to the neighbors by the piggy-back before scheduling, that is, the message sent by hosts carry their next transmission requests for reducing the following transmission and power consumption. However, the mechanism may not be running well if they are not regular transmissions. Mumey et al. (2012) calculate the distances and relations for hosts in the network topology and adjust the situations, including the number of beams and the range of a single beam, of hosts for the network topology. Mehrotra and Bose (2014) calculate and adjust the power of a beam according the situation of radio gain. When the communication neighbors and the relations of neighbors are fixed, it saves the power consumption because hosts can decrease the transmission power in some beams and close useless beams. However, the network topology and the relations of neighbors vary for mobile systems. It is of cost but of no effect to calculate and predict the best beam situations.

This chapter aims to propose approaches for routing and transmission with smart antennas. The proposed approaches consider parallel transmissions and delay constraints of the transmissions. Therefore, the network throughput and transmission guarantee that they will be better than other existing approaches. We do not calculate the transmission and beams of hosts accurately in advance in this research. We predict the routes in advance, but the transmission schedules are decided when the transmission requests are raised. The features of proposed mechanisms are listed as follows.

1. Effective building of the routing paths and transmission schedule based on cluster topology.
2. The number of packets for searching routes in the networks is much less than those of existing approaches.
3. The end-to-end delay of a message sent through the built routing paths is shorter than its delay constraint.
4. Not only the shortest unicasting routing paths, but also shared multicasting paths for the least transmission spending routing are found.
5. The transmission schedules arranged by cluster heads proposed in the research successfully avoid the deafness problem and effectively improve the transmission parallelism.
6. The schedule is arranged according to the viewpoints of both senders and receivers.
7. The maximums of transmissions are shorter than their delay constraints.

2.3 PROBLEM UNDER STUDY AND ITS STATEMENT

Smart antenna systems can support simultaneous transmission (or reception) of multiple packets on different beams using the same channel. However, smart antennas still have interference problem and hardware constraints. Those constraints include a beam mode constraint, data diversity constraint, interference constraint, transmitting rate constraint, and receiving rate constraint. For example, a host cannot communicate with more than one neighboring host in a beam, which is called the diversity constraint. Hence, we should take account of hardware constraints when improving

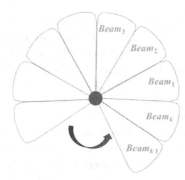

FIGURE 2.1

A host with multibeam smart antenna.

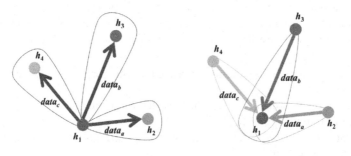

FIGURE 2.2

The transmitting or receiving at the same time.

the network performance with smart antennas. The constraints are described in detail in the problem subsection.

2.3.1 NETWORK ASSUMPTIONS

Let $G = (V, E)$ represent the network topology, where V is the set of n hosts $h_1, h_2, \ldots, h_n$, and E is the set of neighboring relationships between hosts, $E \in V \times V$. Each host has a unique ID and is equipped with a k-beam smart antenna system, which can communicate with k different directions of nodes as shown in Fig. 2.1, where $k = 2^i$ with positive integer i. Thus, with k-beam smart antenna, a host can simultaneously transmit (or receive) multiple packets on different beams at a given time, as shown in Fig. 2.2. The mechanisms proposed in this chapter are based on cluster topology and TDMA technology. A host may be a head or a member in its cluster. While building routes and scheduling the transmissions, the time slots for transmissions of heads and members are alternate. For ease of scheduling, we uni-

Receiver beam / Sender	h_1	h_2	h_3	....	h_n
h_1	0	3	0	....	0
h_2	1	0	1	....	0
h_3	0	3	0	....	0
⋮	⋮	⋮	⋮	....	⋮
h_n	0	0	0	....	0

(a)

Receiver / Sender	h_1	h_2	h_3	....	h_n
h_1	0	0	60	....	0
h_2	12	0	0	....	0
h_3	0	0	0	....	0
⋮	⋮	⋮	⋮	....	⋮
h_n	0	0	0	....	0

(b)

Receiver / Sender	h_1	h_2	h_3	....	h_n
h_1	0	6	12	....	0
h_2	6	0	0	....	0
h_3	12	0	0	....	0
⋮	⋮	⋮	⋮	....	⋮
h_n	0	0	0	....	0

(c)

Receiver / Sender	h_1	h_2	h_3	....	h_n
h_1	0	0	12	....	0
h_2	3	0	0	....	0
h_3	0	0	0	....	0
⋮	⋮	⋮	⋮	....	⋮
h_n	0	0	0	....	0

(d)

FIGURE 2.3

The examples of matrixes. (a) A Neighboring Matrix NE; (b) A Transmission Demand Matrix D; (c) A Transmission Rate Matrix R; (d) A Delay Constraint Matrix TC.

form the direction of each beam for all hosts. The following defines some terms for presenting the problem formulation and our mechanisms.

For ease of presentation, let $c_{i,j}$ represent a communication pair where host h_i intends to send data to host h_j.

Definition 1 (Neighboring Matrix, NE). To manage the locations of neighboring hosts, a neighboring matrix $NE = [ne_{i,j}]_{n \times n}$ stores the beam location where the neighboring hosts is, where $1 \le i, j \le n$. Each element of the matrix $ne_{i,j}$ stores the data representing whether host h_j is located at a beam sector of host h_i or not. If yes, then the value of the element is the number of the located beam. Otherwise, the value of the element is zero. Fig. 2.3(a) shows an example of the neighboring matrix. The data in $ne_{2,1}$ is 1, i.e., h_1 and h_2 are neighbors, and h_1 is located in the beam 1 of h_2.

Property 1. Because we assume that the directions of antennas are uniform, when host h_a is located the in ith beam of h_b, the h_b must be located in the $(i + k/2)$th or $(i - k/2)$th beam of h_a, where k is the number of beam forming, and vice versa.

Definition 2 (Transmission Demand Matrix, D). A Transmission Demand Matrix $D = [d_{i,j}]_{n \times n}$ represents the data volume of transmission requirements in the network. The value of element $d_{i,j}$ denotes the data volume of transmission that intends to transmit host h_i to host h_j. Fig. 2.3(b) is an example of matrix D. The value of $d_{2,1}$ is 12, which means that the host h_2 intends to send data to host h_1 and the volume of the data is 12 MB.

Definition 3 (Transmission Rate Matrix, R). A Transmission Rate Matrix $R = [r_{i,j}]_{n \times n}$ defines the transmission rate in the network. The value of entry $r_{i,j}$ denotes the largest transmission rate for $c_{i,j}$. Fig. 2.3(c) gives an example of matrix R.

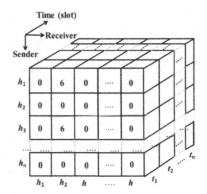

FIGURE 2.4

An example of Scheduling Matrix (S).

The value $r_{1,3}$ is 12, which means the fastest transmission rate of transmission pair $c_{1,3}$ is 12 Mb/s. Usually, the value of $r_{i,j}$ is the same as that of $r_{j,i}$. If h_i is not a neighbor of h_j, then $r_{i,j}$ and $r_{j,i}$ are zero because they cannot transmit to each other directly.

Definition 4 (Delay Constraint Matrix, TC). A Delay Constraint Matrix $TC = [tc_{i,j}]_{n \times n}$ represents the delay constraint of the requirement data in the network. The value of element $tc_{i,j}$ denotes the delay constraint of the data in $c_{i,j}$. Fig. 2.3(d) gives an example of matrix TC. The value of $tc_{2,1}$ is 3, which means that the delay constraint of $c_{2,1}$ is 3 time slots.

Definition 5 (Scheduling Matrix, S). Let T denote the total number of time slots required for accomplishing all transmissions completely. The Scheduling Matrix $S = [s_{i,j,t}]_{n \times n \times T}$ defines the transmission scheduling of hosts during time T. When $c_{i,j}$ is arranged transmitting at time slot t, the element of $s_{i,j,t}$ is the value of transmission rate for transmission $c_{i,j}$ at time slot t, which must be smaller than or equal to $r_{i,j}$. Each element of S is defined as follows:

$$s_{i,j,t} = \begin{cases} r & \text{if the transmission of } c_{i,j} \text{ is arranged at slot } t \text{ in rate } r, \\ 0 & \text{otherwise.} \end{cases}$$

Moreover, when the value of $s_{i,j,t}$ is not equal to 0, $|s_{i,j,t}|$ equals 1; otherwise, the value of $|s_{i,j,t}|$ equals 0. Fig. 2.4 gives an example of the scheduling matrix where time slot t_1 is allocated for $c_{1,2}$ and $c_{3,2}$.

Definition 6 (Efficient routing paths). Let $DC_{k,i}$ represent the delay constraint for a packet transmitted from host h_k to host h_i. In general, there may be more than one path from a source to a destination host. For time-sensitive transmissions, an efficient transmission path should be accomplished in the time less than the delay constraint. For host h_k, let $p_{i,j}$ and $\delta(p_{i,j})$ denote the jth path to host h_i and the delay time of the path, respectively. The set of routing paths $P_{k,i}$ is a set that includes the transmissions whose delay time is less than $DC_{k,i}$. Thus, $P_{k,i}$ is defined as follows:

$$P_{i,j} = \{p_{i,j} | \delta(p_{i,j}) \in (0, DC_{k,j})\},\qquad(2.1)$$

where $p_{i,j}$ is the jth path from h_k to h_i.

Definition 7 (Shared path). For a source host h_k, let $Shared(i, j)$ be the maximal shared path for destinations h_i and h_j. Because these paths start from the same source, the maximal shared paths are obtained simply by verifying the first different hosts of these paths. For example, if the paths to destination h_i and h_j started from h_a are different from the next host of h_k, then the $Shared(i, j)$ is a path from h_a to h_k.

2.3.2 PROBLEM FORMULATION

Given the network $G = (V, E)$ with a set of transmission requirements between hosts, we aim to maximize the network throughput by exploiting the parallel transmission advantage of smart antenna. The k-beam smart antennas system allows a host to communicate with k neighbors in different beams at any given time. However, to avoid the interference and deafness problems, these parallel transmissions should satisfy the constraints of smart antenna system. For ease of representing those constraints strictly, we define the operations as follows. If you are reading for technical review, then the mathematical descriptions may be omitted.

Operation 1 (Logical AND operation ($\odot$)). The operator $\odot$ defines the logical AND operation whose result equals 1 when two operands are equal to 1. Given two integers i and j, where $0 \le i, j \le 1$, it is defined as follows:

$$i \odot j = \begin{cases} 1 & \text{if } i = 1 \text{ and } j = 1, \\ 0 & \text{if } i = 0 \text{ or } j = 0. \end{cases}$$

Operation 2 (Logical NAND operation ($\oplus$)). The operator $\oplus$ defines the logical NAND operation whose result equals 1 when one of the operands equals 0. Given two integers i and j, where $0 \le i, j \le 1$, it is defined as follows:

$$i \oplus j = \begin{cases} 1 & \text{if } i = 0 \text{ or } j = 0, \\ 0 & \text{if } i = 1 \text{ and } j = 1. \end{cases}$$

Operation 3 (Logical XNOR operation ($\otimes$)). The operator $\otimes$ defines the logical XNOR operation whose result equals 1 when both operands have the same values. Given two integers i and j, it is defined as follows:

$$i \otimes j = \begin{cases} 1 & \text{if } i = j, \\ 0 & \text{if } i \ne j. \end{cases}$$

Operation 4 (Continuous Product operation (Π)). The operator Π defines the continuous product operation whose elements are multiplied continuously. Given integers

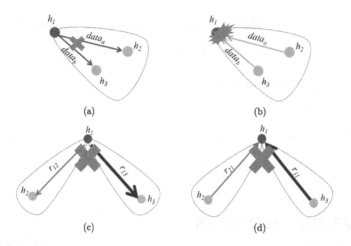

(a)   (b)

(c)   (d)

FIGURE 2.5

Hardware constraints of smart antenna. (a) Data Diversity; (b) Interference Constraint; (c) Transmitting Rate Constraint; (d) Receiving Rate Constraint.

a_p, a_q, and a_r, the operation can be shown as follows:

$$\prod_{i\in\{p,q,r,\dots\}} a_i = a_p \otimes a_q \otimes a_r \cdots .$$

We now introduce the constraints and their mathematical descriptions.

Beam Mode Constraint

At a given time, the operating mode of a host can be either transmitting or receiving. In addition, all beams of a host should be at the same mode, either transmitting or receiving, at any given time. This hardware constraint is referred to as Beam Mode Constraint. That is, we cannot assign a host to send message using one beam and receive message using another beam at a given slot t. Therefore, while scheduling, we should verify the beam mode constraint as follows:

$$|s_{i,j,t}| \odot |s_{k,i,t}| = 0, \ \forall i, j, k, t.$$

Data Diversity Constraint

Each beam of the smart antenna can send only one package at a given time slot. In other words, when h_j and h_k are both located in the same beam of host h_i, then h_i cannot send different packages to h_j and h_k at the same time. This hardware constraint is referred to as Data Diversity Constraint. As shown in Fig. 2.5(a), h_2 and h_3 are located in the same beam of h_1, and h_1 cannot send $data_a$ to h_2 and send $data_b$ to h_3 at the same time. Furthermore, the data diversity constraint can be illustrated by Eq. (2.2). If h_j and h_k are located in the same beam of host h_i, then according to Operation 3, $ne_{i,j} \otimes ne_{i,k}$ equals 1, and communication $c_{i,j}$ and communication $c_{i,k}$ cannot transmit at the same time, that is, according to Operation 2, the value of $|s_{i,j,t}| \odot |s_{i,k,t}|$ should be 0. If h_j and h_k are not located in the same beam of host h_i, then according to Operation 3, $ne_{i,j} \otimes ne_{i,k}$ equals 0, and communication $c_{i,j}$ and communication $c_{i,k}$ can transmit at the same time, that is, because they are not in the same beam, the value of $|s_{i,j,t}| \odot |s_{i,k,t}|$ can be 0 or 1. Therefore, while scheduling, we can verify the data diversity constraint by the equation

$$(|s_{i,j,t}| \odot |s_{i,k,t}|) \oplus (ne_{i,j} \otimes ne_{i,k}) = 1, \ \forall i, j, k, t. \tag{2.2}$$

Interference Constraint

Each beam of the smart antenna can receive one package from one neighboring host at a given time slot, that is, if h_i and h_l are both located in the same beam of host h_j, then h_i and h_l cannot send their own package to h_j at the same time. This hardware constraint is called Interference Constraint. As shown in Fig. 2.5(b), h_2 and h_3 are in the same beam of h_1, and h_1 cannot receive $data_a$ from h_2 and receive $data_b$ from h_3 at the same time. Moreover, the interference constraint can be illustrated by Eq. (2.3). By Property 1, if hosts h_i and h_l are in the same beam of host h_j, then host h_j is in the same beam of hosts h_i and h_l, the value of $ne_{i,j} \otimes ne_{l,j}$ is 1 according to Operation 3, and communication $c_{i,j}$ and communication $c_{l,j}$ cannot transmit at the same time; that is, the result of $|s_{i,j,t}| \odot |s_{l,j,t}|$ should be 0 according to Operation 2. If hosts h_i and h_l are not in the same beam of host h_j, then host h_j is not in the same beam of hosts h_i and h_l, and the value of $ne_{i,j} \otimes ne_{l,j}$ is 0 according to Operation 3. Thus the communication $c_{i,j}$ and communication $c_{l,j}$ can transmit at the same time because they are not in the same beam. Thus the value of $|s_{i,j,t}| \odot |s_{l,j,t}|$ can be 0 or 1 according to Operation 2. Therefore, while scheduling, we can verify the interference constraint by the equation

$$(|s_{i,j,t}| \odot |s_{l,j,t}|) \oplus (ne_{i,j} \otimes ne_{l,j}) = 1, \ \forall i, j, l, t. \tag{2.3}$$

Transmitting Rate Constraint

Each host can transmit messages to its neighboring hosts at a given slot only in one transmitting rate. In other words, if h_x and h_y are the neighboring hosts that are in the different beams of h_i and if the transmission rates of $c_{i,x}$ and $c_{i,y}$, say $r_{i,x}$ and $r_{i,y}$, are different, then h_i cannot transmit data to h_x by rate $r_{i,x}$ and transmit data to h_y by rate $r_{i,y}$ at the same time; it transmits data to h_x and h_y by rate $\min(r_{i,x}, r_{i,y})$ instead. This hardware constraint is called Transmitting Rate Constraint. For example, Fig. 2.5(c) depicts the constraint. Thus, while scheduling, we can verify the transmission rate constraint by Eq. (2.4). In other words, when we schedule $c_{i,p}, c_{i,q}, c_{i,w}$ at time t, i.e., host h_i sends data to h_p, h_q, and h_w, the arranged $r_{i,p}, r_{i,q}, r_{i,w}$ must be the same.

$$\prod_{j \in K} s_{i,j,t} = 1, \ K = \{j| \ \forall \ |s_{i,j,t}| = 1\}, \forall i, t. \tag{2.4}$$

Receiving Rate Constraint

Each host can receive messages from its neighboring hosts at a given slot only in one receiving rate. That is, if h_x and h_y are the neighboring hosts that are in different beams of h_i and if the transmission rates of $c_{x,i}$ and $c_{y,i}$, $r_{x,i}$ and $r_{y,i}$, are different, then h_i cannot receive data from h_x by rate $r_{x,i}$ and receive data from h_y by rate $r_{y,i}$ at the same time. This hardware constraint is called Receiving Rate Constraint. For example, Fig. 2.5(d) depicts the constraint. Thus, while scheduling, we can verify the transmission rate constraint by Eq. (2.5). In other words, when we schedule $c_{p,i}$, $c_{q,i}$, and $c_{w,i}$ at time t, i.e., hosts h_p, h_q, and h_w, send data to host h_i, the arranged $r_{p,i}$, $r_{q,i}$, $r_{w,i}$ must be the same.

$$\prod_{l \in K} s_{l,i,t} = 1, \; K = \{l | \forall |s_{l,i,t}| = 1\}, \forall i, t. \tag{2.5}$$

2.3.3 OBJECTIVES

This research proposes mechanisms for finding efficient routing paths and the maximal transmission throughput. The routing mechanism is used to find the routing when two hosts are in the distance larger than one-hop transmission. Thus the objectives of routing mechanism are to find the shortest path between two hosts and the minimal cost in the efficient paths. The scheduling mechanism is used to arrange transmissions during a cluster. Therefore, the objectives of scheduling are to arrange transmissions that have the maximal transmission parallelism. We further describe the objectives in detail.

2.3.3.1 Routing Objectives

Objective: The shortest paths for unicasting

Let $L_{k,i}$ be a path selected from $P_{k,i}$ (defined in Eq. (2.1)) with its length denoted by $|L_{k,i}|$. In general, the cost of a path depends on the length of the path. For unicasting transmission, the minimal cost of transmissions is the cost of the shortest path of each pair. For multicasting, the cost of transmissions is the summary of selected paths with removed the shared part. If the transmission packet is multicasting and the path has shared with other path except the source host. This work attempts to minimize the cost for the transmissions of all multicast packets. The goal is defined as follows:

$$L_{k,i} = \min\{l_{k,i} | l_{k,i} \in P_{k,i}\}, \; \forall k \text{ in the network.}$$

Objective: The minimal cost for multicasting

Let PKT be the set of transmitted packets in a time period of the network. Then $|PKT|$ and pkt_i denote the numbers of transmitted packets and ith transmitted packet in the set, respectively. In addition, the number of destinations for pkt_i is denoted by $|mc_i|$, and $cost_{i,j}$ represents the transmission cost for pkt_i to its jth destination. The total cost for transmitting the packets, $Cost_{total}$, can be defined by Eq. (2.6), where we aim to minimize the total cost for all packets to their destinations in the network environment.

$$Cost_{total} = \sum_{k=1}^{n} \min(\sum_{m=1}^{n} (\sum_{i=1}^{n} (|L_{k,i}| - Shared(k,m))) + Shared(k,m)). \quad (2.6)$$

2.3.3.2 Scheduling Objectives

Because of data diversity constraint and interference constraint, the schedule arranges the transmissions for the same beam of a host in different time slots. While scheduling, we can verify the constraints by Eqs. (2.2) and (2.3). Moreover, due to transmitting rate constraint and receiving rate constraint, when we arrange two or more transmissions for a host at the same time, the transmission rate should adjust to the same smaller rate. Thus the scheduled time for each communication pair may be longer than it could be. The following represents the relation we just mentioned:

$$\sum_{t=0}^{T_t} s_{i,j,t} = d_{i,j}, \text{ and } \sum_{t=0}^{T_t} |s_{i,j,t}| \geq d_{i,j}/r_{i,j}, \forall i, j \in (1,n). \quad (2.7)$$

Objective: The maximal satisfying transmissions in the minimal time

When all the communications are scheduled at time t, all the values of $s_{i,j,t+k}$ for any positive integer k in Matrix S are 0. Thus, let T_t denote the total number of time slots required for completing all transmissions, which is defined by Eq. (2.8). In addition, let $\omega(s_{i,j})$ and $delay(c_{i,j})$ denote the set of the transmission slots allocated for $c_{i,j}$ and the delay time of $c_{i,j}$, respectively. They can be defined by Eqs. (2.9) and (2.10). Let A denote the set of transmission pairs whose delay constraint is satisfied and which is represented by Eq. (2.11). In this chapter, we attempt to get a larger number of elements in A, consequently resulting in a smaller value of T_t:

$$T_t = \min\{t | \sum_{j=1}^{n} \sum_{i=1}^{n} |s_{i,j,t}| = 0\}, \quad (2.8)$$

$$\omega(s_{i,j}) = \{t \in [0, T_t] | s_{i,j,t} \neq 0\}, \quad (2.9)$$

$$delay(c_{i,j}) = \max\{t \in [0, T_t] | s_{i,j,t} \neq 0\}, \quad (2.10)$$

$$A = \{c_{i,j} | \text{ where } delay(c_{i,j}) < \max(\omega(s_{i,j}))\}. \quad (2.11)$$

The scheduling mechanism attempts to satisfy the maximal number of transmission delay constraints for packages and maximize the network throughput, i.e., minimize the total time required to complete the maximal number of transmissions. Thus the proposed algorithm aims to achieve the following goals:

$$\max(|A|) \text{ and } \min T_t. \quad (2.12)$$

2.4 THE PROPOSED APPROACH

To exploit the smart antenna system efficiently, we propose cluster-based mechanisms for transmission arrangement. The clustering schemes enhance the spatial reuse, which is the same as the advantage of smart antenna system. The well-scheduled clustering schemes reduce the amount of transmissions in the environment and decrease the interference problem. Therefore, we propose mechanisms to determine routing paths and schedule the transmissions according to the clustered smart antenna systems. To perform the transmissions efficiently, we separate the jobs into three steps. The first is to define the clusters and cluster heads in the environment; the second is to determine the routing paths for each transmission pair; and the third is to schedule parallel transmissions. These steps are described in detail in the following subsections.

2.4.1 DEFINING THE CLUSTERS IN THE CONCERNED ENVIRONMENT

The proposed routing and scheduling mechanisms are cluster-based wireless sensor networks. The cluster-based scheme may improve the spatial reuse of communications, reduce the amount of time of transmissions, decrease the communication interference, and save the consumption of transmission energy. Using cluster-based scheme, the cluster heads are selected for managing communications for intracluster and intercluster. Thereby, many studies have proposed different definitions of clusters and heads. We are not going to define any new mechanism for clustering and selecting heads, only introducing some mechanisms to achieve these goals.

The studies that define clusters can be classified into two categories: one is forming clusters before choosing cluster head (Xing et al., 2011; Muthuramalingam et al., 2010; Lai et al., 2009; Yua et al., 2012; Kour and Sharma, 2010; Wang et al., 2013; Sasikumar and Anitha, 2014); the other is selecting cluster heads and then forming the clusters (Zhang et al., 2005; Deshpande and Patil, 2013). While forming clusters before choosing cluster head, the hosts may form their clusters according to their locations, communication ranges, mobility, and environment situations (Hussein et al., 2010; Kour and Sharma, 2010; Wang et al., 2013; Sasikumar and Anitha, 2014). The hosts trend to cluster with closed neighbors because the communications for longer distance consume higher communication cost. Thus the locations of hosts may determine their clusters. In addition, the hosts in the same cluster are able to communicate with each other. Thus the range of a cluster is subject to the communication range of hosts. Moreover, the hosts have similar mobility, or situation may cluster together because their corresponding locations may last a longer time period. After clustering, the heads may be selected according to the remaining energy, connectivity with neighbors, the cost for communications, or the stabilities. The remaining energy of heads is concerned because the cluster heads have to manage and negotiate the intracluster and intercluster communications. The host having higher connectivity and lower communication cost with neighbors represents that it spends less energy consumption for communications with neighbors

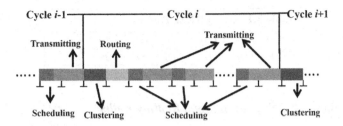

FIGURE 2.6

The cycles in proposed approach.

in the cluster. Thus the connectivity with neighbors and the cost for communicating with neighbors are concerned. By contrast, if selecting cluster heads before forming the clusters, then the hosts having much powerful are deployed or selected to be the heads or candidate heads in the environment (Zhang et al., 2005; Deshpande and Patil, 2013). In addition, if the neighbors are notified by the neighboring head(s), then the neighbors join to an appropriate cluster by notifying the head of the cluster. When the number of heads is larger than that required by the environment, all the heads are candidate heads, and they take turns to be the real heads.

To satisfy the real time situation and prolong the lifetime of wireless sensor networks, the clusters and heads should be reformed and reselected after some time period because the remaining energy and the locations of sensors in the networks may vary.

2.4.2 DETERMINING THE ROUTING PATHS FOR EACH TRANSMISSION PAIR

In this section, we propose a cross-layer routing algorithm based on cluster-based network topology for smart antenna systems. Both the superiorities and constraints of smart antenna are taken into account. In addition, we also consider the delay constraints of packages. Each cluster has one Head and two or more members. When a member belongs to more than one cluster, it is named a Gateway. A cluster head can communicate with the members in one-hop transmission. A gateway can communicate with two or more cluster heads to negotiate one another. A routing path usually crosses two or more clusters. The Heads know the locations of members and the rate of transmission for each transmission pair in their cluster. Based on the frame defined in previous section, the time is divided into many frames, and each frame is divided into many time slots. The routing paths are built in a cycle after clustering, as shown in Fig. 2.6. The proposed algorithm decides the routing path in three phases. The first phase is building the delay-guaranteed routing paths. The second phase is finding shared paths for minimizing cost. The third is adjusting the routing paths according to the transmissions in each cluster. We further describe these phases.

Table 2.1 RREQ/RREP Packets

RREQ	ID_S	ID_P	ID_D	SL	TL	DC
RREP	ID_S	ID_P	ID_D	SL	dt	DU

2.4.2.1 Building Delay-Guaranteed Routing Paths

In this phase, we search all the routing paths that can guarantee the delay constraint of transmissions. Using a smart antenna, cluster heads, and members can take turns to transmit messages. If not transmitting, the host can receive messages. Thus, when Heads transmit packets at $2k$th time slots, where k is a positive integer, the Gateways and members receive at $2k$th time slots; in the other words, the Gateways and members transmit packets and Heads receive packets at $(2k + 1)$th time slots, and vice versa.

For ease of presentation, the packets for sources to search destinations are named Route Request packets (RREQ). By contrast, the packets for destinations to reply sources are named Route Reply packets (RREP). The RREQ consists of fields ID_S, ID_P, ID_D, SL, TL, and DC, as shown in Table 2.1(a). The ID_S, ID_D, and ID_P represent the identifiers of source host, destination hosts, and the packet, respectively. The sequence list SL records the list of hosts that relay the packet from source; the time list TL records the list of time stamps while arriving at each relay host. The DC denotes the delay constraint of the packet from source to destination. On the other hand, the RREP consists of ID_S, ID_P, ID_D, SL, and DU, as shown in Table 2.1(b). The ID_S and ID_P are copied from the corresponding RREQ packet. The ID_D is the identifier of returning destination. The SL denotes the list of hosts that relay the packet from source to destination; the packet will be returned by reversing the list; the dt denotes the time delay during two relay hosts, and DU represents the delay time from Source to Destination. The values of dt and DU are described in detail in later subsections.

The routing path is built when destination host and relaying hosts transmit RREP packets back to the source. When transmitting the RREQ and RREP packets, the related hosts include the sources, destinations, Heads, and Gateways. That is, if a host is neither a Head nor a Gateway, then it participates in the building phases only when it is the source or the destination. The details of building the routing paths with guaranteed delay are described in the following subsections.

A. Creating and Transmitting RREQ Packets

When a host is going to build a routing path, it creates an RREQ packet. While creating an RREQ packet, the source host fills the ID_S, ID_P, ID_D and appends the current time stamp to the TL field of the packet. In addition, the time constraint is calculated and filled in the DC field. When finding multiple destinations, the ID_D is a list having all the identifiers of destinations. After creating and filling information, the source transmits the RREQ packet to its neighbors. If the source host is a member, then the Head of the cluster must receive the packet. Otherwise, a gateway of the

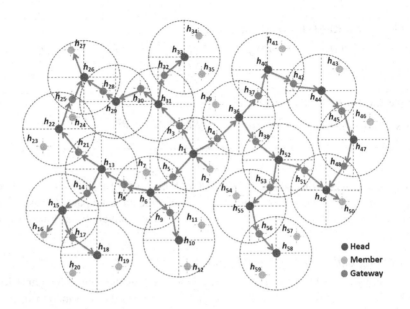

FIGURE 2.7

An example of transmitting RREQ spreading.

same cluster must receive the packet. On receiving an RREQ packet, each host first verifies the ID_D and DC fields. When the current time is later than the DC, the host omits the packet since the packet is out of date. In addition, when the SL of the packet is a superset of that in other existing RREQ, the host omits the packet since the packet is redundant. When the ID_D is the host itself, it replies a corresponding RREP packet when the delay time is earlier than the DC. The replying process is described in detail in the next subsection. Otherwise, when the host is neither the source nor the destination, the host appends its identifier to SL and the current time stamp to TL. After appending to SL and TL, the host stores the RREQ message and relays the packet to its other neighbors. When the relay host is a Gateway, it transmits the packet to Heads of other clusters. When the relay host is a Head, it transmits the packet to Gateways or the destination host in the cluster. All relaying hosts store the RREQ records and periodically eliminate the existing RREQ records when the delay time is more than double of the DC in that packet, representing that the route through this host is not a path appropriate for the source host to destination host.

For example, Fig. 2.7 shows that h_2 is a source host, and h_2 creates an RREQ packet as in Table 2.2. The first broadcasting h_2 transmits the packet to h_1, and then h_1 verifies the DC and modifies the SL and TC of the packet. After that, h_i transmits the packet again. Due to using smart antennas, hosts can transmit only by some needed beams, and the heads and members must transmit packets by turns in our mechanism. Therefore, h_1 transmits the RREQ packet to h_3, h_4, and h_5. After veri-

Table 2.2 The RREQ Packet Created by h_2

ID_S	ID_P	ID_D	SL	TL	DC
h_2	1	h_{16}, h_{27}, h_{50}	$NULL$	0	10

Table 2.3 The RREP Packet Created by h_{50}

ID_S	ID_P	ID_D	SL	dt	DU
h_2	1	h_{50}	$h_2, h_1, \ldots$	1	7

fying and modifying the packet, h_3, h_4, and h_5 transmit the RREQ packets to Heads h_{31}, h_{36}, h_6, and so on. The transmission will continue until the packet arrives at its destination or the time is later than DC. In Fig. 2.7, the destinations are h_{16}, h_{27}, and h_{50}.

B. Creating and Replying RREP Packets

On receiving an RREQ packet, a destination host replies the source with an RREP packet. While creating the RREP packet, the host copies the information in fields of ID_S and ID_P, appends the destination to the SL, and fills the time duration of transmission from source to the destination into field DU. The identifier of the host itself is filled in the field of ID_D. The difference between the last time stamp in RREQ, and the time of the RREQ received by the destination is filled into the dt field. After creating and filling information, the destination host transmits the RREP packet to the neighbor in the last of SL. If the destination host is a member, then the Head of the cluster must receive the packet. Otherwise, the gateways of the same cluster will receive the packet. For example, Table 2.3 shows the RREP packet created by h_{50} in Fig. 2.7.

When a host receives an RREP packet, it may relay the RREP back if it relayed the corresponding RREQ before. In addition, it stores the routing path into its routing table after it relays the RREP. For example, when host h_j receives an RREP packet from its neighbor h_k, it verifies its RREQ records; when the ID_S and ID_P are the same as an RREQ in the records, it relays the RREP to the previous host of it in the SL, say h_i, which can be found either in the RREP packet or in the corresponding RREQ packet. Before relaying the RREP, h_j gets the original value in dt field, say $t_{j,k}$, and modifies the value in dt field of RREP to the time difference between the time stamps of h_i and h_j stored in the RREQ packet. In addition, the original value $t_{j,k}$ is used to calculate the delay constraint, say const, from h_j to h_k using Eq. (2.13). After relaying the RREP, host h_j stores the path from the source host, in ID_S field, to the destination host, in ID_D field, into the routing table and the previous neighbor h_i and the next neighbor h_k in Last field and Next field, respectively. Moreover, the calculated const is stored in the dc field. After relaying the RREP packet and storing the routing path, the host h_j eliminates the corresponding RREQ record:

$$const = DC \times t_{i,j} \div DU. \tag{2.13}$$

Table 2.4 The Routing Table after h_{51} Receives RREP					
ID_S	ID_P	ID_D	$Last$	$Next$	dc
h_2	1	h_{50}	h_{52}	h_{49}	1

In the example from Fig. 2.7, the source host h_2 sends an RREQ packet as listed in Table 2.2, and one of the destination host h_{50} sends an RREP packet as shown in Table 2.3. When host h_{51} receives an RREP packet from host h_{49}, a Routing Record in the Routing Table may be constructed as Table 2.4.

C. Routing Table in each Host

The information in Routing table is used for the host to verify when it helps transferring a packet to the destination of the packet. While receiving a packet, the host verifies the identifier of the destination host. When a record in the routing table has the same destination, the host gets the identifier of next host and the delay constraint in the $Next$ field and dc field of the routing path, respectively. Thus the host knows the receiver of this packet and the constraint of delay transferring. To guarantee the delay constraint of a packet, the delay time of each transmission should be limited in a certificated duration, namely, delay constraint of the transmission. As we have mentioned, a host can have the total delay time and the delay constraint of the transmission from the source to the destination in DU field of RREP and DC field of RREQ, respectively. The delay time of the host to its next neighbor is obtained according to the dt field of RREP. Therefore we calculate the delay constraint if a host receives an RREP and is the source of the RREP packet. It also records the routing Map of the routing path in SL field. After the double of the packets' DC, the source host reorganizes all the routing paths that are illustrated in the next section.

2.4.2.2 Finding Shared Paths for Multicasting

After receiving all the RREP packets from different routing paths, the source host should choose a proper path for its transmission. When it is a unicast transmission, the source decides the routing by choosing the route with the shortest transmission time. When it is a multicast transmission, not only the transmission time but also the shared path should be taken into account. To reduce the power consumption for transmitting, we may take account of shared path first if the transmission delay is less than the delay constraint. The source maintains a Routing Map by all RREP packets. When there are more than one routing paths to a destination, we choose the path with more benefit, which may share a part of the path with others. We define $p_{i,j}$ as the jth path connecting the current host to host h_i. In addition, $\delta(p_{i,j})$ represents the number of hops through the path $p_{i,j}$. When two paths, for example, $p_{i,j}$ and $p_{k,h}$, from a source to different destinations share a part of routes, the part of shared path is denoted by $Shared(p_{i,j}, p_{k,h})$. For example, when the paths to destination h_i and h_j started from h_a are different from the next host of h_k, the shared path is from h_a to h_k. We also define the length of a path $p_{i,j}$ as $|p_{i,j}|$, which is the number of hops

Table 2.5 A Routing Map of h_2

Step	1	2	3	4	5	6	7	8	9	10	11	12	13
$Path_{50,1}$	h_2	h_1	h_4	h_{36}	h_{38}	h_{52}	h_{51}	h_{49}	h_{50}				
$Path_{27,1}$	h_2	h_1	h_3	h_{31}	h_{30}	h_{29}	h_{28}	h_{26}	h_{27}				
$Path_{16,1}$	h_2	h_1	h_5	h_6	h_8	h_{13}	h_{14}	h_{15}	h_{16}				
$Path_{27,2}$	h_2	h_1	h_5	h_6	h_8	h_{13}	h_{21}	h_{22}	h_{25}	h_{26}	h_{27}		
$Path_{50,2}$	h_2	h_1	h_4	h_{36}	h_{37}	h_{40}	h_{42}	h_{44}	h_{45}	h_{47}	h_{48}	h_{49}	h_{50}

on the path $p_{i,j}$ including host i and j. Thus every two paths from the same source to different destinations must satisfy the following rule:

$$\min(|p_{i,j}|, |p_{k,h}|) \geq |Shared(p_{i,j}, p_{k,h})| \geq 0.$$

For each destination, the source host finds the shortest path to the destination and the maximal shared paths with other destinations. Suppose that the maximal cost for each transmission is γ; then the cost for path $p_{i,j}$ is $\gamma \times \delta(p_{i,j})$. The cost for a packet to all destinations is obtained by Eqs. (2.14) and (2.15), and this paper aims to find the minimal cost for all transmissions in the networks.

$$Cost_i = \gamma \times \min(\sum_{j=1}^{|des|}(\delta(p_{i,j}) - \sum_{k=j}^{|des|}\delta(Shared(p_{i,j}, p_{i,k}))), \tag{2.14}$$

$$Cost_{net} = \sum_{i=1}^{n} Cost_i. \tag{2.15}$$

Taking Fig. 2.7 and Table 2.5 as an example, source h_1 receives two RREPs created by h_{27} and one RREP created by h_{16}. Finding the longest shared path of $Path_{27,1}$ and $Path_{16,1}$ and the longest shared path of $Path_{27,2}$ and $Path_{16,1}$, we have that $Path_{27,1}$ and $Path_{16,1}$ share the path from h_2 to h_{13}, and $Path_{27,2}$ and $Path_{16,1}$ share from h_2 to h_1. After finding shared path, we calculate the cost of each path to its destinations and choose a path with minimal cost for each destination. The cost of $Path_{27,1}$, $Path_{27,2}$, and $Path_{16,1}$ are 8, 10, and 8, respectively. However, the total cost of routing to h_{27} and h_{16} through $Path_{27,1}$ and $Path_{27,2}$ are $15(= 8 + 8 - 1)$ and $13(= 8 + 10 - 5)$, respectively, since $Path_{27,1}$ shares link 1 with $Path_{16,1}$ and $Path_{27,2}$ shares link with $Path_{16,1}$. Although the cost of $Path_{27,1}$, i.e., 8, is less than $Path_{27,2}$, i.e., 10, the sharing of the $Path_{27,2}$ and $Path_{16,1}$ reduces the total cost of these two destinations. Hence we choose $Path_{27,2}$ and $Path_{16,1}$ for h_{27} and h_{16}, respectively.

2.4.2.3 Adjusting the Routing Path

In this section, we try to shorten the routing time by finding shorter transmissions in clusters. Heretofore the routing paths are built by Heads and Gateways excluding Source and Destination. However, the transmission time differs from the transmission rate and times. Therefore we try to adjust the routing path by considering other

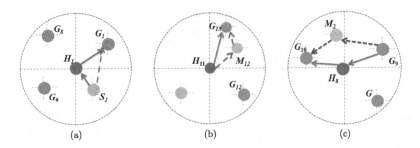

FIGURE 2.8

Examples of cases. (a) An example of Case1; (b) An example of Case2; (c) An example of Case3.

paths in the same cluster. Let $d_{i,j}$ represent the distance between host h_i and host h_j, and $r_{i,j}$ represent the transmission rate for h_i sending to h_j. In any path the built routing in a cluster may be one from the following three cases: a Member to a Gateway through the Head (denoted using $M - H$ and $H - G$), a Gateway to a Member through the Head (denoted using $G - H$ and $H - M$), or a Gateway to other Gateway through the Head (denoted using $G_i - H$ and $H - G_j$). Since having the location and transmission rate for all members, the Heads can adjust the paths by considering Eqs. (2.16)–(2.18), where H, M, and G represent Head, Member, and Gateway, respectively. In the first case, $M - H$ and $H - G$, a Member transmits to the Head, and the Head transmits to a Gateway, when it satisfies Eq. (2.16), the transmission time of $M - G$ is less than the transmission time of $M - H$ plus $H - G$; thus the transmission $M - G$ can replace transmissions $M - H$ and $H - G$. In the second case, $G - H$ and $H - M$, a Gateway transmits to the Head, and the Head transmits to a Member; when it satisfies Eq. (2.17), the transmission time of $G - M$ is less than the transmission time of $G - H$ plus $H - M$; thus the transmission $G - M$ can substitute for transmissions of $G - H$ and $H - M$. In the third case, $G_i - H$ and $H - G_j$, a Gateway G_i transmits to the Head, and the Head transmits to other Gateway G_j; when it satisfies Eq. (2.18), the transmission time of $G_i - G_j$ is less than the transmission time of $G_i - H$ plus $H - G_j$; thus the transmissions of $G_i - H$ and $H - G_j$ can be replaced by transmission $G_i - G_j$.

$$\frac{d_{MG}}{r_{MG}} < \frac{d_{MH}}{r_{MH}} + \frac{d_{HG}}{r_{HG}}, \tag{2.16}$$

$$\frac{d_{GM}}{r_{GM}} < \frac{d_{GH}}{r_{GH}} + \frac{d_{HM}}{r_{HM}}, \tag{2.17}$$

$$\frac{d_{G_iG_j}}{r_{G_iG_j}} < \frac{d_{G_iH}}{r_{G_iH}} + \frac{d_{HG_j}}{r_{HG_j}}. \tag{2.18}$$

For example, Fig. 2.8(a) shows an example of Case1, and it satisfies Eq. (2.16), which means that the transmission time of $S_1 \rightarrow H_1 \rightarrow G_1$ is longer than the transmission time of $S_1 \rightarrow G_1$; therefore the transmission of $S_1 \rightarrow H_1 \rightarrow G_1$ is replaced

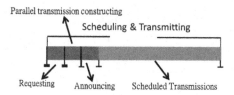

FIGURE 2.9

The periods of scheduling and transmitting subframes.

by the transmission of $S_1 \rightarrow G_1$ for rapid transmission. Fig. 2.8(b) shows an example of Case2 and satisfies Eq. (2.17), which means that the transmission time of $G_{13} \rightarrow H_{11} \rightarrow M_{12}$ is longer than the transmission time of $G_{13} \rightarrow M_{12}$; therefore the transmission of $G_{13} \rightarrow H_{11} \rightarrow M_{12}$ is replaced by the transmission of $G_{13} \rightarrow M_{12}$. Fig. 2.8(c) shows an example of Case3 and satisfies Eq. (2.18), which means that the transmission time of $G_9 \rightarrow H_8 \rightarrow G_{10}$ is longer than the transmission time of $G_9 \rightarrow G_{10}$; therefore the transmission of $G_9 \rightarrow H_8 \rightarrow G_{10}$ is replaced by the transmission of $G_9 \rightarrow G_{10}$.

2.4.3 SCHEDULE PARALLEL TRANSMISSIONS PAIRS

After defining the clusters and determining routing paths, when there are transmission requests, the transmissions are scheduled in order to exploit the smart antenna system effectively. In this section, we propose a scheduling algorithm based on the cluster-based network topology for a smart antenna system. As we mentioned in the last section, because each Head knows the locations of members and the transmission rate of each transmission pair, the transmission scheduling is arranged by the Heads. Therefore, the Heads collect the transmission requirements of members and arrange the transmission sequence for them.

The proposed scheduling algorithm partitions the period of each transmission frame into four phases, shown in Fig. 2.6, including those dealing with request collection, parallel transmission construction, scheduling and announcement, and data transmission, shown in Fig. 2.9. The scheduling algorithm consists of the former three-four phases. The details of the algorithm are described in the following subsections. For ease of presentation, the hosts in our examples are equipped with 4-beam smart antennas, unless otherwise stated.

2.4.3.1 Collecting Transmission Requests

In request collection phase, Heads first announce the time slots for request transmission to their members according to the locations (or beams) of these members around the Head. The member located in different directions of Headers' beams can transmit their requests at the same time. In other words, when two or more members are in the same beam of head, their requests should transmit in different time slots because

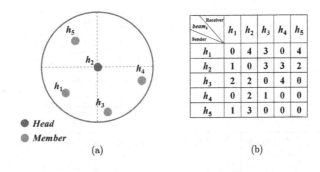

Sender \ Receiver beam	h_1	h_2	h_3	h_4	h_5
h_1	0	4	3	0	4
h_2	1	0	3	3	2
h_3	2	2	0	4	0
h_4	0	2	1	0	0
h_5	1	3	0	0	0

● Head
● Member

(a)   (b)

FIGURE 2.10

(a) A cluster in the network environment with head h_2 and (b) the Matrix NE of h_2.

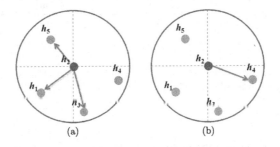

(a)   (b)

FIGURE 2.11

The transmission in time slots: (a) slot i and (b) slot $i + 1$.

of interference problem. The Heads arrange the requesting time by the Neighboring Matrixes NEs. The value in each field of NE is the beam number of the neighbor located. Thus the neighbors located in the same beam number are not arranged at the same time to transmit requests. After that, members transmit their requests to their Heads in their allowed time slots. Each request is transmitted among the hosts in the cluster, either head or member. For example, Fig. 2.10(b) is the neighboring matrix NE of the network depicted in Fig. 2.10(a). Since the hosts h_4 and h_3 are in the same beam of h_2, h_4, and h_3 cannot transmit their request in the same slot. Hence, h_2 may announce that h_1, h_3 and h_5 can transmit their requirement in time slot i, and h_4 can transmit its requirement in time slot $i + 1$ as shown in Fig. 2.11.

After announcing request time, the Heads collect the requirements of their members and store them in Matrixes D and TC. Each requirement includes the volume and delay constraint of the requested transmission. Matrix D stores the requesting volumes of each transmission requirement, and Matrix TC stores the delay constraints of each transmission requirement. As mentioned in the last section, the delay constraint is stored in the routing table built while determining the routing path. Taken Fig. 2.10 as an example, Fig. 2.12(a) represents the demand transmission pairs in

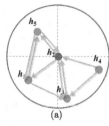

(a)

Receiver Sender	h_1	h_2	h_3	h_4	h_5
h_1	0	0	12	0	12
h_2	12	0	12	0	24
h_3	0	12	0	0	0
h_4	0	12	12	0	0
h_5	12	24	0	0	0

(b)

Receiver Sender	h_1	h_2	h_3	h_4	h_5
h_1	0	0	60	0	36
h_2	12	0	24	0	48
h_3	0	24	0	0	0
h_4	0	48	12	0	0
h_5	12	48	0	0	0

(c)

Receiver Sender	h_1	h_2	h_3	h_4	h_5
h_1	0	0	5	0	12
h_2	11	0	13	0	9
h_3	0	7	0	0	0
h_4	0	8	14	0	0
h_5	6	10	0	0	0

(d)

FIGURE 2.12

Transmissions and matrixes in a cluster. (a) Transmission demand pairs in the cluster; (b) The Transmission Rate Matrix R; (c) The Transmission Demand Matrix D; (d) The Delay Constraint Matrix TC.

the cluster. After collecting the requirement, the Head h_1 saves the request volume of each transmission pair, say $c_{i,j}$, in $d_{i,j}$. Fig. 2.12(c) is the produced Matrix D. Moreover, h_1 saves the transmission delay constraint of each transmission pair $c_{i,j}$ in $t_{i,j}$. Fig. 2.12(d) is the produced Matrix TC. Note that the Matrix R, shown in Fig. 2.12(b), stores the transmission rate of each transmission pair.

2.4.3.2 Constructing Parallel Transmissions

To improve the parallel transmission by multiple beams of each host, this phase constructs parallel transmission groups using both sender-oriented and receiver-oriented procedures. These groups are the basis of our scheduling algorithm. The sender-oriented parallel transmission procedure (SOP) groups the transmissions that can be arranged at the same time period by the view of senders. By contrast, the receiver-oriented parallel transmission procedure (ROP) groups the transmissions that can be arranged at the same time period by the view of receivers. The following describes the steps of SOP.

Step 1. Group the communication pairs that can transmit in the same time period by each host. Thus we have groups $gs_1, gs_2, \ldots, gs_i$, where n, n is the number of hosts in the cluster. The elements in each group are the pairs of communicating hosts. For example, $gs_1 = \{c_{1,2}, c_{1,3}, c_{1,5}\}$ represents the transmission requirements including transmissions from host h_1 to hosts h_2, h_3, and h_5.

Step 2. Sort the groups by the urgencies of groups in decreasing sequence. The urgency of a group is determined by the most urgent communication in the group. The most urgent communication is the transmission having the least delay constraint time. Thus the group having the most urgent communication is the first group after sorting.

Step 3. Verify the transmission requests of latter groups. If any communication can transmit with the former group in a parallel manner without violating the minimal delay constraint and the characteristics of smart antenna, then the transmission request is moved to the former group.

The procedure of ROP is similar to that of SOP; the only difference is that sender viewpoint is changed into receiver viewpoint in the first step. For example, the group gr_1 is defined as $gr_1 = \{c_{2,1}, c_{3,1}, c_{6,1}\}$, which represents the transmission requirements including transmissions from hosts h_2, h_3, h_6 to host h_1.

Taken Fig. 2.12(a) as an example, the Step 1 finds five parallel transmission groups, where $gs_1 = \{c_{1,3}, c_{1,5}\}, gs_2 = \{c_{2,1}, c_{2,3}, c_{2,5}\}, gs_3 = \{c_{3,2}\}, gs_4 = \{c_{4,2}, c_{4,3}\}, gs_5 = \{c_{5,1}, c_{5,2}\}$. Step 2 arranges the group in the order of gs_1, gs_5, gs_3, gs_4, and gs_2. In Step 3, we found that transmission requests $c_{4,2}$ and $c_{4,3}$ can communicate with gs_1 at the same time, that is, they are moved to gs_1. Thus the procedure finds three parallel transmission groups $gs_1 = \{c_{1,3}, c_{1,5}, c_{4,2}, c_{4,3}\}, gs_5 = \{c_{5,1}, c_{5,2}, c_{3,2}\}$, and $gs_2 = \{c_{2,1}, c_{2,3}, c_{2,5}\}$.

The Receiver-oriented Parallel Transmission Procedure (ROP) is the same as SOP but by receiver view. Taken the same example, the result groups of ROP are ordered as $gr_1 = \{c_{2,1}, c_{5,1}\}, gr_2 = \{c_{3,2}, c_{4,2}, c_{5,2}\}, gr_3 = \{c_{1,3}, c_{2,3}, c_{4,3}\}, gr_5 = \{c_{1,5}, c_{2,5}\}$. After parallel transmission construction phase, the cluster Head schedules the transmission pairs based on the result of SOP and ROP, Matrix R, and Matrix TC. The details are described in the next subsection.

2.4.3.3 Scheduling According to Receivers and Senders

In this phase, we schedule the transmissions according to their transmission delay constrains, transmission rates, and transmission volume. To avoid against the constraints of smart antenna communication, we may adjust the transmission rate of some communication pairs to increase the parallelism and throughput. The procedure is based on the groups of SOP and ROP. The steps are shown as follows.

Step 1. Select the transmission requirement with the smallest transmission delay constraint. The information is stored in the Delay Constraint Matrix TC.

Step 2. Find the group including the selected communication in the result of SOP, usually in the first group, and arrange the group into schedule. The arranged time is the same as the selected communication need. When the arranged communications have the same senders or receivers, the transmission rate should coordinate with the less one because of the characteristics of smart antennas. If the coordination is against the delay constraint of selected communication, then the transmission of selected communication do not coordinate with other communications.

Step 3. Modify the information in Matrix TC and Matrix D. When the transmission requirement of a communication has been satisfied, the related information in Matrixes TC and D is removed. When some data have been transmitted but not

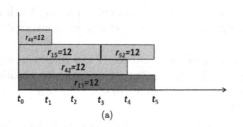

Receiver \ Sender	h_1	h_2	h_3	h_4	h_5
h_1	0	0	0	0	0
h_2	12	0	24	0	48
h_3	0	24	0	0	0
h_4	0	0	0	0	0
h_5	12	24	0	0	0

(a)   (b)

FIGURE 2.13

The schedule of former 5 time slots. (a) The arranged transmission; (b) the changed Transmission Demand Matrix D.

satisfied the volume of the requirement, the remaining volume in the Matrix D is modified.

Step 4.   Modify the groups according to Matrix D. If there is no volume in D, then the transmission request is removed from its group. Moreover, the modified groups should be adjusted by Step 2 and Step 3 of SOP again.

Step 5.   Repeat from Step 1 to Step 4 until there is no volume in Matrix D; that is, when there is no transmission request in Matrix D, the scheduling based on the group of SOP is finished.

Step 6.   Use ROP instead of SOP and repeat from Step 1 to Step 5. Then we have another schedule based on the groups of ROP.

Step 7.   Compare the total time of result in Step 5 with that in Step 6; then choose the smaller one to be our scheduling result.

For ease of description, we continue the example used in the last subsection. The matrixes and network environment are shown in Fig. 2.12. The process is as follows.

1. From Matrix TC we have that $c_{1,3}$ is the pair with the smallest transmission delay constraint. Hence, $c_{1,3}$ is selected.

2. Since the groups of SOP are ordered as $gs_1 = \{c_{1,3}, c_{1,5}, c_{4,2}, c_{4,3}\}$, $gs_5 = \{c_{5,1}, c_{5,2}, c_{3,2}\}$, and $gs_2 = \{c_{2,1}, c_{2,3}, c_{2,5}\}$, we select gs_1 that contains $c_{1,3}$ and arrange the group into the schedule. Note that the time of the group in schedule depends on the transmission time of $c_{1,3}$. Since $c_{1,3}$ and $c_{1,5}$ are transmitted at the same time, from Matrix R we let the transmission rate of h_1 be 12 MB/s. In addition, from Matrix D we know that the transmission time for $c_{1,3}$ is 5 time slots. Thus the first 5 time slots are arranged for the gs_1. Note that when $c_{1,5}$ finishes transmission, the transmission $c_{5,2}$ can be arranged in group gs_1 because $c_{5,2}$ does not collide with the other transmissions in gs_1. However, because the transmissions $c_{1,3}$ only needs 5 time slots, the transmission $c_{5,2}$ can only be arranged from the third slot to the fifth time slot.

3. Since $c_{1,3}, c_{1,5}, c_{4,2}$, and $c_{4,3}$ have been transmitted during the former 5 time slots, modify the requirement information in Matrix D and TC. Fig. 2.13 shows the schedule and matrixes at present.

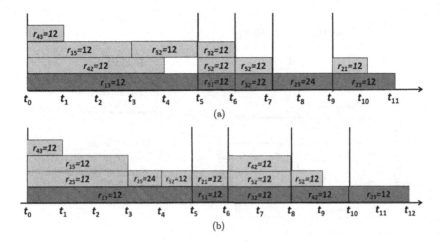

FIGURE 2.14

The scheduling results. (a) Schedule from SOP; (b) Schedule from ROP.

4. Since now $d_{1,3}, d_{1,5}, d_{4,2},$ and $d_{4,3}$ of D are 0, remove these transmission requests from the groups. After adjusting these groups by SOP, the groups are ordered as $gs_5 = \{c_{5,1}, c_{5,2}, c_{3,2}\}$ and $gs_2 = \{c_{2,1}, c_{2,3}\}$.

5. Repeating the above procedures, we have the schedule shown in Fig. 2.14(a), and the total time for these transmissions is 11 time slots.

6. Then we make another schedule by the result of ROP. The schedule is shown in Fig. 2.14(b), and the total time for these transmissions is 12 time slots.

7. Under the maximal satisfying transmissions, because the schedule from SOP is in the length less than that from ROP, we select the result from SOP to be our schedule.

2.5 IMPLEMENTATION

For evaluations, we may implement some modules in the simulations environment as shown in Fig. 2.15. The modules include Smart Antenna Module, Evaluators, and Smart Antenna Host. Those functions of the modules are listed as follows.

1. Smart Antenna Module: A module constructs smart antenna hosts for different situations.

2. Smart Antenna Host: A smart antenna host includes Antenna Module, Cluster Module, QoS Module, Routing Module, Scheduling Module, and Transmission Module.

   a. The Antenna Module is used for setting and referred the status of antenna.

   b. The Cluster Module is used to build the cluster and arrange the heads and gateways of each cluster.

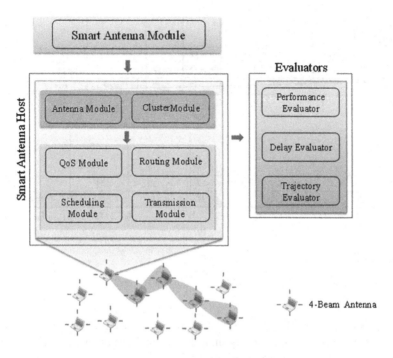

FIGURE 2.15

The architecture of simulation environment.

   c. The QoS Module is used to vary the delay constraints of each transmission.

   d. The Routing Module is designed according to routing mechanism used to find the routes for each communication pair.

   e. The Scheduling Module is designed according to scheduling mechanism to arrange the transmissions in the networks.

   f. The Transmission Module is designed to simulate the transmissions according to the Scheduling Module, Routing Schedule, Cluster Module, and QoS Module.

3. Evaluator:

   a. The Performance Evaluator records the running output to calculate the performance and throughput of mechanisms.

   b. Delay Evaluator records the time delay of packets to calculate and show the delay time of packets and their delay constraints.

   c. Trajectory Evaluator shows the routing situations of packets to verify if the routing path satisfies the designs.

The evaluation environment is built by means of implementing the designed modules. The variety of evaluations is built by inputting different parameters in Evaluators.

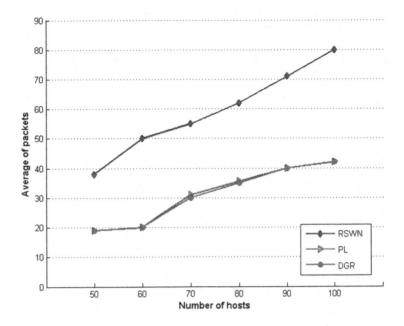

FIGURE 2.16

The number of packets for building routing path.

2.6 EVALUATION

In this section, we evaluate the performance of the proposed mechanisms. The simulations are built using MATLAB 2014a. To focus the efficiency of each mechanism, we compare the related routing mechanisms and scheduling mechanisms. For our simulation, a number of hosts are randomly distributed in a 500×500 meters square area. Each host is equipped with 8-beam smart antenna with 20-m communication range. Moreover, the transmission delay constraint of each package is $(512\beta/2\alpha) + \gamma$, where α, β are positive constants, and γ is a random variable. In Section 2.6.1, we investigate the performance of the proposed mechanism DGR with existing RSWN (Jawhar et al., 2010) and PL (Liu and Lin, 2005). In Section 2.6.2, we investigate the performance of the proposed JRS with existing ESR (Chang et al., 2013) and HTS (Wong and Jia, 2013).

2.6.1 ROUTING EVALUATIONS

First, we evaluate the number of packets in the networks for building routes according to these mechanisms. The PL is a cluster-based algorithm, and the transmissions in each cluster are scheduled. But it does not arrange the transmissions among clusters. The RSWN finds the routing by flooding the searching packets around the network. Fig. 2.16 shows the comparison of the amount of transmitted packets between the

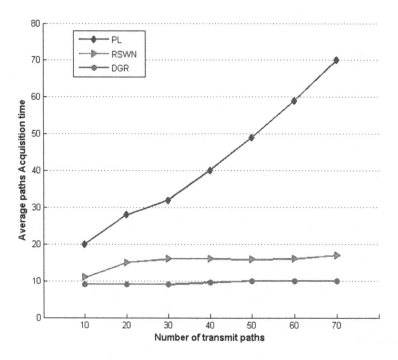

FIGURE 2.17

The comparison of finding satisfied path.

proposed algorithm and others. The RSWN chooses relay hosts by flooding, and thus all the neighboring hosts relay the request packet. While using cluster-based mechanism, only some hosts in the neighboring relay the request packet. Therefore, the number of packets for building path by RSWN is much larger than those by PL and DGR.

Fig. 2.17 shows the comparison of the searching period of the proposed algorithm with RSWN and PL. Because the transmissions among different clusters are complicated, when the number of hosts or transmitting paths is greater, the collisions for PL are more serious, and the time for finding routes is much longer. To avoid packets collision, the RSWN arranges the time slots of neighbors for transmission. Thus the average time for finding paths is less than that for PL. Using DGR, only Heads and Gateways help to relay the packets, and the time slots for Heads and Gateways are alternate. Therefore, the time for DGR is much less than those for RSWN and PL.

Fig. 2.18 shows the comparison of power consumption. The DGR reduces the number of relaying hosts, and the routes for which the transmission delays are longer than the delay constraint time will be omitted. Therefore, the power consumption of DGR is much less than those of PL and RSWN.

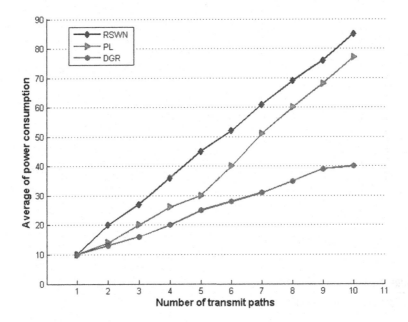

FIGURE 2.18

The comparison of power consumption.

2.6.2 SCHEDULING EVALUATIONS

First, we evaluate the throughput for mechanisms. Fig. 2.19 shows the results of comparisons. In general, when the numbers of hosts are less than 250, each cluster has the number of members less than 10, and the network throughputs of these mechanisms are increasing when the numbers of hosts are increasing. However, since the parallelism is bounded by the number of beams, the network throughput will not increase when the numbers of hosts are greater than 250. In Fig. 2.19, the proposed JRS and ESR outperform HTS because of taking parallel transmission into account. Moreover, due to considering the parallelism from sender and receiver viewpoints, the proposed JRS is better than ESR.

Second, we compare the number of transmissions satisfying their delay constraint, as shown in Fig. 2.20. Because JRS aims to deal with the transmission delay constraint first, the proposed JRS is more satisfactory than ESR and HTS.

Third, Fig. 2.21 shows the result for the impact of beam numbers on network throughput. Because the parallelism is increased with the number of beams, the mechanisms in Fig. 2.19 have better network throughputs. The JRS and ESR outperform the HTS since they consider the maximized parallelism. In comparison, the JRS is better than the ESR because we consider the difference of transmission rate between transmitting along and parallel transmitting with other transmission pairs.

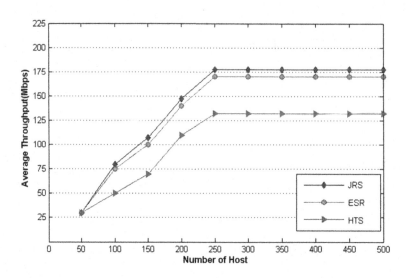

FIGURE 2.19

The comparison of network throughput.

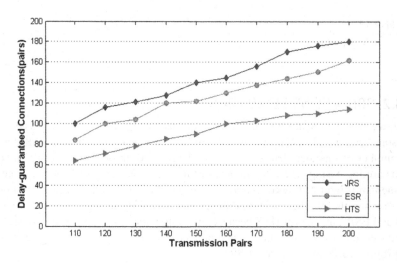

FIGURE 2.20

The comparison of transmission delay guaranteed.

2.7 CONCLUSIONS

In this chapter, we introduced a series of mechanisms for efficient exploiting smart antenna systems. These mechanisms are used for routing searching and transmission scheduling. While routing searching, for reducing the number of transmission packets

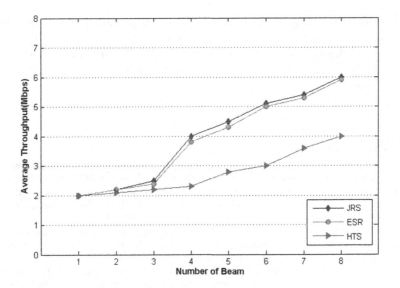

FIGURE 2.21

The throughput varies from different number of beams.

in the network, the cluster-based mechanisms are proposed. The cluster-based routing mechanism eliminates the deafness and interference problems. The DGR builds not only the shortest unicasting routes but also the maximal shared multicasting routes for smart antenna systems in a short time period. The cluster-based scheduling mechanism schedules the required transmissions with maximal parallelism from sender and receiver viewpoints. The mechanism can schedule the maximal number of transmission requests and satisfies their delay constrains. The results of evaluations show that our proposed routing mechanism and scheduling mechanism outperform other existing mechanisms in terms of throughputs, power consumption, and delay guaranteed. The lessons learned in the chapter are as follows.

a. Successful exploiting the advantage of smart antenna systems and keep off the hardware constraint of smart antenna.
b. Proposing mechanisms to guarantee the transmission delay by means of guaranteeing the transmission delay in each transmission path.
c. Increasing the parallelism of transmissions is a proper approach to exploit smart antenna efficiently.

2.8 FUTURE WORKS AND CHALLENGES

The next step and on-going work within the research is an extension of the generalization of the mechanisms. For ease of presentation, the mechanisms assume that the number of beams and the directions of all the antennas are unified. The flexibilities of the mechanisms in terms of the directions and the number of beams are not proved. The proof of flexibilities is the first step to generalize the mechanisms. Moreover, the hardware of smart antennas is the critical challenge for implementation of the mechanisms.

REFERENCES

Bao, L., Garcia-Luna-Aceves, J., 2005. Receiver-oriented multiple access in ad hoc networks with directional antennas. Wireless Networks 11 (1–2), 67–79. http://dx.doi.org/10.1007/s11276-004-4747-8.

Bazan, O., Jaseemuddin, M., 2011. A conflict analysis framework for QoS-aware routing in contention-based wireless mesh networks with beamforming antennas. IEEE Transactions on Wireless Communications 10 (10). http://dx.doi.org/10.1109/TWC.2011.081611.101328.

Chang, C.-T., Chang, C.-Y., Wang, T.-L., Lu, Y.-J., 2013. Throughput enhancement by exploiting spatial reuse opportunities with smart antenna systems in wireless ad hoc networks. Computer Networks 57 (13), 2483–2498. http://dx.doi.org/10.1016/j.comnet.2013.03.018.

Chen, S., Peng, M., Zhang, H., Wang, C., 2015. Investigation of service success probability for downlink heterogeneous cellular networks with cell association and user scheduling. In: IEEE Wireless Communications and Networking Conference. IEEE, pp. 1434–1439.

Chou, Z.-T., Huang, C.-Q., Chang, J.M., 2014. QoS provisioning for wireless LANs with multi-beam access point. IEEE Transactions on Mobile Computing 13 (9), 2113–2127. http://dx.doi.org/10.1109/TMC.2013.85.

Deshpande, V.V., Patil, A.R.B., 2013. Energy efficient clustering in wireless sensor network using cluster of cluster heads. In: International Conference on Wireless and Optical Communications Networks, pp. 1–5.

Gossain, H., Cordeiro, C., Agrawal, D.P., 2005. MDA: an efficient directional MAC scheme for wireless ad hoc networks. In: The IEEE Global Telecommunications Conference. IEEE, pp. 3633–3637.

Hung, L.-L., Chang, C.-Y., Wu, S.-H., 2011. Maximizing throughput for delay-constraint transmissions with smart antenna systems in WLANs. In: The International Workshop on Mobile Commerce, Cloud Computing, Network and Communication Security, pp. 583–588.

Hussein, A., Yousef, S., Al-Khayatt, S., Arabeyyat, O., 2010. An efficient weighted distributed clustering algorithm for mobile ad hoc networks. In: International Conference on Computer Engineering and Systems, pp. 221–228.

Jain, V., Gupta, A., Lal, D., Agrawal, D., 2005. A cross layer MAC with explicit synchronization through intelligent feedback for multiple beam antennas. In: IEEE Global Telecommunications Conference. IEEE, pp. 3196–3200.

Jawhar, I., Wu, J., Agrawal, D.P., 2010. Resource scheduling in wireless networks using directional antennas. IEEE Transactions on Parallel and Distributed Systems 21 (9), 1240–1253. http://dx.doi.org/10.1109/TPDS.2009.171.

Ko, Y.-B., Shankarkumar, V., Vaidya, N., 2000. Medium access control protocols using directional antennas in ad hoc networks. In: The IEEE International Conference on Computer Communications. IEEE, pp. 13–21.

Kour, H., Sharma, A.K., 2010. Hybrid energy efficient distributed protocol for heterogeneous wireless sensor network. International Journal of Computer Applications 4 (6), Article 7. http://dx.doi.org/10.5120/828-1173.

Lai, W.-K., Shieh, C.-S., Lee, Y.-T., 2009. A cluster-based routing protocol for wireless sensor networks with adjustable cluster size. In: International Conference on Communications and Networking in China, pp. 1–5.

Li, Y., Luo, X., 2010. Performance analysis of QoS multicast routing in mobile ad hoc networks using directional antennas. International Journal of Computer Network and Information Security 2 (2), 26–32.

Liu, J.-S., Lin, C.-H.R., 2005. Energy-efficiency clustering protocol in wireless sensor networks. Ad Hoc Networks 3 (3), 371–388. http://dx.doi.org/10.1016/j.adhoc.2003.09.012.

Mavromoustakis, C.X., Dimitriou, C.D., Mastorakis, G., 2013. On the real-time evaluation of two-level BTD scheme for energy conservation in the presence of delay sensitive transmissions and intermittent connectivity in wireless devices. International Journal on Advances in Networks and Services 6 (34), 148–162.

Mehrotra, R., Bose, R., 2014. An integrated framework for optimizing power consumption of smart antennas. In: National Conference on Communications, pp. 1–5.

Mumey, B., Judson, I., Tang, J., Xing, Y., 2012. Topology control in multihop wireless networks with multi-beam smart antennas. In: International Conference on Computing, Networking and Communications, Wireless Communications Symposium, pp. 1020–1024.

Muthuramalingam, S., RajaRam, R., Pethaperumal, K., Devi, V., 2010. A dynamic clustering algorithm for MANETs by modifying weighted clustering algorithm with mobility prediction. International Journal of Computer and Electrical Engineering 2 (4), 709–714.

Nasipuri, A., Ye, S., You, J., Hiromoto, R., 2000. A MAC protocol for mobile ad hoc networks using directional antennas. In: IEEE Wireless Communications and Networking Conference. IEEE, pp. 1214–1219.

Rafique, M.I., 2013. Exploiting smart antennas for spatial reuse and multiplexing in wireless mesh networks. In: IEEE International Symposium and Workshops on a World of Wireless, Mobile and Multimedia Networks. IEEE, pp. 1–7.

Raman, B., Chebrolu, K., 2005. Design and evaluation of a new MAC protocol for long-distance 802.11 mesh networks. In: The ACM Annual International Conference on Mobile Computing and Networking. ACM, pp. 156–169.

Sasikumar, M., Anitha, R., 2014. Performance evaluation of heterogeneous-HEED protocol for wireless sensor networks. International Journal of Advanced Research in Computer and Communication Engineering 3 (2), 5555–5558.

Wang, J., Zhu, X., Cheng, Y., Zhu, Y., 2013. A distributed, hybrid energy-efficient clustering protocol for heterogeneous wireless sensor network. International Journal of Grid and Distributed Computing 6 (4), 39–49. http://dx.doi.org/10.5120/828-1173.

Wendt, S., Samhat, A.E., Renzo, F.D., 2011. Cross-layer resource allocation in WIMAX multi-hop networks using smart antennas. In: International Workshop on Cross Layer Design, pp. 1–5.

Wong, G.K.W., Jia, X., 2013. An efficient scheduling scheme for hybrid TDMA and SDMA systems with smart antennas in WLANs. Wireless Networks 19 (2), 259–271. http://dx.doi.org/10.1007/s11276-012-0464-x.

Xing, Z., Gruenwald, L., Phang, K., 2011. A robust clustering algorithm for mobile ad hoc networks. In: Next Generation Networks and Ubiquitous Computing.

Yua, J., Qia, Y., Wangb, G., Gua, X., 2012. A cluster based routing protocol for wireless sensor networks with nonuniform node distribution. International Journal of Electronics and Communications 66 (1), 54–61. http://dx.doi.org/10.1016/j.aeue.2011.05.002.

Zhang, Z., Ma, M., Yang, Y., 2005. Energy efficient multi-hop polling in clusters of two-layered heterogeneous sensor networks. In: The 19th IEEE International Parallel and Distributed Processing Symposium. IEEE, p. 81b.

ACRONYMS AND GLOSSARY

List of acronyms with explanation

DC	the delay constraint of the packet from source to destination
DU	the delay time from source host to destination host
G	gateway, a host located in two or more clusters
H	head, a host located in the center of a cluster
ID_D	the identifier of destination host
ID_P	the identifier of the packet

ID_S	the identifier of source host
M	member, a host located in a cluster
MB	megabyte
Matrix D	transmission demand matrix
Matrix NE	neighboring matrix
Matrix R	transmission rate matrix
Matrix S	transmission scheduling matrix
Matrix TC	delay constraint matrix
ROP	receiver-oriented parallel transmission procedure
RREQ	route request packet
RREP	route reply packet
SL	sequence list, a list of hosts that relay the packet from source
SOP	sender-oriented parallel transmission procedure
TDMA	time division multiple access
TL	a list of time stamps while arriving at each relay host

Glossary of terms with explanation

Beam mode constraint the constraint that the operating mode of a host can be either transmitting or receiving at a given time.

Data diversity constraint the constraint that each beam of the smart antenna can send only one package at a given time slot.

Delay guaranteed the time period of routing the path is guaranteed to be shorter than the delay constraint.

Destination the destination host of a message or a packet.

Efficient routing path the path that a packet can be sent to the destination in the time duration less than its delay constraint.

Gateway a host located in two or more clusters.

Head a host located in the center of a cluster.

Host a device that can connect with other devices via wireless communication.

Interference constraint the constraint that each beam of the smart antenna can receive one package from one neighboring host at a given time slot.

Member a host located in a cluster.

Multicasting the transmission that a packet is sent to two or more destinations.

One-hop neighbor a neighbor located in the range of direct communication.

Receiving rate constraint the constraint that each host can receive messages from its neighboring hosts at a given slot only in one receiving rate.

ROP a procedure that groups the transmissions by the view of receivers.

Routing path the path that lists the hosts which a packet visited.

Shared(p_1, p_2) the shared path of paths p_1 and p_2.

SOP a procedure that groups the transmissions by the view of senders.

Source the source host of a message or a packet.

Transmitting rate constraint the constraint that each host can transmit messages to its neighboring hosts at a given slot only in one transmitting rate.

Unicasting the transmission that sends a packet to only one destination.

A DTN-BASED MULTI-HOP NETWORK FOR DISASTER INFORMATION TRANSMISSION

3

Shinya Kitada, Goshi Sato, Yoshitaka Shibata

Iwate Prefectural University, Japan

3.1 INTRODUCTION

From the previous researches (Shibata et al., 2014), it is clear that the wireless and mobile networks are very effective as disaster information communication means. However, it is difficult to use cellular network in case of serious disasters because the cellular base stations are damaged by the secondary disaster such as tsunami and landslide. In addition, the network traffic of a cellular network which is concentrated to the base station is congested and cannot be transmitted to the destination.

In this chapter, in order to resolve those problems, we propose a DTN based Multi-hop network which can respond to a severely damaged network environment. In our system, the message issued from the source mobile node can be finally reached to the destination node or the gateway to Internet by multi-hopping via the intermediate mobile nodes. At the same time, we consider vehicle-to-vehicle (or walker) communication where vehicle speed is not constant, eventually the vehicle moves out of communication range while transmitting data. So we use acceleration sensors mounted smart devices. In our system, by reducing redundant message transmission, battery energy can be saved.

The rest of this paper is organized as follows. The related works to our system are summarized in Section 3.2. The network system configuration of our suggested system in disaster areas is shown in Section 3.3. The data flow of multi-network based on DTN is explained in Section 3.4. A prototyped system is precisely explained in Section 3.5. Finally, conclusions and future works are summarized in Section 3.6.

Smart Sensors Networks. DOI: 10.1016/B978-0-12-809859-2.00004-8

3.2 RELATED WORKS

In the Great East Japan on March 11th in 2011, most of the residents in the disaster areas could not leverage mobile phones (Shibata et al., 2014). It is strongly required to develop the data transmission method even through serious disasters happened. Multi-hop network can be deployed in the disaster areas without constructing network infrastructure. However, since the people move with mobile terminal in disaster area, it is difficult to constantly keep connection and transmission of information data. Delay/Disruptions Tolerant Network, simply DTN uses store and forward transport method (Vahdat and Becker, 2000). The mobile node stores the data to its storage device if the communication link to the neighbor mobile node is not existed. If the communication link to the neighbor mobile node can be found, then the mobile node sends the stored data to the neighbor mobile nodes. There are many researches concerned with designing the functions of DTN protocol and running them on network simulators, but there is few researches running on a prototyped system using actual network devices.

DTN is a relay transmission technology to achieve end-to-end transmission under poor and unstable communication environment. By using store-and-forward transmission method, reliable end-to-end communications can be possible. There are several works using such as sensor data and metadata to achieve higher performance (Ramanathan et al., 2007; Tian et al., 2014; Therese et al., 2012). In those works, a DTN Protocol is developed on the network simulator and demonstrate in variety of parameter environments.

On designing a DTN protocol, it is important to consider the following issues such as unstable data communication period between mobile nodes, amount of data transmission to send/receive, and limited battery energy resource. In addition, the development of a DTN protocol on the actual prototype system depends on network hardware and OS specification. Under those conditions, protocol overhead in a limited environment has to be reduced in the system design. There are only a few related works with DTN protocol using the actual prototype system.

In the work Ito et al. (2014), the proposed function of the system combining MANET and DTN can switch between MANET and DTN as necessary to avoid network resources consumption. DTN MapEx (Trono et al., 2015) integrates a DTN protocol and distributed computing function and generates maps for disaster information to share data for decision-making.

3.3 THE PROPOSED SYSTEM

In this section, we propose a DTN based multi-hop network which can realize vehicle-to-walker communication using smart devices. On the occurrence of disasters, since many residents evacuate with smart devices, the limitation of battery

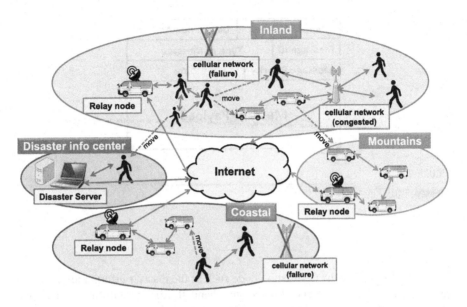

FIGURE 3.1

System configuration.

energy resource should be considered. We manage the node list and acceleration sensor data to reduce the redundant transmission and communication.

Fig. 3.1 shows our proposed network system configuration at the time of disaster where inland, coastal and mountains areas are assumed in the following conditions; 1) cellular network is congested or cannot be functioned in the disaster areas, 2) most of the residents move carrying with smart devices in the disaster area, 3) the relay nodes are deployed each area just after occurrence of disaster.

On the above conditions, DTN based Multi-hop network is configured using the smart devices by residents. The data from a smart device as a mobile node are transmitted to the neighbor nodes by a DTN protocol. This transmission is repeated until arriving at the destination node or the relay node as a network gateway to Internet. The relay node is assumed the following conditions; 1) the relay node by wireless cognitive access network with long distance can connect to Internet, 2) the relay node has a mobile server that collects the received data from other mobile nodes. We regard the transmitted data as *Message* in our proposed network. Fig. 3.2 shows a message structure in our system.

A Message consists of ID and disaster information, sender address, destination address, type of *Message*, *Message* priority. We identify each device as *Address*. Disaster victims create *Message* and decide urgent level using *Message* priority. A *Message* is transmitted based on this priority. The relay node can receive those

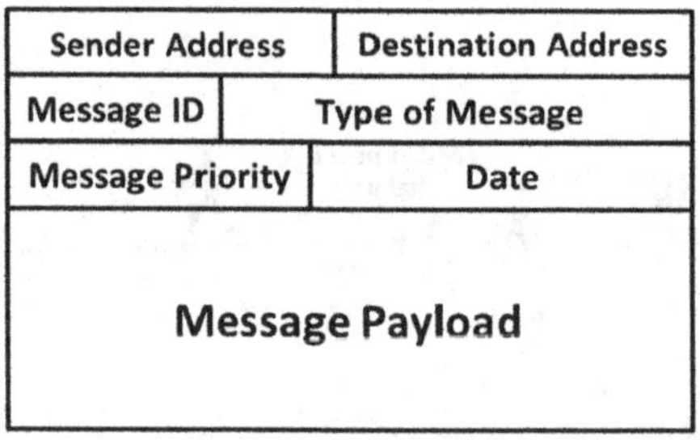

Sender Address	Destination Address	
Message ID	Type of Message	
Message Priority	Date	
Message Payload		

FIGURE 3.2

Message structure.

messages and send to the destination node such as the disaster information center in the other areas through Internet. On the other hand, the disaster information center is assumed in the following conditions; 1) the disaster information center is allocated at the place where there is no damage of tsunami and landslide, 2) access to Internet is always possible. Thus, in our proposed system, the communication between the mobile node and the relay node, and the mobile modes at the different areas can be performed in the same manner.

Fig. 3.3 shows our proposed conceptual network model. When the destination node exists in the same area as the source node, the source node spreads *Messages* to the neighbor nodes, n1, n2, and n3. The received neighbor nodes spread the received *Messages* to the other neighbor nodes. By repeating this process, the *Messages* eventually arrive at the destination node.

In the case in which the source node, n1 and n3 are walkers, their moving speeds are slow. In this case, the source node transmits the *Messages* and can expect that all of the *Messages* in his device can arrive at those neighbor nodes. If the source node has many *Messages*, it transmits higher priority *Messages* first. In the case where the node n2 is a vehicle and its moving speed is faster, the source node may not transmit all of the *Messages* in his device storage and may derive to link down. To prevent the link down while transmission of *Messages*, the source node selects proper *Messages* and transmits based on the relative moving speed between the source node and n2, and *Messages* priority. Thus proper *Message* can arrive at the destination node 1 through node n2.

On the other hand, when we need to send *Messages* to the destination node 2 across Internet, the relay node 1 performs the role of temporary destination as a gateway to Internet. The relay node 1, after receiving *Message* from node n5, transmits to relay node 2 through Internet. Then relay node 2 transmits them to node n7. Thus the *Messages* from the source node can be finally arrived at the destination node 2 across Internet.

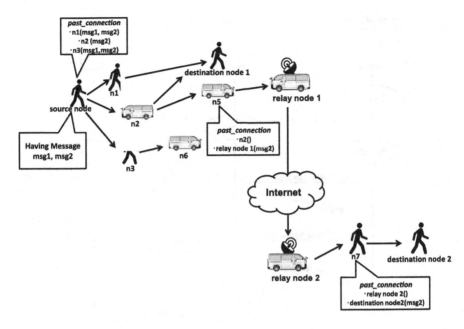

FIGURE 3.3

Conceptual network model.

3.4 NETWORK PROTOCOL

The network routing protocol of our proposed system is based on flooding method such as Epidemic routing [3]. In our routing protocol, each node manages connection nodes using *neighbor_list* to keep neighbor nodes and *past_connection* list to keep past connection state. Fig. 3.4 shows a flow of discovery process of neighbor nodes. First, the transmission node and neighbor nodes exchange Probe Request and Probe Response while changing *Listen* mode and *Search* mode mutually. If the mode is *Search*, the node sends *Probe Request*. Also if the mode is *Listen*, the node waits for *Probe Request* and replies *Probe Response*. When the transmission node received *Probe Response*, then adds the responded node to *neighbor_list*. Next, the transmission node checks *past_connection_list* whether the list is empty or not.

Fig. 3.5 shows an example of *past_connection* and *neighbor_list*. In this example, the transmission node has *Message*1 (msg1) and msg2 to the neighbor n1. First, the transmission node checks the neighbor node1 (n1) which is the top of *neighbor_list* whether there is n1 in *past_connection*. In this example, since n1 exists in *past_connection*, the transmission node passes n1 and goes to n2 which is the second of *neighbor_list*. In this case, since n2 exists in *past_connection*, but the transmission node had not transmitted all *Messages* in the storage. So the transmission node decides to connect n2 and transmits msg2. If all of the nodes in *neighbor_list* exist in

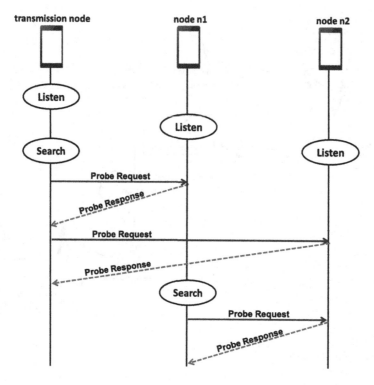

FIGURE 3.4

Discovery process of neighbor node.

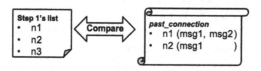

FIGURE 3.5

Compare lists example.

past_connection and the transmission node had transmitted all *Messages* to all nodes in *past_connection*, then the transmission node continues to discover a new neighbor node to update *neighbor_list*. This step is repeated for all of the existing nodes.

Next, both nodes select their own acceleration data form acceleration and gyro sensors, and exchange those data. Both nodes determine the proper *Message* from the storage based on the relative speed and message propriety. Finally, the transmission node transmits the selected *Message*. Then node n2 repeats the message transmission to other nodes in the same manner.

Thus, our algorithm uses the *neighbor list* to manage the neighbor nodes, the *past_connection* to manage the *Messages* whether those are transmitted or not, moreover exchanging acceleration and gyro sensor data before transmitting data. Thus, our algorithm can save duplicated Messages and reduce battery energy consumption by the unnecessary communication.

Algorithm: Managing list to routing.

If *past_connection* is empty;
    The source node sends connection request to the first node in *neighbor_list*.
    Else;
        The source node compares *neighbor_list* with the **past_connection** and focuses on the *i*-th node in *neighbor_list*.
If node *i* in *neighbor_list* $\notin$ *past_connection*;
    The source node sends connection request to the node *i*
    Else if the source node had not transmitted all of *Messages* in the storage;
        The source node sends connection request to the node
    Else;
        When the node *i* is existed and the source node had transmitted all of *Messages*,
        then tries the next node $i + 1$.

3.4.1 SYSTEM ARCHITECTURE

Fig. 3.6 shows the network system architecture which is realized by an application module and network system module on top of network interface in a smart device. Moreover, the network system module is consisted of a routing module, DTN module, and TCP (or UDP)/IP protocol. The application module consists of Message Storage, Message Selector, Message Exchanger, and Sensor Module. Routing module comprises Discovery, Decision, Connect, and *past_connection* modules which can manage the connection state between the own node and the neighbor nodes. The source node discovers the neighbor nodes or relay nodes by Discovery module in Routing module. Next, Decision module maintains the connection to another nodes by referring to *past_connection*. After completion of the connection to the neighbor nodes, the *Message* in Message Storage is transmitted using Message Exchanger. Since TCP protocol is used in Message Exchanger, *Message* can be received/send from/to the other nodes bi-directionally. Then the received *Message* is stored in the Message Storage.

The relay nodes are consisted of several different devices including smart devices, mobile servers, and an Internet access gateway with long distance wireless networks. The smart device in the relay node securely transmits *Messages* to the mobile server. It is desirable to avoid message duplication in our network system. For this reason,

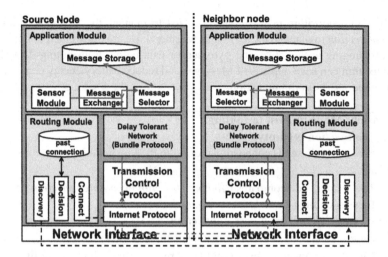

FIGURE 3.6

System architecture.

we implemented a module to temporarily check message duplication. The destination nodes such as Disaster Server checks message duplication by referring Sender address and Message ID Storage and removes if the duplication is found. When the destination nodes are located across the Internet, the Internet gateway server can perform message duplication in the same manner. Thus by removing the message duplication, the redundant message transmission and unnecessary energy consumption can be reduced.

3.5 PROTOTYPE SYSTEM AND PERFORMANCE

In order to verify the effect of our proposed system, we construct a prototype system using commercially available smart devices as shown in Fig. 3.7 and evaluate its functional and performance. The prototype is consisted of multiple different devices including smart terminals for human carrying communication devices, relay node with smart device and Mobile server, and a disaster server as a destination server.

Table 3.1 shows the details of hardware system of our prototype system. We develop our DTN function module and routing function module for as an application of the Android OS based smart terminal using Android SDK. We construct a Multi-hop network by those smart terminals using WLAN IEEE 802.11n which is mounted as common wireless network devices. For Mobile server and Disaster server, mobile PCs operated by Ubuntu OS are introduced. JAVA and MySQL are used for software system development of our architecture. All of the *Message*s are managed by SQLite in the message storage. Fig. 3.8 shows a test Graphical User Interface (GUI) of our

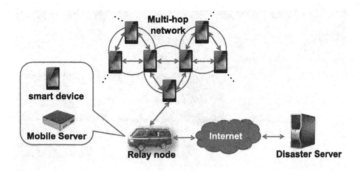

FIGURE 3.7

Prototype system.

Table 3.1 Hardware Specification of Prototype	
Smart Devices	Nexus 7 (2013)
	OS: Android 5.0
	CPU:APQ8064 QuadCore1.5 GHz
	MEM: RAM 2 GB
	Develop. SDK.: Android SDK
	NIC: IEEE 802.11n
Mobile Server	OS: Ubuntu 14.04
	CPU: Core i3 4010U
	MEM: RAM 4 GB
	Develop. SDK: JDK7
Disaster Server	OS: Ubuntu 14.04
	CPU: Core 2 Extreme
	MEM: RAM 4 GB
	Develop. SDK: JDK7

application. This GUI shows a view at the time of the first login on the application of
the smart device.

3.5.1 PERFORMANCE EVALUATION AND DISCUSSIONS

The preliminary test is executed to evaluate the performance of the basic application
system on the smart terminals. Table 3.2 shows summary of the test network parameters and their values. As performance index, the throughput and the packet loss rate
are measured at the outdoor of our university by changing the transmission distance
between the two smart terminals.

Fig. 3.9 shows the result of the performance of the throughput and the packet loss
rate for the basic application system between the smart terminals. As shown, at the
transmission distance from 0~40 m, both the throughput and the packet loss rate are

FIGURE 3.8

Test graphical user interface.

Table 3.2 Test Network Parameters and Their Values	
Parameter	**Values**
Packet size	1 KB
Message size	1 MB
Message transmission protocol	TCP/IP
DTN method	Epidemic
Distance between nodes	0~150 m
Test area	Outdoor (rural area)
Number of test	5 times

almost stable and constant, about 7 Mbits/sec and 10%, respectively. At the transmission distance from 40~80 m, the packet loss rate gradually increases and throughput decreases, and finally all of the packets are almost lost and throughput reached to 0. From this result, it is founded that the possible transmission distance is within 80 m and the throughput is under 7 Mbits/sec. Therefore, in the actual disaster situation, if the people are located outdoor with smart terminals within 0~80 m area, the message transmission can be possible in the area without DTN function. If people are mutually far away, messages are temporally stored in his storage by DTN function. Those messages can be delivered to the neighbor smart terminal by mutually moving.

Next, we examined the performance evaluation of multi-hop network environment. Table 3.3 shows summary of the test network parameters and their values for the performance evaluation. The end-to-end transmission time from the source smart terminal to the destination disaster server was measured for three different message sizes when the number of the hops is increased. Each smart terminal node was placed

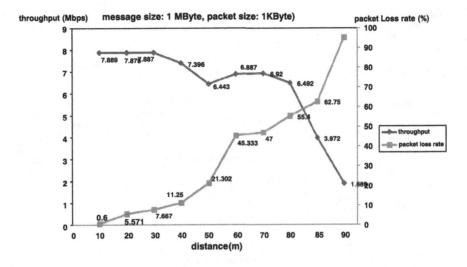

FIGURE 3.9

Throughput and packet loss rate.

Table 3.3 Test Network Parameters and Their Values for Multi-Hop Network	
Parameter	**Value**
Number of hops	1 to 5
Distance between nodes	3 m
Packet size	1 KB
Message size	1 MB, 512 KB, 1 KB
Message transmission protocol	TCP/IP
DTN method	Epidemic
Test area	Indoor: 150×5 (m^2)
Number of test	5 times

every 3 m at distance. The test was repeated 5 times and the observed end-to-end transmission time were to derive the average values.

Fig. 3.10 shows the result of end-to-end transmission time from the source smart terminal to the destination disaster server for three different message sizes when the number of the hops is increased from 1 to 5 depending on the size of the messages. Therefore by considering this result, it is possible to estimate the end-to-end transmission time even when the number of the hops is increased to more than 5.

On the other hand, the differences of the end-to-end transmission times for three message sizes are small. This reason is that the packet processing, such as packet connection establishment on TCP on the smart terminal node is dominant for the end-to-end transmission time compared with the number of the packet transmission.

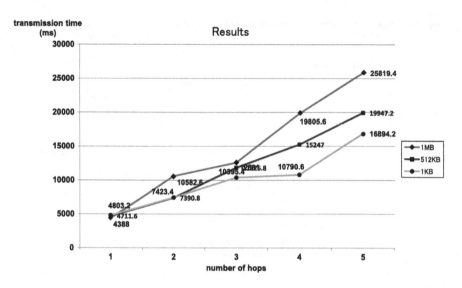

FIGURE 3.10

Result of measure test.

As result, the message transmission by TCP should be improved by using more simple reliable transmission protocol or introducing the larger packet size more than 1 Kbyte to relatively reduce the influence of the protocol processing time.

3.6 CONCLUSIONS

In this paper, we proposed an effective message transmission method by DTN based Multi-hop network for the case of infrastructure failure at time of disaster. It is possible to consider the message transmission and communication interruption using DTN protocol function. Also we used the neighbor list and the *past_connection_list* to manage routing, and acceleration sensor data to reduce link down while transmitting data. Furthermore, communication between disaster areas also becomes possible by using relay node capable of Internet access. In order to verify effect of the proposed system, we built a prototype system and experimented the performance evaluation. Through the measurement test with actual devices, we found that the current prototype provides reasonable performance but need to improve message transmission time when the size of the multi-hop is larger.

As our future works, we will examine the message transmission on vehicle-to-vehicle communication. Then we will design more effective message transmission method using acceleration sensor data. Next, we plan to improve DTN protocol which can ensure to reach to the relay node, and attempt the experimental test with various metrics (duplicated data ratio, delivery ratio etc.) while we compare existing protocols. Also since performance changes due to the mobility of nodes, we will work for an experimental test for dynamic change of the nodes.

ACKNOWLEDGMENT

The research was supported by SCOPE (Strategic Information and Communications R&D Promotion Program) Grant Number 142302010 by Ministry of Internal Affairs and Communications in Japan.

REFERENCES

Ito, M., Nishiyama, H., Kato, N., 2014. A novel communication mode selection technique for DTN over MANET architecture. In: International Conference on Computing, Networking and Communications, pp. 551–555.

Ramanathan, R., Hansen, R., Basu, P., Rosales-Hain, R., Krishnan, R., 2007. Prioritized epidemic routing for opportunistic networks. In: Proceedings of the 1st International MobiSys Workshop on Mobile Opportunistic Networking, pp. 62–66.

Sato, G., Uchida, N., Shibata, Y., 2015. Performance evaluation of software defined and cognitive wireless network based disaster resilient system. In: The 29th IEEE International Conference on Advanced Information Networking and Applications, pp. 741–746.

Shibata, Y., Uchida, N., Shiratori, N., 2014. Analysis and proposal of disaster information network from experience of the great East Japan earthquake. IEEE Communications Magazine 52 (3), 44–48.

Therese, J., Fajard, B., Yasumoto, K., Shibata, N., Sun, W., Ito, M., 2012. DTN-based data aggregation for timely information collection in disaster areas. In: The 13th IEEE International Conference on Wireless and Mobile Computing, Networking and Communications, pp. 333–340.

Tian, C., Ci, L., Cheng, B., Li, X., 2014. A 3D location-based energy aware routing protocol in delay tolerant networks. In: The 13th IEEE International Conference on Dependable, Autonomic and Secure Computing, pp. 485–490.

Trono, E.M., Arakawa, Y., Tamai, M., Yasumoto, K., 2015. DTN MapEx: disaster area mapping through distributed computing over a delay tolerant network. In: Mobile Computing and Ubiquitous Networking, pp. 179–184.

Vahdat, A., Becker, D., 2000. Epidemic Routing for Partially-Connected Ad Hoc Networks. Technical Report CS-2000-06 Department of Computer Science, Duke University.

ACRONYMS AND GLOSSARY

List of acronyms with explanation

CPU   central processing unit
DTN   delay/disruption tolerant network
DTN MapEx a distributed computing system for disaster map generation that operates over a delay/disruption tolerant network
GUI   graphical user interface
MANET   mobile adhoc network
MEM   main memory
NDN   never die network
OS   operating system

PCs	personal computers
RAM	random access memory
SDK	software development kit
SDN	soft defined network
TCP	transmission control protocol
UDP	user datagram protocol
WLAN	wireless local area network

Glossary of terms with explanation

Challenged network a network environment where network connections are often disconnected and unstable such as the network state just after large scale disaster.

Destination node a node that receives data at the end.

Disaster information center the location of the central information center where all of the disaster information in this local area are collected and distributed.

Disaster server a server which collects and delivers the disaster related information in the local area. Usually the disaster server is set to the disaster-response headquarter in the local area.

Epidemic routing a basic DTN routings which is based on the flooding.

Flooding transmitting packets to all of the neighbor nodes.

Gateway a host which is connected to another network such as Internet.

IEEE802.11n one of IEEE standard local area networks standard.

Internet gateway a network equipment that transmits/receives the packets to/from Internet environment.

Link down the state where the connection between the sending node and the destination node is disabled.

Message a set of data that is consisted of multiple packets.

Message ID a unique identification number to describe and identify the data packet.

Message priority descriptor indicating priority level.

Mobility of nodes speed of mobility such as vehicles or pedestrians.

Multi-hop network wireless network where the data are transmitted through the multiple number of intermediate mobile nodes between the source and destination nodes.

Relay node intermediate node which transmits the data from the source to the destination nodes.

Sensor data physical data which can be observed from various devices such as smartphones, tablet terminals and vehicles.

Source node a node that generates data.

Type of message descriptor indicating what kind of messages is included.

Vehicle-to-vehicle communication a communication method between a moving car with wireless communication devices to directly exchange to others in adhoc manner.

Vehicle-to-walker communication a communication method where a moving car with wireless communication device to directly exchange to the pedestrian who carry a smartphone or tablet terminal.

INTELLIGENT ENERGY MANAGEMENT FOR ENVIRONMENTAL MONITORING SYSTEMS

Petr Musilek*,†, **Michal Prauzek**†, **Pavel Krömer**†, **James Rodway***, **Tomáš Bartoň***

**Department of Electrical and Computer Engineering, University of Alberta, Edmonton, AB, Canada* †*Faculty of Computer Science, VŠB Technical University of Ostrava, Ostrava, Czech Republic*

4.1 INTRODUCTION

Environmental monitoring involves collecting data over time and across volumes of space that exhibit significant internal variation (Culler et al., 2004). The emergence of new sensors and monitoring devices brings new possibilities for research and operational monitoring of many different environments. Measuring and recording physical and biotic parameters aids understanding of the environment and its changes. Examples include ecosystem classification or monitoring changes in land use and ecosystems, such as forest succession (Sanchez-Azofeifa et al., 2013).

The development of environmental monitoring systems (EMSs) requires a strong partnership between environmental scientists and computer engineers (Benson et al., 2010). Different environments (e.g. tropical, arctic or underground) not only call for different types of measurements, but also pose different requirements on the performance of environmental monitoring devices. These requirements include the number and type of input/output interfaces, sampling frequency, data processing capabilities, computing performance, battery operation, communication interfaces, device size, and operating temperature range. EMSs are commonly installed in remote or inaccessible locations. This brings even more stringent requirements of higher reliability and energy independence. Energy availability also affects the duty cycle of monitoring device operation (Dewan et al., 2014).

This contribution uses monitoring of terrestrial ecosystems as an example to develop a general structure of EMS and to exemplify typical sensors used for measuring environmental conditions and their nominal energy requirements. With energy harvesting, monitoring systems can operate indefinitely, albeit with varying fidelity and subject to common failures. However, it is how energy is provided and used that determines how long a monitoring system can operate without the need for replenish-

Smart Sensors Networks. DOI: 10.1016/B978-0-12-809859-2.00005-X

ment. The way energy is used can be controlled by a number of power management techniques. Sequences and combinations of these techniques then form energy management strategies with the goal to minimize energy consumption while keeping the quality of service at an acceptable level. The complexity of energy-related tasks requires the use of non-trivial approaches offered, for example, by methods of computational intelligence considered in this work.

This chapter is organized in six sections. Section 4.2 describes the structure of EMS and lists some sensors commonly used for monitoring terrestrial ecosystems. Following section 4.3 concentrates on EMS power supplies and energy harvesting. Section 4.4 provides an overview of energy management techniques and develops methodology for design, testing, and evaluation of energy management strategies. Recent contributions of the authors in the area of intelligent energy management are reviewed in section 4.5. Finally, major conclusions and trends in intelligent energy management are laid out in section 4.6.

4.2 ENVIRONMENTAL MONITORING SYSTEMS

The main purpose of EMSs is to obtain information needed for decision support. In its broadest sense, the term "environmental monitoring" may refer to surveillance of indoor or outdoor spaces, natural or man-made environments, as well as special tasks such as air pollution or water quality monitoring. Approaches to environmental monitoring range from remote sensing using aircraft and satellites, through laboratory analysis of field-collected samples, to in-situ monitoring devices and wireless sensor networks. In terms of monitored modalities, an EMS can consider physical, chemical, biological or population-related parameters of surveilled environments.

To facilitate the description of EMSs and their energy-related operation, we will concentrate our attention on the systems for monitoring terrestrial ecosystems. These ecosystems include the entire spectrum of natural environments ranging from tropical rain-forests to ice-free zones of polar deserts. They also represent a wide range of environmental energy regimes: from rainy seasons to polar nights. The principles and conclusions developed in the terrestrial context can be extended to other environments as well as to more specific monitoring systems, e.g. systems for monitoring indoor or outdoor air quality. A typical suite of variables monitored in the context of terrestrial ecosystems is illustrated in Fig. 4.1.

4.2.1 STRUCTURE OF ENVIRONMENTAL MONITORING SYSTEMS

An EMS contains three main functional modules and three ancillary modules (shown as solid and dashed blocks, respectively, in Fig. 4.2). The *sensor interface* provides signal conditioning (filtering, amplification, isolation, and analog-to-digital conversion) for analog sensors, and communication protocols (e.g. RS-232, inter-integrated Circuit (I^2C), serial peripheral interface (SPI), TII) for digital sensors (Zhou and Mason, 2002). The *data processing and control unit* uses a microcontroller unit (MCU)

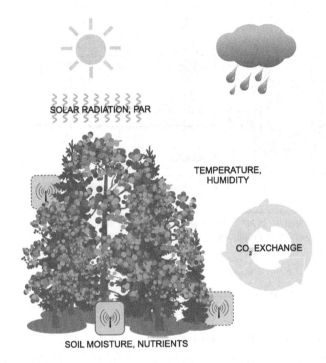

SOLAR RADIATION, PAR

TEMPERATURE, HUMIDITY

CO_2 EXCHANGE

SOIL MOISTURE, NUTRIENTS

FIGURE 4.1

A terrestrial ecosystem with its main processes and characteristics, and stand-alone (solid outline) and networked (dashed outlines) monitoring devices.

to process the digital data from the sensor interface, and to coordinate the time keeping, data storage and transmission operations of the system. Data is provided to a data management system or the end user through the *data interface*.

The *memory* module serves for temporary or permanent storage of collected data, and can be implemented using a non-volatile memory device (e.g. EEPROM or FLASH memory). The *time tracking module* establishes the basis for precise time-keeping of the recorded values or events. The *power supply module* provides any voltages required by the other components of the system derived from an energy source such as battery pack or mains connection, or harvested from the environment.

4.2.2 SENSORS FOR ENVIRONMENTAL MONITORING

EMSs depend on sensors to measure the parameters of interest and convert them into signals which can be used by the data processing and control unit. The large variety of parameters that can be measured in terrestrial environments requires the use of disparate sensing elements. This section identifies

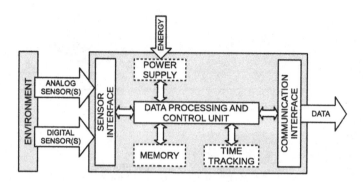

FIGURE 4.2

A general block diagram of an environmental monitoring system. The main functional modules are shown as white blocks with solid outline, while the ancillary modules have dashed outline.

several major sensors and describes their operating principles and energy demands.

4.2.2.1 Temperature Measurement

Temperature can be measured using sensors based on a range of different technologies: resistance temperature detectors (RDTs), thermistors, thermocouples, bipolar junction transistors (BJTs), and micro-electro-mechanical systems (MEMS) resonators (Roozeboom et al., 2013). Each technology provides different degree of accuracy, temperature range, and response time. Due to their construction, temperature sensors do not require a significant amount of power for their operation.

4.2.2.2 Light Monitoring

Photosynthesis, one of the most important processes in terrestrial ecosystems, is driven by light (Huemmrich et al., 1999; Lee, 1998; Pastorello et al., 2011; Kooistra et al., 2009). Corresponding measurements are performed using pyranometers (for solar radiation) and quantum sensor instruments (for photosynthetically active radiation, PAR) that differ mainly in their spectral response. While pyranometers are sensitive to the entire spectrum of solar radiation, PAR sensors use filters that limit their response to the desired (narrow) band of wavelengths. The actual sensing elements are based on thermopile or photovoltaic devices (Cotfas et al., 2011) that generate voltage or current upon illumination (Cox, 2001). Due to their active mode of operation, light monitoring sensors do not significantly contribute to the overall energy consumption of an EMS.

4.2.2.3 Soil Monitoring

Measurement of soil properties is important for understanding land-atmosphere interactions (Mittelbach et al., 2011), ecosystem classification, and assessment of seasonal changes (Zhao et al., 2013). The most important parameters are soil temperature, moisture, and pH. In many cases, soil properties are measured simultaneously at several depths (Vargas et al., 2010).

Soil moisture sensors are electromagnetic devices exploiting the dependence between soil water content and other properties, such as electrical conductivity or permittivity (Robinson et al., 2008). A typical electromagnetic soil moisture sensor consumes about 100 mW of power, placing a significant burden on the power supply of a remote EMS, especially when the use of multiple sensors is required.

Soil temperature is usually measured using thermistors. With typical nominal resistance of 10 kΩ–100 kΩ (Campbell, 2013), soil temperature sensors are not significant energy loads. In addition, temperature sensing is often included with soil moisture devices.

Recently developed integrated, multimodal sensors combine measurements of soil temperature, electrical conductivity, and pH (Futagawa et al., 2012). Such semiconductor-based devices not only expand the variety of measured modalities, but also reduce power consumption.

4.2.2.4 Gas Analysis

Concentrations or fluxes of some gases (e.g. CO, CO_2, CH_4) can be used to characterize land-atmosphere interactions (Vargas et al., 2010), or to provide warnings of ecosystem disturbances. Some gases are detected by their characteristic absorption in the atmosphere, through nondispersive infrared (NDIR) light spectroscopy. The main disadvantage of these sensors is their relatively high power consumption of 100 to 500 mW.

Semiconductor sensors, which are based on observable gas/semiconductor interactions, require a relatively high operating temperature (200–1,000 °C). Implementation of a heating element in such sensor devices (Capone et al., 2003) implies high power consumption limiting their use to EMSs powered from the mains supply.

Electrochemical sensors produce current or voltage proportional to the amount of a target gas in the atmosphere. Their power consumption is the lowest among all gas sensors. They also have good linearity and selectivity, and excellent accuracy and repeatability (Kumar et al., 2013).

4.3 POWER SUPPLIES FOR TERRESTRIAL ENVIRONMENTAL MONITORING SYSTEMS

Although power supplies have only a supporting role in the functioning of EMSs, their efficiency dramatically affects the requirements for system maintenance (e.g. recharging or replacement of batteries). This section provides a description of power supply components and an overview of energy harvesting.

4.3.1 COMPONENTS OF POWER SUPPLIES

4.3.1.1 Energy Sources

EMSs considered in this article are usually deployed at remote locations, precluding the use of mains power supply. Primary (non-rechargeable) batteries are typically the first option for powering such a field-installed EMS. They have many advantages, including high capacity (e.g. modern lithium-based batteries) and temperature stability. Their biggest disadvantage is the impossibility of recharging and the need of periodic, manual replacement.

Energy harvesting devices (Mateu and Moll, 2005) offer a good alternative (or supplement) to primary batteries in EMS power supplies. Harvesting sources are usually combined with energy storage devices, described in the following subsection. Combinations of primary batteries and different energy harvesting sources can also be used to power an EMS.

4.3.1.2 Energy Storage Devices

Storage devices can stockpile energy for use when environmental energy is not available. EMSs can be deployed in various climates characterized by diverse conditions and different patterns of ambient energy availability. For example, tropical climates have long periods of warm, sunny weather alternating with rainy seasons. Polar regions, on the other hand, have low ambient temperatures and experience extended periods of no insolation during polar nights. For these reasons, storage devices must be carefully selected, both from the perspective of their operating conditions and storage capacity. Other important characteristics include peak available power, durability (cycling capacity), and self-discharge rate. Specific to environmental monitoring applications are operational constraints related to environmental protection and impact (Ibrahim et al., 2008).

The lead-acid batteries are used most often. Their popularity is due in part to their low cost, high reliability, and efficiency. However, they have relatively low cycling capacity ($5 \cdot 10^2$–10^3) and poor performance at temperatures below $-20\,°C$ and above $+60\,°C$ (Chen et al., 2009). The nickel-cadmium (NiCd) batteries have a higher specific energy (50–60 Wh/kg), but have a comparatively low cycle life (10^3–$2 \cdot 10^3$) and higher cost. Compared to lead acid batteries, they can withstand more extreme conditions ($-40\,°C$ to $+60\,°C$). They can be also fully discharged without significant loss of capacity, lifetime, or efficiency (Gonzalez et al., 2004). In the context of EMSs, the main problem of NiCd batteries is the presence of cadmium which is extremely toxic. The lithium-ion batteries have high efficiency, power density (75–200 Wh/kg), and cell voltage. However, as they use lithium, these batteries may cause fires when exposed to moisture. This limits their use in EMSs, along with their relatively high cost. Most batteries, except lithium-ion, do not perform well in cold temperatures due to the increase of their internal resistance that leads to loss of capacity. The opposite holds true for their operation at elevated temperatures, but at the cost of a significant shortening of their service life or even permanent damage.

Supercapacitors have great cycling capacity (10^4–10^6) and can operate over a wide range of temperatures ($-55\,°C$ to $+85\,°C$) without significant loss of capacity. Their operational lifetime can reach up to 10 years before their capacity is reduced to 80% (Simjee and Chou, 2008). This makes them ideal for applications in EMSs, along with little to no negative environmental impacts (Bradbury, 2010; Chen et al., 2010). Their main disadvantages are their low specific energy (5–15 Wh/kg) and high leakage current in comparison to lead, nickel or lithium-based cells (Yang and Zhang, 2013). Their self-discharge rate is about 5% per day, which makes them unsuitable for long term storage of energy, unless they are continuously recharged (Bradbury, 2010). In addition, supercapacitors have highly nonlinear behavior, demanding the use of advanced modeling techniques for energy storage estimation (Yang and Zhang, 2013; Weddell et al., 2011). Supercapacitors can be also combined with rechargeable batteries to form a hybrid storage system which extends operational time (Ongaro et al., 2012).

4.3.1.3 Power Converters

Both battery-powered and energy harvesting systems often need to boost low input voltage or convert a high voltage to a level required for device operation (Nymand and Andersen, 2010). This can be accomplished by an integrated DC/DC converter – a circuit that changes direct current (DC) voltage at one level to a DC voltage at a different level. The converters are also responsible for providing a constant supply voltage to other EMS components using energy from batteries or other storage devices. The efficiency factors associated with the converter plays a major role in determining the operational lifetime of systems they power (Raghunathan et al., 2002).

4.3.1.4 Loads

An EMS contains a number of functional and ancillary modules (see Fig. 4.2) that act as electrical loads. These include sensors, MCU, memory devices, and communication interfaces. Many loads can operate in different modes (e.g. standby and power off), and complex loads like the MCUs can operate in several power saving modes (e.g. idle, power down, power save, standby, and extended standby).

Energy management starts with low power design of all system components. The length of available operational time largely depends on loads, and thus the energy consumption that is designed in the sensor platform cannot be significantly reduced through power management or optimization (Raghunathan et al., 2006). Therefore, the most effective low-power design method is simply to make low power consumption a key objective in the design process, because a significant amount of power can be saved by avoiding waste (Horowitz et al., 1994).

4.3.2 ENERGY HARVESTING SYSTEMS

Energy harvesting devices generate electric energy from the environment using direct energy conversion techniques (Mateu and Moll, 2005). The character of energy harvesting sources (Harb, 2011) has a direct impact on the energy management strategy (Pimentel and Musilek, 2010). Some management strategies make use of harvesting source models to predict future energy availability (Vigorito et al., 2007). Other characteristics to consider during design are power density, required form of the output power (DC or alternating current (AC)), voltage/current curve (for maximum power point operation), and periodicity. Energy is most commonly harvested from the sun, but wind and vibration energy is often harvested as well; examples of these techniques are provided in Pimentel et al. (2010), Tan and Panda (2011) and Dallago et al. (2012).

In order to operate indefinitely, a system cannot consume, on average, more power than the harvested source can provide. Otherwise, if the consumption exceeds the production, the system will sooner or later deplete its energy stockpile and stop working due to the empty energy reservoir and absence of environmental energy. This leads to undesirable system performance.

4.3.2.1 Topologies of Energy Harvesting Systems

There are three main topologies for energy harvesting systems: autonomous, hybrid autonomous, and battery-supplemented. Depending on the configuration, energy management strategies with different design goals are required (Pimentel and Musilek, 2010).

Autonomous harvesting systems

Autonomous harvesting systems fully satisfy their energy needs from ambient sources, without batteries (Pirapaharan et al., 2012; Yi et al., 2008). Autonomous systems can only operate when the energy source is available, but their lifetime and performance are not limited by storage inefficiencies (e.g. round trip efficiency, self-discharge and aging). These systems are inherently governed by the so-called energy neutrality principle since they can never consume more energy than their harvesting device can deliver (Kansal et al., 2007).

Autonomous hybrid harvesting systems

Autonomous hybrid harvesting systems are the most common type of energy harvesting systems. They have an energy reservoir implemented using a secondary battery or supercapacitor. The harvesting device collects energy for system operation and recharging of storage (Raghunathan et al., 2005). This arrangement can dramatically increase the operational lifetime of the system. With proper energy management, this topology can achieve 0% dead time operation. The battery and the energy harvesting device must be sized so that they satisfy the energy needs of the system, possibly using the energy-neutrality principle (Kansal et al., 2004). The system can sometimes consume more energy than the harvesting source provides (using battery reserves), but the production/consumption rates have to be balanced over the long run.

Battery-supplemented harvesting systems

These systems usually have a battery as the main source of energy and a harvesting device that plays an important, but secondary, role. The goal of energy management in such systems is to limit battery energy usage and to increase the system's lifetime, e.g. by making external recharging or replacement of batteries less frequent (Kansal et al., 2007). This system can use primary or secondary batteries. Harvested energy can directly or indirectly power the load or its specific parts. An example can be found in Janek et al. (2007). This approach greatly increases system reliability and allows extended data acquisition, processing or transfer. As long as the primary batteries have some useful charge left, the system can continue to operate in situations when secondary storage is depleted and environmental energy is not available for harvest.

4.4 ENERGY MANAGEMENT STRATEGIES

Energy management can be applied at three different levels: the MCUs, node, and network (Kompis and Sureka, 2010). MCU-level techniques are driven by optimizing microcontroller power dissipation and involve selection of the MCUs unit and possibly the use of dynamic voltage and frequency scaling. Node-level techniques focus on adapting the device performance depending on temporal variations of energy availability and sensing, computing, and data communication. Finally, network-level techniques concentrate on optimizing node density, data routing, and synchronization of communication events across networked sensing systems.

A successful energy management strategy is not trivial, as it must take into account all aspects of the system (Pimentel and Musilek, 2010). It starts with low power design of all system components and the use of efficient power conversion techniques. It involves a number of power management techniques, such as power scheduling, duty cycling, etc. Common techniques found in the literature are summarized in Table 4.1. The use of these techniques is tied to the hardware configuration. For example, the commonly-used duty cycling technique requires that the underlying hardware supports sleep modes. Individual power management techniques, described in the following subsection, can be combined to form energy management strategies outlined at the end of this section.

4.4.1 POWER MANAGEMENT TECHNIQUES

4.4.1.1 Maximum Power Point Tracking

Maximum power point tracking (MPPT) is a technique that allows the maximum available energy to be transferred from a transducer (Kimball et al., 2009). Energy transducers, such as solar panels, have nonlinear characteristic power curves; the optimal load for maximum power output depends on the operating point for given conditions (Esram and Chapman, 2007). This technique requires monitoring the incoming energy, determining the optimal operation point, and adapting the load (Pimentel and

Table 4.1 Power Management Techniques

System Strategies	
Maximum power point tracking (MPPT)	Dynamic voltage scaling (DVS)
(Adaptive) duty cycling	Dynamic frequency scaling (DFS)
Dynamic voltage and frequency scaling (DVFS)	
Strategies for Peripherals (Moser et al., 2007; Arms et al., 2005)	
Adaptive sensing rate	Adaptive memory management
Sensor power gating	Signal conditioning power gating
Event-driven sampling	Sample rate reduction/minimization
Sleep between samples	Scalable fidelity
Reduction of wireless data transmission	Transceiver duty cycle reduction
Strict power management	Event-driven transmission strategy

Musilek, 2010). The optimal load can be determined using a number of different techniques (Esram and Chapman, 2007).

An MPPT power management system was proposed in Shao et al. (2007) for solar energy harvesting devices powering wireless sensor networks. The system used a charge pump with the operating frequency adjusted to maximize the power output to the battery or MCU. Similar work was presented in Yi et al. (2008), applying MPPT to vibration energy harvesting. A power management unit activated different operation modes, matching the load to the available energy level.

4.4.1.2 Duty Cycling

Duty cycling periodically turns individual loads off (or to low power modes) when not needed, thus lowering the overall power consumption of the system. Sleep modes, supported by most MCUs and communication modules, make duty cycling a straightforward approach; it is very commonly used in environmental monitoring and energy harvesting systems (Kansal et al., 2007; Moser et al., 2007, 2010).

The simplest, static implementation of this technique is to set a fixed duty cycle so that the average energy consumption matches the average energy production (Pimentel and Musilek, 2010). However, this approach suffers from serious shortcomings (Pimentel, 2012). If an energy source provides abundant energy (more than is consumed), a system with fixed duty cycle will waste excess energy once the storage element is fully charged. Conversely, if energy consumption exceeds production, the system will eventually deplete the storage reserves and stop working.

Indeed, the variable nature of environmental energy availability requires the use of dynamic duty cycling. It can make better use of the harvested energy by increasing the duty cycle when abundant energy is available and decreasing it when there is a shortage. This follows from the fact that energy storage is not ideal, and thus harvested energy should be used directly rather than stored first (Kansal et al., 2007). In general, such adaptive duty cycling approaches match energy production and consumption in real time.

4.4.1.3 Dynamic Voltage and Frequency Scaling

While duty cycling can provide significant energy savings in idle states, additional energy can be conserved by optimizing system performance in the active state (Sinha and Chandrakasan, 2001). There are two effective techniques for reducing processor energy consumption: dynamic voltage scaling (DVS) and dynamic frequency scaling (DFS). Their basic premise is that the MCU operating voltage and frequency can be adjusted to match instantaneous processing requirements.

DVS lowers the operating voltage of the MCU (or other circuits) which causes a quadratic reduction of the switching power. Decreasing the operating frequency through DFS reduces the switching power linearly and thus does not affect the total energy consumed per task (Sinha and Chandrakasan, 2001). However, it can be used to extend energy consumption over longer periods in times of low energy availability. In both cases, the MCU operates at a slower pace but remains active. Both techniques can also be combined to implement dynamic voltage and frequency scaling (DVFS) (Liu et al., 2012).

The scaling techniques allow the system to keep operating at a slow pace, but without compromising the execution of important tasks. In comparison, duty cycling completely stops all tasks during the sleep mode (Pimentel, 2012). DVS-based power management for battery powered systems is described in Zhong and Xu (2007), which also includes a protocol to statistically guarantee real-time performance. An example of a DVFS scheme for the dynamic power management of an embedded MCU is described in Ok et al. (2008).

4.4.1.4 Power Management for Peripherals

A summary of power reduction strategies that can be applied to peripherals (such as sensors and RF transceivers) is presented in Arms et al. (2005) and compiled in Table 4.1.

The need for sensor power management was identified in Raghunathan et al. (2006) since some sensors consume considerable amounts of energy. Examples include sensors that need high rate and high resolution analog to digital (A/D) converters, sensor arrays, active transducers, and sensors for detecting biological and chemical agents. To reduce the energy consumption of those sensors, the authors suggest applying methods of scalable fidelity and shutdown, similar to those used for processors. This can be achieved by acquiring "a measurement sample only if needed, when needed, where needed, and with the right level of fidelity" (Raghunathan et al., 2006). These sensor-specific techniques are analogous to the DVFS and duty cycling for MCU.

Power management for communication devices usually exploits tradeoffs between the use of local memory and data transmission (Moser et al., 2007). This technique, called buffer management (Niyato et al., 2007), stores incoming data and schedules it for future transmission.

4.4.2 FROM TECHNIQUES TO STRATEGIES

While individual power management techniques can reduce the rate at which energy is consumed by specific subsystems or during certain operating periods of EMSs, they are usually combined into energy management strategies. These strategies must be designed and optimized from a system perspective, taking into account all components of EMSs, their roles in system operation, and their interactions. Even from a component perspective, multiple configurations should be considered and the most suitable selected for a particular environment or situation. For example, while some techniques are designed to maximize the output power of solar cells, it is the power transfered from the power converter that drives the energy losses of the power supply (Shao et al., 2007). Similarly, rather than using storage with inevitably limited round trip efficiency, using solar energy directly can significantly improve the overall energy efficiency of the system (Raghunathan et al., 2005). These examples clearly show the need of a systematic approach to the development of energy management strategies. In general, it starts with identification of the various issues involved in the efficient use of stored and/or harvested energy (Kansal et al., 2007). The results of this analysis are then used for specification of practical energy management strategies.

4.4.2.1 Design

There are many different approaches for design of energy management strategies. However, any successful system strategy should manage device activities without compromising performance quality (Kompis and Sureka, 2010) or system reliability. Sometimes just a reformulation of the problem or its deeper analysis can lead to a new solution with lower energy demands (Horowitz et al., 1994). System components must be designed with strategy, environment and sensors considerations at the same time. The size of energy storage devices (e.g. secondary batteries or supercapacitors) should be selected using a statistical sizing strategy (Ongaro et al., 2012) that ensures a specific percentage of EMS activity per year. The energy storage requirements can be reduced by introducing a delay between data sampling and transmission using a buffer for the sampled data (Levron et al., 2011).

Many systems are dependent on the amount of energy present in the environment. As most natural energy sources are intermittent, their energy availability should be predicted to be used efficiently. Kansal (Kansal et al., 2007) presented a three part harvesting aware energy management strategy: 1) prediction of energy availability using an energy generation model, 2) computation of optimal duty cycles based on the predictions, and 3) dynamic, real-time adaptation of the duty cycle according to the observed energy generation profile.

An important aspect of any energy harvesting strategy is the design of wake-up models. As energy storage in harvesting systems can only be recharged during periods of low activity (i.e. various low power or sleep modes), a discontinuous operation model must be used (Mateu and Moll, 2005). There are two basic wake-up schemes (Zhu and Ni, 2007). A synchronous mode, where the system plans next wake up, and an asynchronous mode, where the system wakes up dynamically after meeting certain monitored or external conditions (Dewan et al., 2014). There are different asynchronous wake-up methods, based on internal triggers or external impulses (e.g. (RFID) tag) (Janek et al., 2007). Synchronous wake-up usually has its own scheduling algorithm. A basic scheme for task scheduling is presented in Liu et al. (2012): (i) generate initial schedule, (ii) balance workload, (iii) check energy availability for each scheduled task and tune up the schedule, (iv) speed up task execution when the system detects an overflow of incoming environmental energy.

4.4.2.2 Development and Testing

Long-term functional testing of EMSs would be lengthy and hence expensive, rendering such procedures impractical. Instead, long-term EMS operation can be modeled using sophisticated simulations taking into account energy availability and other environmental parameters. Simulation results can provide useful information about energy management strategy performance, in a short time and at a low cost. Simulation scenarios should be planned as complex processes where high level goals are accomplished by execution of sequences of simple tasks. An example clearly demonstrating the importance of simulations for environments where actual functional testing is not possible is mission scheduling for the Mars rover (Diaz et al., 2013). Similar approaches can be used for EMS intended for deployment in inaccessible environments, e.g. in the Arctic (Prauzek et al., 2014b) or in tropical regions (Watts et al., 2014).

4.4.2.3 Evaluation

A study by the Sensors and Instrumentation Knowledge Transfer Network (Kompis and Sureka, 2010) identified a framework for benchmarking low power systems employing power management techniques. This framework considers three primary concerns: (i) actual power consumption under typical conditions, (ii) system operability and usability, and (iii) impact of power management on system reliability. Other studies look at energy strategies directly as attempts to resolve the trade-off between the energy consumption and quality of service (Niyato et al., 2007) or fidelity (Hsu et al., 2006; Raghunathan et al., 2006) provided by the system. Watts (Watts et al., 2013) termed this trade-off the "energy-for-data exchange." Indeed, when only limited energy is available for system operation, its performance will be affected. Thus, the strategies can be seen as choices of which aspects of system performance should be sacrificed.

4.5 COMPUTATIONAL INTELLIGENCE IN EMS ENERGY MANAGEMENT

This section provides an overview of several state-of-the-art approaches to EMS energy management, based on various methods of computational intelligence. They range from decision trees and neural networks to clustering and evolutionary computing approaches, applied to energy availability prediction, data compression, rule-based control of individual sensor nodes, and network-wide sampling coordination. All presented techniques have a single common goal to reduce energy consumption of EMS and thus to extend their deployment lifetime and reduce their maintenance requirements. To achieve this goal, they must satisfy so called efficiency condition: the performance of EMSs with intelligent energy management must be better than that of their conventional counterparts.

Most methods described in this contribution do not require networked environment to operate and are applicable to both, standalone and networked sensor nodes. Network level techniques, such as E-BACH outlined in section 4.5.4, benefit from a plurality of sensor nodes that share information on the current state of their energy stores and the environment to distribute node activity in a way that reduces network-wide energy consumption. Although their performance generally depends on the number of nodes, there is no requirement for a minimum network size.

4.5.1 PRESSURE-BASED FORECASTING OF SOLAR ENERGY AVAILABILITY

Forecasting of solar energy availability is an important part of a number of applications, including predictive energy management in EMS. Weather forecasting techniques range from simple persistence, to the use of sophisticated numerical weather prediction models (Lynch, 2008). Weather is described using various atmospheric parameters, including pressure, temperature, wind, precipitation, and hundreds of other meteorological variables across large geographic areas and vertical expanses. Processing these huge datasets requires the use of some of the most powerful supercomputers (Lynch, 2008).

Crude prediction of storms and cloud cover can be performed using a single variable of atmospheric pressure, as its changes can signify changes in weather. Large drops are associated with storms and increasing cloud cover, while rises indicate improving weather conditions and clearing skies (Stull, 2000). These principles have been used in home weather stations for a number of decades (Brown, 1999), with accuracy in the neighborhood of 75% (Vizard, 1994).

Using a single variable for forecasting has both benefits and drawbacks. While the accuracy of the forecast may suffer, the use of few variables allows for the application of simple prediction algorithms that can be easily implemented with very limited computing resources. Additionally, fewer variables translates into fewer required sensors and associated energy costs. Moreover, atmospheric pressure sensors are not prone to errors caused by improper sensor placement, e.g. temperature sensors placed in direct sunlight (WMO, 2008). However, when other relevant variables are measured, they may be integrated into the forecast to improve its accuracy.

Using atmospheric pressure measurements, the amount of solar energy expected at a location during a day may be predicted (Rodway et al., 2015). Due to differing day lengths during a year, there is a degree of seasonality present in the actual energy time series. To alleviate this problem, a basic estimate can be produced solely on the location and elevation of the site, for each day of the year. Plots of measured and

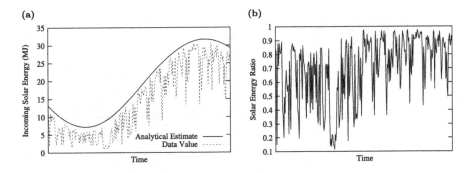

FIGURE 4.3

Example solar energy values. (a) Solar energy; (b) Ratio of measured/estimated solar energy.

analytical values of solar energy for a site in the northwestern United States is shown in Fig. 4.3a. The actual daily solar energy is divided by the analytical estimate to provide a percentage of energy lost to cloud cover and this percentage is the target for development of an actual, site-specific predictor (Fig. 4.3b). The final amount of solar energy is obtained by multiplying the predicted energy by the analytical estimate for that day.

The predictors can be implemented using a number of learning methods. Regression trees (Loh, 2014) are an attractive option because of their simple implementation on the limited hardware of an EMS. In order to create a tree predicting the percentage of daily solar energy in relation to the analytically calculated estimate, measurements of atmospheric are taken at specific times during a day. These measurements and differences between them then form the inputs to the regression tree during its formation (Rodway et al., 2015). To identify natural times to evaluate and act on the expected amounts of future energy, the measurements are taken at points relative to the local sunrises and sunsets. This way, regression trees may be created for each desired node location, or a more generic forecasting tree may be created for the entire network, at the expense of increased prediction error. Different forecasting horizons are also possible, however looking further ahead also incurs larger prediction error. Results reported by Rodway et al. (2015) confirm validity of this approach for energy availability prediction in EMS. The authors have already successfully included pressure-based predictors in simulation studies involving an entire EMS wireless sensor network (Rodway et al., 2016).

4.5.2 ENERGY MANAGEMENT IN EMS USING FUZZY CONTROL

The dependence of environmentally-powered EMS devices on the ambient energy brings forward the possibility to use energy availability prediction (e.g. the pressure-based approach introduced in Section 4.5.1) to estimate the state of charge of EMS energy storage elements (Castagnetti et al., 2012), to improve their performance (Piorno et al., 2009), or to reconfigure them (Li et al., 2014).

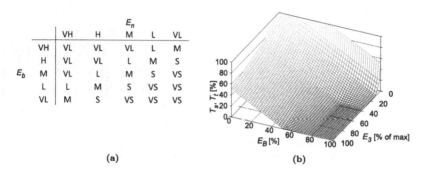

		E_n			
	VH	H	M	L	VL
E_b VH	VL	VL	VL	L	M
H	VL	VL	L	M	S
M	VL	L	M	S	VS
L	L	M	S	VS	VS
VL	M	S	VS	VS	VS

(a)   (b)

FIGURE 4.4

The rule base (a) and the sensing and transmission cycle surfaces of the basic controller before optimization (b). The fuzzy terms for inputs E_b, E_n are very low (VL), low (L), medium (M), high (H), and very high (VH), and for outputs T_s, T_t very short (VS), low (S), medium (M), long (L), and very long (VL).

In this section, such approach is illustrated through the design process of predictive fuzzy controller. During the development, the controller performance can be evaluated using a series of simulation experiments based on an accurate hardware model of the EMS device and energy availability data measured at the site intended for its deployment. Such simulation-based development approach can provide useful information about the performance of given energy management system, at a low cost and in a short time.

The structure of the controller is first devised manually, considering the nature of the energy management problem. For the purpose of energy management, one can define the following linguistic variables and corresponding fuzzy sets. The input variables are *normalized energy buffer level*, E_B, and *energy harvesting outlook* for n hours ahead, E_n. The fuzzy controller has two outputs *sensing cycle*, T_s (the time between two consecutive measurements) and *transmission cycle*, T_t (the time between two wireless data transmissions). Behavior of the controller can be described using several scenarios involving its desired response to specific values of the input variables. For example, when the forecast indicates high amount of available energy and the energy buffer is full, the device should take measurements as often as desired and transmit collected data without delay. Conversely, when the forecast is close to zero and the buffer level is low, the device should only use a minimal acceptable sensing rate and delay data transmissions to conserve energy. A fuzzy rule base and control surface of a controller implementing these scenarios is shown in Fig. 4.4.

Although the manually designed controller may operate correctly, its performance can be tuned through appropriate modifications of its parameters. The lack of an effective learning mechanism in fuzzy control system can be overcome using evolutionary computing. Differential evolution (DE) was selected as the method of choice for tuning the controller parameters, based on the literature review. DE evolves a population of candidate solutions by their iterative modification through differential mutation and crossover. In each iteration, mutation is applied to the current popu-

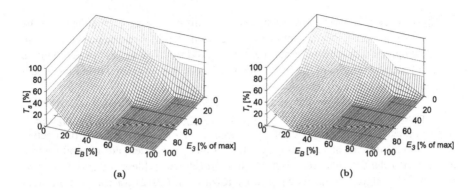

FIGURE 4.5

Controller surfaces of the optimized controller: sensing (a) and transmission cycle (b).

lation to form so called trial vectors. These vectors are further modified by various crossover operators. At the end of each iteration, the trial vectors compete with existing candidate solutions (controllers) for survival in the population.

The controllers were evolved and examined through a series of computational experiments using an enhanced version of a recently developed software simulator of the wireless sensor devices (Krömer et al., 2014a, 2014b). The operations considered by the simulator include analog and digital sensor readings, transfer of data between internal memories, and wireless data transmission between the node and a base station. At each simulation step, the simulator records a number of device attributes that are used for evaluating fitness function and for analyzing node performance.

The optimized controller provides consistent, predictable power management that yields data with a well spread temporal distribution and small number of failures (Prauzek et al., 2016). The control system is designed and optimized offline but fuzzy rule processing is calculated online (Prauzek et al., 2014a). The control surfaces for this controller are shown in Fig. 4.5. In contrast to static control strategies, the harvesting-aware approach substantially reduces the number of monitoring failures during which the phenomena of interest is not sensed for more than a maximum acceptable period of time. This enables the use of complex dynamic sensing and monitoring strategies powered by ambient energy that can respond to various application requirements. The resulting fuzzy control strategies are products of pre-deployment, site-specific optimization, and thus they generally do not require online tuning. As a result, they can be implemented in the form of a compiled fuzzy controller, e.g. as a look-up table. This allows their use for monitoring tasks that require real-time performance.

4.5.3 IN-NODE DATA COMPRESSION

Since wireless transmissions are very energy expensive, it is desirable to reduce the amount of data a wireless sensor node has to transmit. A possible approach is to use a compression algorithm to process the measured data. Time series data often exhibit strong correlation between subsequent samples. This unnecessary redundancy can be removed to minimize the amount of messages to transmit. Another possibility is to use a data prediction schema. In this case, measured samples are first used to build a model of the phenomena of interest. Data queries are then answered using the model instead of transferring all data samples.

An example of an algorithm suitable for in-node data compression is derivative-based prediction (DBP) (Raza et al., 2012, 2015; Bogliolo et al., 2014). The algorithm builds a piece-wise linear model of the measured data and transmits the parameters of the model instead of raw values of data points. As models are generally more concise then the data they represent, such approach can achieve a significant compression. The algorithm has been used by Raza et al. (2012) for transmitting light intensity data in a tunnel, achieving 97% suppression of the number of transmitted messages. Another deployment of the algorithm, by Potsch et al. (2014), achieved 95% suppression of messages of temperature measurements.

DBP is a simple, lossy data compression algorithm that approximates a time series with a sequence of linear models. It continuously accumulates data samples into a circular buffer that is initially empty. After the buffer is filled, a straight line is fitted to the buffered data: the first and the last data points are used to compute the slope of the time series. Together with the current data point, this slope forms a linear model which is then transmitted to the base station. This way, both devices share the same model.

After the first model is computed, both the base station and the wireless node draw samples from the model. The base station uses these data points to answer user queries, instead of real measured samples. The node continues to measure the real values and compares them against the model. If a significant deviation between the values, the node computes a new slope, updates the model and transmits the parameters to the base station. It is worth noting that the algorithm works in real time: the model is transmitted immediately after it has been created or updated as necessary.

An advantage of this particular algorithm is that it gives a guarantee on the data quality – a new model is transmitted if the real value exceeds a threshold, which sets a limit on the maximal absolute error. Another important factor is the computational simplicity of the algorithm – the slope of the linear model is computed simply by subtracting the first and last sample in the buffer and dividing by the time difference. The memory requirement is also small as the buffer size is typically less than 30 data samples.

Bartoň and Musilek (2016) introduced two improvements of this algorithm. If the real-time property of the algorithm is not required (i.e. a short delay in data delivery is acceptable), additional 40% of messages can be reduced. In this approach, called *delayed DBP*, the slope is not computed from past–current data points, but from past–future data point pair. Obviously, the algorithm cannot access future data and has to wait for it to arrive. Thus the algorithm is no longer real-time. An example of how DBP and delayed DBP approximate a time series with a sequence of linear models is shown in Fig. 4.6. An alternative approach, called *DBP with look ahead*, removes the need for delay using a time-series prediction approach when few future

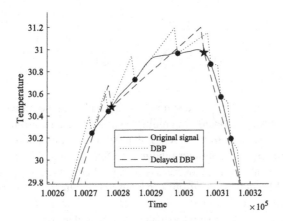

FIGURE 4.6

An example of how DBP approximates the time series with a sequence of linear models. The solid line is a temperature time series which has been approximated by DBP (dotted) and Delayed DBP (dashed). We can see, that whenever the distance between the model and the data gets over certain threshold a new model is generated – depicted as a full circle or a star. Reprinted, with permission from Bartoň and Musilek (2016). © 2016 IEEE.

points of the time series are predicted by a recurrent artificial neural network, and the slope is computed using this forecast.

The practicality of DBP approach and its derivatives is hindered by the need of transmitting service packets to maintaining the wireless network routing. Raza et al. (2012) identified these auxiliary packets as an important source of the remaining wireless transfers. As a result, a device will not benefit from more advanced data compression as long as the amount of network routing packets is not reduced, e.g. through more efficient network routing protocols (Krömer and Musilek, 2015). Alongside compression-based data traffic reduction, there are other approaches based on data fusion that gather results from multiple nodes and aggregate them in-network (Dasgupta et al., 2003).

4.5.4 ENTROPY-BASED CLUSTERING HIERARCHY

High-level clustering of EMS nodes can be performed for a number of different purposes, including fault-tolerance, load balancing, network connectivity, and other energy, and network-related objectives (Abbasi and Younis, 2007; Afsar and Tayarani-N, 2014). Organization of sensor nodes into clusters can be based on various criteria including information properties of collected data (Musilek et al., 2015).

Entropy is a measure of regularity/predictability vs. disorder/randomness (Shannon, 1948; Yentes et al., 2013). This general concept has been used in a number of environmental applications including EMS, e.g. in networks of hydrological measurement stations described by Singh (2013).

Marginal entropy of a random variable, $X = \{x_1, x_2, \ldots, x_n\}$, is defined as

$$H(X) = -\sum_i P(x_i) \log_2 P(x_i), \tag{4.1}$$

where $P(x_i)$ is the probability of a symbol, x_i, appearing in the sequence, X. Derived entropy-based measures, considering multiple variables and their mutual relationship, include e.g. *conditional entropy* $H(X|Y)$, that defines the amount of randomness in a variable, X, with respect to another random variable, Y, and *information gain* which measures how the entropy of X decreases with the knowledge Y (Quinlan, 1993). In the domain of EMS, *marginal entropy* can be used to evaluate the information content of a single variable (i.e. a sensor), *conditional entropy* describes the dependency of two variables, while *joint entropy* evaluates the whole data set.

Nevertheless, it is not easy to analyze time series-like data using entropies. The reasons include unknown probability distributions and inability to bin sequential (streaming) data. One possible approach is to use the approximate entropy and sample entropy (Yentes et al., 2013) that are often used in physiological data analysis (Richman and Moorman, 2000; Yentes et al., 2013; Baumert et al., 2013; Peng et al., 2014). They can be also used to evaluate relationships between two variables through *cross entropy* (Richman and Moorman, 2000).

Algorithmic (Kolmogorov) complexity is a measure of information contained in a sequence (Li and Vitányi, 2008). It is defined as the length of the shortest program that is needed to generate the sequence. Kolmogorov complexity cannot be directly computed but can be efficiently approximated by LZ76 – a data compression technique proposed by Lempel and Ziv in 1976 (Lempel and Ziv, 1976) which is known to be related to the entropy rate (Lempel and Ziv, 1976) and has been used to approximate entropy and algorithmic complexity before (Baumert et al., 2013; Peng et al., 2014; Henriques et al., 2013; Weijs et al., 2013). Kolmogorov marginal complexity approximated by LZ76, $\hat{K}(x) = LZ76(x)$, can be used to evaluate the *joint* Kolmogorov complexity of two time series, x and y (Li et al., 2004)

$$\hat{K}(xy) = \frac{\hat{K}(x, y) + \hat{K}(y, x)}{2},$$

where x, y stands for the concatenation of sequences x and y. Because the order of the concatenation may affect the result, mean of both possible concatenations is used. *Conditional* Kolmogorov complexity $\hat{K}(x|y)$ is defined as the shortest program that can regenerate x from y. $\hat{K}(x)$ is a special case of $\hat{K}(x|\lambda)$ where λ is the empty sequence. The intuition in the context of approximation via compression is that if the two series are similar, then the data in y can be used to improve the compression which will result in lower complexity. It can be estimated as the difference of the unconditional complexity estimates, i.e. $\hat{K}(x|y) = \hat{K}(xy) - \hat{K}(y)$ (Li et al., 2004).

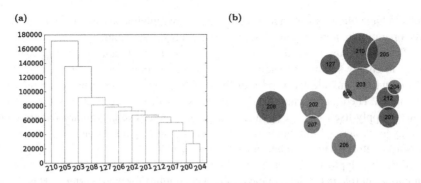

FIGURE 4.7

Illustration of entropy-based clustering hierarchy (E-BACH). (a) Hierarchical agglomerative clustering of the Tumbarumba dataset using weighted average linkage, performed on joint Kolmogorov complexity, $\hat{K}(xy)$. (b) E-BACH partition and node entropy ranking for the Tumbarumba site (circle diameter corresponds to relative entropy of the node). Reprinted, with permission from Musilek et al. (2015). © 2015 IEEE.

The validity of Kolmogorov complexity for environmental data in the area of EMS has been demonstrated experimentally. An analysis of a real-world data set collected by an EMS deployed in Tumbarumba, NSW, Australia, has shown a strong positive correlation between SampEn and LZ76 (Musilek et al., 2015).

Data entropy and its approximations can be used to identify clusters of sensors based on their role in the monitored environment. Depending on particular measures used, results of such clustering outline a hierarchy of nodes according to their information contents, and identify groups of sensor nodes that provide similar data. This can be further exploited to streamline EMS operations with respect to various high-level objectives such as spatial and temporal coverage of the environment, volume of energy harvested and consumed, data availability and immediacy, and others.

An example of an entropy-aware node clustering strategy is the entropy-based clustering hierarchy (E-BACH) algorithm (Musilek et al., 2015). It represents a strictly data-centric approach that considers the properties of collected data as the primary criteria for wireless sensor network (WSN) partitioning, as well as for cluster head (CH) selection and node scheduling. Entropy measures and their approximations are in this approach treated as information distances between sensor data. The corresponding distance matrix is used to perform hierarchical agglomerative clustering (Johnson, 1967). Clustering results for the best combination of entropy measures and linkage criteria are shown in Fig. 4.7.

4.6 CONCLUSIONS AND FUTURE WORK

This chapter provides an overview of energy management for environmental monitoring systems and reports on corresponding state-of-the-art methods using computational intelligence. Due to the nature of their construction and deployment, EMS devices have often very stringent requirements that include autonomy, maintenance-free operation in harsh conditions, and strict energy constraints.

The power supply is a key component of every EMS. It converts energy obtained from an energy source and supplies it to the load components of the system. Most common loads include data processing and control units, sensors, and communication interfaces. There is a variety of diverse sensors with varying energy demands. The power supply itself can use a variety of power sources (e.g. primary batteries and energy harvesters), different energy storage elements (e.g. rechargeable batteries and supercapacitors), and a number of energy converters.

Given the typical deployment conditions of EMS, energy harvesting often plays an important role in the development of power supplies for these systems. Harvesting can endow EMS with energy neutrality, in theory allowing the systems to operate indefinitely, without any additional external source of energy. Harvesting sources can be integrated in power supplies in several ways, leading to three main topologies of energy harvesting systems (autonomous, autonomous hybrid, and battery-supplemented).

Management of energy in EMS is necessary to strike a balance between their performance and endurance. It can be applied at three different levels: the MCU, node and network. In any case, a successful energy management starts with low-power design, and involves application of a number of power management techniques. The most common techniques include maximum power point tracking, duty cycling, dynamic voltage and frequency scaling, and a variety of scheduling mechanisms for peripherals.

Energy management strategies combine power management techniques to achieve performance objectives of EMS under energy availability constraints. In general, these objectives involve continuous operation of EMSs and fidelity of monitoring results they provide. The use of energy harvesting brings new possibilities through application of harvesting-aware energy management. This approach can take full advantage of environmental energy by predicting its availability and adapting the system performance accordingly, in a look-ahead fashion. This and other techniques and strategies are facilitated by modern electronic components that feature a variety of low-power and sleep modes. However, the complexity of the underlying tasks, combined with the limited computing resources of typical EMS, require application of simple yet powerful algorithms. A number of such approaches based on methods of computational intelligence have been recently introduced and are briefly described in this work.

With growing number, scale, and complexity of environmental monitoring applications, it is no longer possible to develop and test all EMS on their actual deployment sites. Simulation studies allow to perform these tasks in a short period of time and for a fraction of the cost of field testing. At the same time, they allow for testing of realistic operating scenarios that take into account historical or model data describing the actual conditions at the deployment sites. The use of simulations can also provide new insights about EMSs design methods and dramatically reduce their time to market.

REFERENCES

Abbasi, A.A., Younis, M., 2007. A survey on clustering algorithms for wireless sensor networks. Computer Communications 30 (14–15), 2826–2841.

Afsar, M.M., Tayarani-N, M., 2014. Clustering in sensor networks: a literature survey. Journal of Network and Computer Applications 46, 198–226.

Arms, S., Townsend, C., Churchill, D., Galbreath, J., Mundell, S., 2005. Power management for energy harvesting wireless sensors. In: Proceedings of SPIE, the International Society for Optical Engineering, vol. 5763, pp. 267–275.

Bartoň, T., Musilek, P., 2016. Derivative based prediction with look ahead. In: 2016 International Joint Conference on Neural Networks. IJCNN 2016, Vancouver, BC.

Baumert, M., Voss, A., Javorka, M., 2013. Compression based entropy estimation of heart rate variability on multiple time scales. In: EMBC, 35th Annual Int. Conf. of the IEEE, pp. 5037–5040.

Benson, B., Bond, B., Hamilton, M., Monson, R., Han, R., 2010. Perspectives on next-generation technology for environmental sensor networks. Frontiers in Ecology and the Environment 8 (4), 193–200.

Bogliolo, A., et al., 2014. Towards a true energetically sustainable wsn: a case study with prediction-based data collection and a wake-up receiver. In: 2014 9th IEEE Int. Symp. on Ind. Embedded Syst. (SIES), pp. 21–28.

Bradbury, K., 2010. Energy Storage Technology Review. Tech. rep., Duke University Energy Initiative. http://people.duke.edu/~kjb17/tutorials/Energy_Storage_Technologies.pdf (last accessed 01-October-2016).

Brown, A., 1999. Severe Weather Detector and Alarm. US5978738 A, US Patent Office.

Campbell, 2013. 107 and 108 Temperature Probes Brochure. Tech. rep., Campbell Scientific, Inc. http://www.campbellsci.com/107-l (last accessed 01-October-2016).

Capone, S., Forleo, A., Francioso, L., Rella, R., Siciliano, P., Spadavecchia, J., Presicce, D.S., Taurino, A.M., 2003. Solid state gas sensors: state of the art and future activities. Journal of Optoelectronics and Advanced Materials 5 (5), 1335–1348.

Castagnetti, A., Pegatoquet, A., Belleudy, C., Auguin, M., 2012. An efficient state of charge prediction model for solar harvesting wsn platforms. In: 2012 19th International Conference on Systems, Signals and Image Processing. IWSSIP 2012, pp. 122–125.

Chen, H., Cong, T.N., Yang, W., Tan, C., Li, Y., Ding, Y., 2009. Progress in electrical energy storage system: a critical review. Progress in Natural Science 19 (3), 291–312.

Chen, H., Wei, B., Ma, D., 2010. Energy storage and management system with carbon nanotube supercapacitor and multidirectional power delivery capability for autonomous wireless sensor nodes. IEEE Transactions on Power Electronics 25 (12), 2897–2909.

Cotfas, D., Cotfas, P., Borza, P., Ursutiu, D., Samoila, C., 2011. Wireless system for monitoring the solar radiation. Environmental Engineering and Management Journal 10 (8), 1133–1137.

Cox, J.F., 2001. Fundamentals of Linear Electronics: Integrated and Discrete. Cengage Learning.

Culler, D., Estrin, D., Srivastava, M., 2004. Overview of sensor networks. Computer 37 (8), 41–49.

Dallago, E., Danioni, A., Marchesi, M., Venchi, G., 2012. An autonomous power supply system supporting low-power wireless sensors. IEEE Transactions on Power Electronics 27 (10), 4272–4280.

Dasgupta, K., Kalpakis, K., Namjoshi, P., 2003. An efficient clustering-based heuristic for data gathering and aggregation in sensor networks. IEEE Wireless Communications and Networking Conference 3, 1948–1953.

Dewan, A., Ay, S., Karim, M., Beyenal, H., 2014. Alternative power sources for remote sensors: a review. Journal of Power Sources 245, 129–143.

Diaz, D., Cesta, A., Oddi, A., Rasconi, R., R-Moreno, M., 2013. Efficient energy management for autonomous control in rover missions. IEEE Computational Intelligence Magazine 8 (4), 12–24.

Esram, T., Chapman, P.L., 2007. Comparison of photovoltaic array maximum power point tracking techniques. IEEE Transactions on Energy Conversion 22 (2), 439–449.

Futagawa, M., Iwasaki, T., Murata, H., Ishida, M., Sawada, K., 2012. A miniature integrated multimodal sensor for measuring pH, EC and temperature for precision agriculture. Sensors (Switzerland) 12 (6), 8338–8354.

Gonzalez, A., Gallachir, B., McKeogh, E., Lynch, K., 2004. Study of Electricity Storage Technologies and Their Potential to Address Wind Energy Intermittency in Ireland. Tech. rep., Sustainable Energy Authority of Ireland. http://www.seai.ie/uploadedfiles/FundedProgrammes/REHC03001FinalReport.pdf (last accessed 01-October-2016).

Harb, A., 2011. Energy harvesting: state-of-the-art. Renewable Energy 36 (10), 2641–2654.

Henriques, T., Goncalves, H., Antunes, L., Matias, M., Bernardes, J., Costa-Santos, C., 2013. Entropy and compression: two measures of complexity. Journal of Evaluation in Clinical Practice 19 (6), 1101–1106.

Horowitz, M., Indermaur, T., Gonzalez, R., 1994. Low-power digital design. In: IEEE Symposium on Low Power Electronics, pp. 8–11.

Hsu, J., Zahedi, S., Kansal, A., Srivastava, M., Raghunathan, V., 2006. Adaptive duty cycling for energy harvesting systems. In: Proceedings of the International Symposium on Low Power Electronics and Design, Vol. 2006, pp. 180–185.

Huemmrich, K.F., Black, T.A., Jarvis, P.G., McCaughey, J.H., Hall, F.G., 1999. High temporal resolution NDVI phenology from micrometeorological radiation sensors. Journal of Geophysical Research D: Atmospheres 104 (D22), 27935–27944.

Ibrahim, H., Ilinca, A., Perron, J., 2008. Energy storage systems – characteristics and comparisons. Renewable and Sustainable Energy Reviews 12 (5), 1221–1250.

Janek, A., Steger, C., Preishuber-Pfluegl, J., Pistauer, M., 2007. Power management strategies for battery-driven higher class UHF RFID tags supported by energy harvesting devices. In: 2007 IEEE Workshop on Automatic Identification Advanced Technologies – Proceedings, pp. 122–127.

Johnson, S.C., 1967. Hierarchical clustering schemes. Psychometrika 2, 241–254.

Kansal, A., Hsu, J., Zahedi, S., Srivastava, M.B., 2007. Power management in energy harvesting sensor networks. ACM Transactions on Embedded Computing Systems 6 (4). http://dx.doi.org/10.1145/1274858.1274870.

Kansal, A., Potter, D., Srivastava, M., 2004. Performance aware tasking for environmentally powered sensor networks. Performance Evaluation Review 32 (1), 223–234.

Kimball, J., Kuhn, B., Balog, R., 2009. A system design approach for unattended solar energy harvesting supply. IEEE Transactions on Power Electronics 24 (4), 952–962.

Kompis, C., Sureka, P. (Eds.), 2010. Power Management Technologies to Enable Remote and Wireless Sensing. ESP Central Ltd (the Electronics, Sensors and Photonics KTN).

Kooistra, L., Bergsma, A., Chuma, B., de Bruin, S., 2009. Development of a dynamic web mapping service for vegetation productivity using Earth observation and in situ sensors in a sensor web based approach. Sensors 9 (4), 2371–2388.

Krömer, P., Musilek, P., 2015. Bio-Inspired Routing Strategies for Wireless Sensor Networks. Intelligent Systems Reference Library, vol. 85. Springer, pp. 155–181.

Krömer, P., Prauzek, M., Musilek, P., 2014a. Harvesting-aware control of wireless sensor nodes using fuzzy logic and differential evolution. In: Workshop on Energy Harvesting Communications. IEEE SECON 2014, Singapore, pp. 51–56.

Krömer, P., Prauzek, M., Musilek, P., Bartoň, T., 2014b. Optimization of wireless sensor node parameters by differential evolution and particle swarm optimization. In: IBICA, pp. 13–22.

Kumar, A., Kim, H., Hancke, G., 2013. Environmental monitoring systems: a review. IEEE Sensors Journal 13 (4), 1329–1339.

Lee, X., 1998. On micrometeorological observations of surface-air exchange over tall vegetation. Agricultural and Forest Meteorology 91 (1), 39–49.

Lempel, A., Ziv, J., 1976. On the complexity of finite sequences. IEEE Transactions on Information Theory 22 (1), 75–81.

Levron, Y., Shmilovitz, D., Martínez-Salamero, L., 2011. A power management strategy for minimization of energy storage reservoirs in wireless systems with energy harvesting. IEEE Transactions on Circuits and Systems I: Regular Papers 58 (3), 633–643.

Li, M., Chen, X., Li, X., Ma, B., Vitanyi, P., 2004. The similarity metric. IEEE Transactions on Information Theory 50 (12), 3250–3264. http://dx.doi.org/10.1109/TIT.2004.838101.

Li, M., Vitányi, P.M.B., 2008. An Introduction to Kolmogorov Complexity and Its Applications, third edition. Texts in Computer Science. Springer.

Li, Y., Jia, Z., Xie, S., 2014. Energy-prediction scheduler for reconfigurable systems in energy-harvesting environment. IET Wireless Sensor Systems 4 (2), 80–85.

Liu, S., Lu, J., Wu, Q., Qiu, Q., 2012. Harvesting-aware power management for real-time systems with renewable energy. IEEE Transactions on Very Large Scale Integration (VLSI) Systems 20 (8), 1473–1486.

Loh, W.-Y., 2014. Fifty years of classification and regression trees. International Statistical Review 82 (3), 329–348. http://dx.doi.org/10.1111/insr.12016.

Lynch, P., 2008. The origins of computer weather prediction and climate modeling. Journal of Computational Physics 227 (7), 3431–3444.

Mateu, L., Moll, F., 2005. Review of energy harvesting techniques and applications for microelectronics. In: Proceedings of SPIE – the International Society for Optical Engineering, PART I, Vol. 5837, pp. 359–373.

Mittelbach, H., Casini, F., Lehner, I., Teuling, A., Seneviratne, S., 2011. Soil moisture monitoring for climate research: evaluation of a low-cost sensor in the framework of the Swiss soil moisture experiment (SwissSMEX) campaign. Journal of Geophysical Research D: Atmospheres 116 (5).

Moser, C., Thiele, L., Brunelli, D., Benini, L., 2007. Adaptive power management in energy harvesting systems. In: Proceedings – Design, Automation and Test in Europe. DATE, pp. 773–778

Moser, C., Thiele, L., Brunelli, D., Benini, L., 2010. Adaptive power management for environmentally powered systems. IEEE Transactions on Computers 59 (4), 478–491.

Musilek, P., Kromer, P., Barton, T., 2015. E-bach: entropy-based clustering hierarchy for wireless sensor networks. In: 2015 IEEE/WIC/ACM International Conference on Web Intelligence and Intelligent Agent Technology (WI-IAT), Vol. 3, pp. 231–232.

Niyato, D., Rashid, M.M., Bhargava, V.K., 2007. Wireless sensor networks with energy harvesting technologies: a game-theoretic approach to optimal energy management. IEEE Wireless Communications 14 (4), 90–96.

Nymand, M., Andersen, M., 2010. High-efficiency isolated boost dcdc converter for high-power low-voltage fuel-cell applications. IEEE Transactions on Industrial Electronics 57 (2), 505–514.

Ok, S., Kim, J., Yoon, G., Chu, H., Oh, J., Kim, S.W., Kim, C., 2008. A DC-DC converter with a dual VCDL-based ADC and a self-calibrated DLL-based clock generator for an energy-aware EISC processor. In: Proceedings of the Custom Integrated Circuits Conference, pp. 551–554.

Ongaro, F., Saggini, S., Mattavelli, P., 2012. Li-ion battery-supercapacitor hybrid storage system for a long lifetime, photovoltaic-based wireless sensor network. IEEE Transactions on Power Electronics 27 (9), 3944–3952.

Pastorello, G.Z., Sanchez-Azofeifa, G. Arturo, Nascimento, M.A., 2011. Enviro-net: from networks of ground-based sensor systems to a web platform for sensor data management. Sensors 11 (6), 6454–6479.

Peng, Z., Genewein, T., Braun, D.A., 2014. Assessing randomness and complexity in human motion trajectories through analysis of symbolic sequences. Frontiers in Human Neuroscience 8, 168. http://dx.doi.org/10.3389/fnhum.2014.00168.

Pimentel, D., 2012. Energy Management for Automatic Monitoring Stations in Arctic Regions. Ph.D. thesis University of Alberta.

Pimentel, D., Musilek, P., 2010. Power management with energy harvesting devices. In: 2010 23rd Canadian Conference on Electrical and Computer Engineering (CCECE), pp. 1–4.

Pimentel, D., Musilek, P., Knight, A., Heckenbergerova, J., 2010. Characterization of a wind flutter generator. In: 2010 9th Conference on Environment and Electrical Engineering. EEEIC 2010, pp. 81–84.

Piorno, J., Bergonzini, C., Atienza, D., Rosing, T., 2009. Prediction and management in energy harvested wireless sensor nodes. In: 1st International Conference on Wireless Communication, Vehicular Technology, Information Theory and Aerospace and Electronic Systems Technology, 2009. Wireless VITAE 2009, pp. 6–10.

Pirapaharan, K., Gunathillake, W., Lokunarangoda, G., Nissansani, M., Palihena, H., Hoole, P., Aravind, C., Hoole, S., 2012. Design of a battery-less micro-scale RF energy harvester for medical devices. In: 2012 IEEE-EMBS Conference on Biomedical Engineering and Sciences. IECBES 2012, pp. 270–272.

Potsch, T., et al., 2014. Model-driven data acquisition for temperature sensor readings in wireless sensor networks. In: 2014 IEEE Ninth Int. Conference on Intelligent Sensors, Sensor Networks and Inform. Process. (ISSNIP), pp. 1–6.

Prauzek, M., Krömer, P., Rodway, J., Musilek, P., 2016. Differential evolution of fuzzy controller for environmentally-powered wireless sensors. Applied Soft Computing Journal 48, 193–206.

Prauzek, M., Musilek, P., Watts, A., 2014a. Fuzzy algorithm for intelligent wireless sensors with solar harvesting. In: IEEE SSCI 2014 – 2014 IEEE Symposium Series on Computational Intelligence – IES 2014: 2014 IEEE Symposium on Intelligent Embedded Systems, Proceedings, pp. 1–7.

Prauzek, M., Musilek, P., Watts, A.G., Michalikova, M., 2014b. Powering environmental monitoring systems in Arctic regions: a simulation study. Elektronika Ir Elektrotechnika 20 (7), 34–37. http://dx.doi.org/10.5755/j01.eee.20.7.8020.

Quinlan, J.R., 1993. C4.5: Programs for Machine Learning. Morgan Kaufmann Publishers Inc., San Francisco, CA, USA.

Raghunathan, V., Ganeriwal, S., Srivastava, M., 2006. Emerging techniques for long lived wireless sensor networks. IEEE Communications Magazine 44 (4), 108–114.

Raghunathan, V., Kansal, A., Hsu, J., Friedman, J., Srivastava, M., 2005. Design considerations for solar energy harvesting wireless embedded systems. In: 2005 4th International Symposium on Information Processing in Sensor Networks, Vol. 2005. IPSN 2005, pp. 457–462.

Raghunathan, V., Schurgers, C., Park, S., Srivastava, M., 2002. Energy-aware wireless microsensor networks. IEEE Signal Processing Magazine 19 (2), 40–50.

Raza, U., et al., 2012. What does model-driven data acquisition really achieve in wireless sensor networks? In: 2012 IEEE Int. Conference on Pervasive Computing and Commun. (PerCom), pp. 85–94.

Raza, U., et al., 2015. Practical data prediction for real-world wireless sensor networks. IEEE Transactions on Knowledge and Data Engineering 27 (8), 2231–2244. http://dx.doi.org/10.1109/TKDE.2015.2411594.

Richman, J.S., Moorman, J.R., 2000. Physiological time-series analysis using approximate entropy and sample entropy. American Journal of Physiology. Heart and Circulatory Physiology 278 (6), H2039–H2049.

Robinson, D.A., et al., 2008. Soil moisture measurement for ecological and hydrological watershed-scale observatories: a review. Vadose Zone Journal 7 (1), 358–389.

Rodway, J., Krömer, P., Karimi, S., Musilek, P., 2016. Differential evolution optimized fuzzy controller for wireless sensor network energy management. In: 2016 IEEE International Conference on Fuzzy Systems (FUZZ–IEEE), pp. 352–358.

Rodway, J., Musilek, P., Lozowski, E., Prauzek, M., Heckenbergerova, J., 2015. Pressure-based prediction of harvestable energy for powering environmental monitoring systems. In: 2015 IEEE 15th International Conference on Environment and Electrical Engineering (EEEIC), pp. 725–730.

Roozeboom, C., Hopcroft, M., Smith, W., Sim, J., Wickeraad, D., Hartwell, P., Pruitt, B., 2013. Integrated multifunctional environmental sensors. Journal of Microelectromechanical Systems 22 (3), 779–793.

Sanchez-Azofeifa, A., Powers, J.S., Fernandes, G.W., Quesada, M. (Eds.), 2013. Tropical Dry Forests in the Americas: Ecology, Conservation, and Management. CRC Press, Boca Raton, FL.

Shannon, C., 1948. A mathematical theory of communication. The Bell System Technical Journal 27 (3), 379–423. http://dx.doi.org/10.1002/j.1538-7305.1948.tb01338.x.

Shao, H., Tsui, C.-Y., Ki, W.-H., 2007. A micro power management system and maximum output power control for solar energy harvesting applications. In: Proceedings of the International Symposium on Low Power Electronics and Design, pp. 298–303.

Simjee, F., Chou, P., 2008. Efficient charging of supercapacitors for extended lifetime of wireless sensor nodes. IEEE Transactions on Power Electronics 23 (3), 1526–1536.

Singh, V., 2013. Entropy Theory and Its Application in Environmental and Water Engineering. Wiley.

Sinha, A., Chandrakasan, A., 2001. Dynamic power management in wireless sensor networks. IEEE Design & Test of Computers 18 (2), 62–74.

Stull, R.B., 2000. Meteorology for Scientists and Engineers. Brooks/Cole.

Tan, Y., Panda, S., 2011. Optimized wind energy harvesting system using resistance emulator and active rectifier for wireless sensor nodes. IEEE Transactions on Power Electronics 26 (1), 38–50.

Vargas, R., Detto, M., Baldocchi, D., Allen, M., 2010. Multiscale analysis of temporal variability of soil CO_2 production as influenced by weather and vegetation. Global Change Biology 16 (5), 1589–1605.

Vigorito, C., Ganesan, D., Barto, A., 2007. Adaptive control of duty cycling in energy-harvesting wireless sensor networks. In: 2007 4th Annual IEEE Communications Society Conference on Sensor, Mesh and Ad Hoc Communications and Networks. SECON, pp. 21–30.

Vizard, F., 1994. Storm warnings at home. Popular Mechanics 72.

Watts, A.G., Musilek, P., Wyard-Scott, L., 2013. Managing the energy-for-data exchange in remote monitoring systems. In: Electrical Power and Energy Conference 2013.

Watts, A.G., Prauzek, M., Musilek, P., Pelikan, E., Sanchez-Azofeifa, A., 2014. Fuzzy power management for environmental monitoring systems in tropical regions. In: World Congress on Computational Intelligence. WCCI 2014, Beijing, China.

Weddell, A., Merrett, G., Kazmierski, T., Al-Hashimi, B., 2011. Accurate supercapacitor modeling for energy harvesting wireless sensor nodes. IEEE Transactions on Circuits and Systems II: Express Briefs 58 (12), 911–915.

Weijs, S.V., van de Giesen, N., Parlange, M.B., 2013. Data compression to define information content of hydrological time series. Hydrology and Earth System Sciences 17 (8), 3171–3187. http://dx.doi.org/10.5194/hess-17-3171-2013.

WMO, 2008. Guide to Meteorological Instruments and Methods of Observation, 7 ed.. WMO (Series). Secretariat of the World Meteorological Organization, Geneva, Switzerland.

Yang, H., Zhang, Y., 2013. Analysis of supercapacitor energy loss for power management in environmentally powered wireless sensor nodes. IEEE Transactions on Power Electronics 28 (11), 5391–5403.

Yentes, J.M., Hunt, N., Schmid, K.K., Kaipust, J.P., McGrath, D., Stergiou, N., 2013. The appropriate use of approximate entropy and sample entropy with short data sets. Annals of Biomedical Engineering 41 (2), 349–365. http://dx.doi.org/10.1007/s10439-012-0668-3.

Yi, J., Su, F., Lam, Y.-H., Ki, W.-H., Tsui, C.-Y., 2008. An energy-adaptive MPPT power management unit for micro-power vibration energy harvesting. In: Proceedings – IEEE International Symposium on Circuits and Systems, pp. 2570–2573.

Zhao, X., Xu, H., Zhang, P., Fu, J., Bai, Y., 2013. Soil water, salt, and groundwater characteristics in shelterbelts with no irrigation for several years in an extremely arid area. Environmental Monitoring and Assessment 185 (12), 10091–10100.

Zhong, X., Xu, C.-Z., 2007. Energy-aware modeling and scheduling for dynamic voltage scaling with statistical real-time guarantee. IEEE Transactions on Computers 56 (3), 358–372.

Zhou, J., Mason, A. Andrew, 2002. Communication buses and protocols for sensor networks. Sensors 7 (2), 244–257.

Zhu, Y., Ni, L., 2007. Probabilistic wakeup: adaptive duty cycling for energy-efficient event detection. In: MSWiM'07: Proceedings of the Tenth ACM Symposium on Modeling, Analysis, and Simulation of Wireless and Mobile Systems. Springer, pp. 360–367.

ACRONYMS AND GLOSSARY

List of acronyms with explanation

A/D analog to digital

AC	alternating current
BJT	bipolar junction transistor
CH	cluster head
DBP	derivative-based prediction
DC	direct current
DE	differential evolution
DFS	dynamic frequency scaling
DVFS	dynamic voltage and frequency scaling
DVS	dynamic voltage scaling
E-BACH	entropy-based clustering hierarchy
EMS	environmental monitoring system
$\mathbf{I^2C}$	inter-integrated Circuit
MCU	microcontroller unit
MEMS	micro-electro-mechanical systems
MPPT	maximum power point tracking
NDIR	nondispersive infrared
RDT	resistance temperature detector
RFID	radio frequency identification
SPI	serial peripheral interface
WSN	wireless sensor network

Glossary of terms with explanation

Duty cycling switching between power-on and power-save modes of a device.

Energy neutrality principle device can never consume more energy than their harvesting part can deliver.

Energy storage a device to capture produced energy for use at a later time.

Energy harvesting process by which electrical energy is extracted from ambient sources.

Energy independence independence of a device on energy supply from power grid.

Environmental monitoring system monitoring device or network, usually installed in a remote or inaccessible location.

Evolutionary computing a family of metaheuristic algorithms based on adopting Darwinian principles of evolution to perform optimization tasks.

Fuzzy control system a rule-based control system based on fuzzy logic.

In-node data compression reducing of the amount of data a wireless sensor node has to transmit.

Maximum power point tracking a technique used to maximize power extraction from various sources under all conditions.

Power converter a device converting electric energy from one form to another.

Power management functionality of electronic devices to adjust their operation to achieve low power consumption.

Pressure-based predictors algorithms using atmospheric pressure to predict changes in weather.

Sensor nodes devices that can acquire data from analog or digital sensors and transmit it.

DATA STREAMING, PROCESSING, AND ANALYSIS

2

SMART SENSOR DATA STREAM DELIVERY TECHNOLOGIES

5

Tomoya Kawakami*, **Yoshimasa Ishi**[†], **Tomoki Yoshihisa**[†], **Yuuichi Teranishi**[‡,†]

*Graduate School of Information Science, Nara Institute of Science and Technology, Japan
[†]Cybermedia Center, Osaka University, Japan [‡]National Institute of Information and Communications Technology, Japan*

5.1 INTRODUCTION

In recent years, various types of applications such as video delivery and environmental monitoring have been possible, and therefore, sensor data stream delivery where sensor data are periodically gained and delivered continuously has been attracting great attention. As for this sensor data stream delivery, it is possible for the same sensor data stream to have different delivery cycles depending on the receivers.

Examples of such demands include:

- Pictures from camera sensor must be delivered in high frame rate such as 10 fps to detect fast moving objects (cars, motorcycles, etc.) by image processing whereas low frame rate such as 2 fps to detect slow moving object (bicycles, pedestrians, etc.).
- Wearable vibration sensor data must be delivered high rate such as 10 Hz to identify users using characteristic walking motion whereas low rate such as 2 Hz to implement a simple pedometer application.
- When a user is continuously checking the amount of rain to decide the timing to go out in the rainy day, the data are delivered once per second to a smart phone if it connects to a power source. However, the data are delivered once per minute in order to reduce power consumption if it does not connect to a power source.

It is general in sensor data stream delivery that sensor data gained by one sensor is shared by a large number of users. Currently, various P2P-based techniques for dispersing the communication load of the deliverer (source node) have been studied in the data streaming (Zhang et al., 2005; Liao et al., 2006; Magharei and Rejaie, 2007; Yu et al., 2011; Sakashita et al., 2010; Banerjee et al., 2002; Tran et al., 2003; Jin et al., 2007; Silawarawet and Nupairoj, 2011; Le and Nguyen, 2012). In these researches, in the case where the same sensor data stream is delivered to a number of terminals (destination nodes), the communication load of the source node is dispersed when the destination nodes send the received data to other destination nodes. When the delivery cycle is different, the sensor data stream whose delivery cycle is a common divisor of the required cycles can be delivered to all of the destination nodes if the delivery cycles are in a multiple relationship or can be approximated as having a multiple relationship. However, the destination nodes receive redundant data which are not included to the times of each required cycle.

Therefore, we proposed smart sensor data stream delivery technologies to distribute communication loads (Kawakami et al., 2014a, 2014b). Our proposing techniques assume a P2P model or cloud model with different delivery cycles. In this chapter, we introduce our proposing methods and those experimental results in simulations.

In the following, the P2P-based technologies are described in Section 5.2. The technologies on the cloud are described in Section 5.3. We describe the related work in Section 5.4. Finally, the conclusion of the chapter is presented in Section 5.5.

5.2 P2P-BASED TECHNOLOGIES
5.2.1 ADDRESSED PROBLEMS
5.2.1.1 Assumed Environment
The purpose of this study is to disperse the communication load in the sensor stream deliveries that have different delivery cycles. The source nodes have sensors so as to gain sensor data periodically. The destination node searches for the corresponding source node and requires a sensor data stream with a desired delivery cycle in a P2P model. Upon reception of the query from the destination node, the source node determines the delivery path of the sensor data stream. The queries are received during delivery, and whenever a query is received, the delivery path is changed. The sensor data stream is delivered along the determined path, and when a destination node sends a sensor data stream to another destination node, a query is received by the destination node to which the sensor data stream is to be sent. The delivery path changes whenever the source node receives a query for delivering a sensor data stream.

5.2.1.2 Input Setting
The source node of sensor data is S and n destination nodes are D_i $(i = 1, \cdots, n)$. In addition, the delivery cycle of S is C_0 and the delivery cycle required by D_i is C_i. The sensor data that has not been gained by the source node cannot be delivered, and therefore, C_i is a multiple of C_0, which can be represented by $C_i = c_j C_0$ using a certain integer c_j $(= 1, 2, \cdots)$.

In Fig. 5.1, each node indicates a source node and the branches indicate delivery paths for the sensor data streams. Concretely, they indicate communication links in an application layer. The branches are indicated by dotted lines because there is a possibility that the branches may not deliver a sensor data stream depending on the

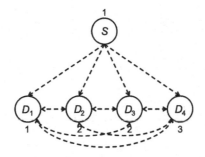

FIGURE 5.1

An example of input setting.

Time	1	2	3	4	5	6	7	...
Table 5.1 An Example of the Sensor Data Delivery								
D_1 (Cycle=1)	o	o	o	o	o	o	o	...
D_2 (Cycle $= 2$)		o		o		o		...
D_3 (Cycle $= 2$)		o		o		o		...
D_4 (Cycle $= 3$)			o			o		...

delivery method. The source node S is at the top and the four destination nodes D_1, $\cdots$, D_4 ($n = 4$) are at the bottom. The figure in the vicinity of each destination node indicates the delivery cycle, and $C_0 = 1$, $C_1 = 1$, $C_2 = 2$, $C_3 = 2$, and $C_4 = 3$. This corresponds to the case where a live camera acquires an image once every second, and D_1 views the image once every second, D_2 and D_3 view the image once every two seconds, and D_4 views the image once every three seconds, for example. Table 5.1 shows the delivery cycle of each destination node and the sensor data to be received in the example in Fig. 5.1.

5.2.1.3 Objective Function

The communication load of S is L_0 and the communication load of D_i is L_i. The communication load SL of the entirety of the system is given by the following equation:

$$SL = \sum_{i=0}^{n} L_i. \tag{5.1}$$

In addition, the following fairness index (FI) is often used as an index for "equality" of specified values (Jain et al., 1984):

$$FI = \frac{\left(\sum_{i=0}^{n} L_i\right)^2}{n \sum_{i=0}^{n} L_i^2} \tag{5.2}$$

where $0 \leq FI \leq 1$, and when $FI = 1$, $L_0 = \cdots = L_n$ and the loads of nodes are perfectly fair. It is indicated that the closer FI is to 1, the more the load is dispersed.

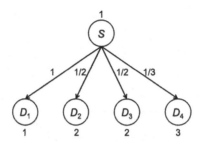

FIGURE 5.2

A case of delivering the sensor data stream directly from the source node.

Another purpose of the present study is to disperse the communication load to the destination nodes while suppressing the communication load of the entirety of the system. Therefore, the objective function is SL and $1 - FI$, and the delivery path is determined to make these values minimum. In the present problem, the received sensor data stream can be sent to another destination node, and each destination node determines the sensor data to be sent.

5.2.1.4 Definition of a Load

The communication load of the source node and the destination nodes is given as the total of the load due to the reception of the sensor data stream and the load due to the transmission. The communication load due to the reception is referred to as the reception load, the reception load of D_i is I_i and the reception load of S is I_0. The communication load due to the transmission is referred to as the transmission load, the transmission load of D_i is O_i and the transmission load of S is O_0.

In many cases, the reception load and the transmission load are proportional to the number of sensor data pieces per unit hour of the sensor data stream to be sent and received. The number of pieces of sensor data per unit hour of the sensor data stream that is to be delivered by D_p to D_q ($q \neq p$; $p, q = 1, \cdots, n$) is $R(p, q)$, and the number delivered by S to D_q is $R(0, q)$.

In the this study, the load with which one piece of sensor data can be received and sent per unit hour is normalized to 1, and the communication load L_r of D_r is given in the following equations:

$$L_r = I_r + O_r = \alpha \sum_{i=0}^{n} R(i, r) + \beta \sum_{i=0}^{n} R(r, i) \tag{5.3}$$

where α and β are loads with which one piece of sensor data is received and sent, respectively.

Fig. 5.2 shows a case where $\alpha = \beta$ and the sensor data is delivered directly from the source node in the example in Fig. 5.1. The figures in the vicinity of the branches are the number of sensor data pieces per unit hour of the sensor data stream. In this example, the sensor data stream is delivered only from the server, and therefore, $R(0, q) = 1/C_q$ and $R(p, q) = 0$. Thus, $R(0, 1) = 1$, $R(0, 2) = 1/2$, $R(0, 3) = 1/2$, and $R(0, 3) = 1/3$. The load at each end is $L_0 = R(1, 0) + R(2, 0) + R(3, 0) + R(4, 0) + R(0, 1) + R(0, 2) + R(0, 3) + R(0, 4) = R(0, 1) + R(0, 2) + R(0, 3) + R(0, 4) = 2.33$, $L_1 = 1.00$, $L_2 = 0.500$, $L_3 = 0.500$, and $L_4 = 0.333$. In this case, $SL = 4.667$ and $FI = 0.617$.

5.2.2 LOAD DISTRIBUTION METHOD

Currently, we have proposed a load distribution method named LLF-H (lowest load first considering hops) (Kawakami et al., 2014b). LLF-H method estimates loads of each nodes and the lowest load node at points in each time sends the data to another destination node. The communication load is dispersed when the destination nodes of which the assumed load is small sends the communication load to the sensor data of the same delivery time that is included in different sensor data streams. In addition, a value that can be calculated from any model is defined as a load, for example, α and β are given to the number of transmissions and receptions of the sensor data as coefficients, in order to determine the delivery path for the purpose of dispersing various values. Moreover, the number of hops until each destination node receives data is important because the number of hops affects the delay of data delivery. In the proposed method, the maximum number of hops allowable to receive data is specified, and delivery paths satisfying the limitation of hops are determined.

In LLF-H method, the load is estimated and the delivery path is determined before the start of delivery, and the following contents are delivered to each destination node after the delivery path has been determined. The contents are about the sensor data sent and received by each destination node at each point in time, and each destination node sends the received sensor data to another destination node according to the timetable. In the timetable, the time is set as the least common multiple of the delivery cycles of all the destination nodes after time 1. After that, the time returns to time 1 and transmissions following the timetable are repeated.

In order to generate the timetable, the amount of calculation becomes enormous due to a large number of combinations, and therefore, in LLF-H method, a restriction is set so that the sensor data stream is sent from a node having a longer cycle to a node having a shorter cycle. As a result of this restriction, the source node first sends the sensor data stream to the destination node having the longest cycle at each time. Likewise, the destination node having the longest cycle at each time sends the sensor data stream to the destination node having the second longest cycle. From the above description, the following loads can be assumed on the basis of the delivery cycles of the destination nodes so as to be used for the preparation of the timetable of the delivery path.

- Reception load within the least common multiple cycle of each destination node
- Transmission load from the source node to the destination node having the longest cycle at each time

Input:
1: *cycles*: Array of delivery cycles of the nodes sorted in ascending order (cycle of source node is −1 and at index 0)
2: *allowableHopCount*: Allowable maximum number of hops
Output:
_____
3: *loads*;   ▷ Array of loads of each node in *cycles*
4: *hopCounts*;   ▷ Array of the number of hops until each node in *cycles* receives data
5: *cycLcm* ← *calculateLCM(cycles)*;   ▷ The least common multiple of the delivery cycles of the destination nodes
6: **for** *i* ← 1 to *cycles.length* **do**
7:     *loads*[*i*] ← α ∗ *cycLcm*/*cycles*[*i*];   ▷ The reception load of each destination node
8: **end for**
9: **for** *i* ← 1 to *cycLcm* **do**
10:     *longNode1st* ← *getLongestNodeIndex(cycles, i)*; ▷ The index of destination node having the longest cycle at time *i*
11:     **if** *longNode1st* ≠ 0 **then**
12:         *requestToSend(0, longNode1st, i)*; ▷ Delivery path from the source node to the destination node having the longest cycle node
13:         *loads*[0] ← *loads*[0] + β;   ▷ The transmission load is added to the source node
14:     **end if**
15:     *longNode2nd* ← *get2ndLongestNodeIndex(cycles, i)*;   ▷ The index of destination node having the second longest cycle at time *i*
16:     **if** *longNode2nd* ≠ *longNode1st* **then**
17:         *requestToSend(longNode1st, longNode2nd, i)*;   ▷ Delivery path from the destination node having the longest cycle to the destination node having the second longest cycle
18:         *loads*[*longNode1st*] ← *loads*[*longNode1st*] + β; ▷ The transmission load is added to the destination node having the longest cycle
19:     **end if**
20: **end for**

FIGURE 5.3
_____
Pseudocode for load estimation and delivery path determination (1/2).

- Transmission load from the destination node having the longest cycle to the destination node having the second longest cycle

On the basis of the approach, the delivery procedure in LLF-H method is described below. Figs. 5.3 and 5.4 show the algorithm of the load estimation and the determination of the delivery paths. In this chapter, the example in Fig. 5.1 and Table 5.1 is used. The transmission and reception load per sensor data piece is $\alpha = \beta = 1$ and the maximum number of hops allowable in LLF-H is 3. Figs. 5.5(a) and 5.5(b) show delivery paths by LLF-H method in the example of Fig. 5.1 and $\alpha = \beta = 1$, the maximum number of hops allowable is three. Fig. 5.5(a) shows the delivery paths determined by delivery cycles and Fig. 5.5(b) shows the delivery paths determined for D_2 at time 6. The numbers in small circles near the delivery path in Fig. 5.5(a) indicate the sensor data at that time. Fig. 5.6 shows the final timetable in the example in Fig. 5.5. The solid arrows in Fig. 5.6 are the paths that have been determined by the delivery cycle of each destination node, and the broken arrows are the paths selected through the load estimation. As a result of LLF-H method in the example of Fig. 5.5, $SL = 4.667$ and there are no redundant communication loads compared with SD method. On the other hand, $FI = 0.992 > 0.876$ and communication loads are more distributed than SD method.

```
21: for i ← 1 to cycLcm do
22:    hopCounts[0] ← 0;                                      ▷ The number of hops of the source node is set to 0
23:    longNode1st ← getLongestNodeIndex(cycles, i);
24:    hopCounts[longNode1st] ← hopCounts[0] + 1;      ▷ The number of hops of the longest cycle node is set to 1
25:    longNode2nd ← get2ndLongestNodeIndex(cycles, i);
26:    hopCounts[longNode2nd] ← hopCounts[longNode1st] + 1; ▷ The number of hops of the second longest cycle
    node is set to 2
27:    for j ← longNode2nd − 1 to 1 do
28:       if i mod cycles[j] = 0 then
29:          minNode ← longNode1st;
30:          minLoad ← loads[minNode];
31:          for k ← longNode2nd to j + 1 do
32:             tmpLoad ← loads[k];
33:             if i mod cycles[k] = 0 and hopCounts[k] < allowableHopCount and tmpLoad < minLoad then
34:                minLoad ← tmpLoad;
35:                minNode ← k;
36:             end if
37:          end for
38:          requestToSend(minNode, j, i);    ▷ Delivery path from the destination node having the minimum load
    to D_j at time i
39:          loads[minNode] ← loads[minNode] + β;    ▷ The transmission load is added to the destination node
    having the minimum load
40:          hopCounts[j] ← hopCounts[minNode] + 1;    ▷ The number of hops from the lowest load node is set
41:       end if
42:    end for
43: end for
```

FIGURE 5.4

Pseudocode for load estimation and delivery path determination (2/2).

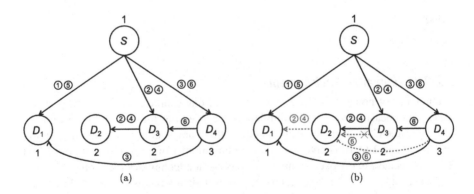

FIGURE 5.5

An example of LLF-H method. (a) Delivery paths determined by delivery cycles; (b) Delivery path determination for D_2 at time 6.

5.2.3 EVALUATION

In this section, we describe the evaluation of the proposed LLF-H method in Section 5.2.2 by simulation.

Time	1	2	3	4	5	6
D_1 (Cycle=1)						
D_2 (Cycle=2)						
D_3 (Cycle=2)						
D_4 (Cycle=3)						
S (Cycle=1)						

FIGURE 5.6

Timetable for sensor data delivery.

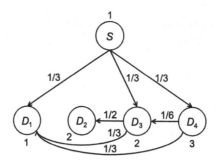

FIGURE 5.7

An example of LCF method.

5.2.3.1 Simulation Environment

The number of nodes n is 2^1, 2^2, $\cdots$, 2^{10}, and one source node is included to the nodes. The measurement of the simulation is carried out for each value. As for the delivery cycle C_i of each destination node without any source node, a random delivery cycle from 1 to 10 is given. The maximum number of hops allowable in LLF-H is $\log_2 n$ because many overlay network construction techniques mention.

According to the existing techniques, sensor data streams having different delivery cycles are delivered as different data streams, and therefore, sensor data is directly delivered from the server. These existing techniques are collectively referred to as Fig. 5.2. We call these techniques SD (source direct) method and compared with LLF-H method. In addition, we compared with the approach that the longest cycle node in each time receives the sensor data from the source node and redelivers to other destination nodes. When some nodes have the same longest cycle, a specific node of them sends to other nodes. We call this technique LCF (longest cycle first) method. Fig. 5.7 shows a case by LCF method in the same example. In the case by LCF method, $SL = 4.667$ and $FI = 0.992$, and thus, the load is dispersed without changing the load of the entirety of the system. However, a destination node having a small load due to reception during a long cycle sends many sensor data streams to another destination node, and therefore, there is a possibility of the transmission load of the destination node with a long cycle increasing in the case where there are many destination nodes at the same time. In addition, the delivery path is determined on the basis of the number of transmissions and receptions of data, and therefore, the effects of the load distribution are low in an environment where the load cannot be simply represented by the number of transmissions and receptions, for example, $\alpha \neq \beta$. Moreover, in order to confirm an influence by guaranteeing the maximum number of hops in LLF-H method, we compared an approach based on load estimation like LLF-H method but not considering hops. We call this technique LLF (lowest load first) method. LLF method is an algorithm that is removed elements related hops from the algorithm of LLF-H method.

In this simulation, each method is tried 20 times with the respective numbers of nodes, and the average value of the results is calculated.

5.2.3.2 Total System Loads

Fig. 5.8(a) shows the communication load SL of the entirety of the system. The longitudinal axis is the communication load of the entirety of the system, and the lateral axis is the number of destination nodes. In the present simulation, a data set is prepared with the cycle and the order at random, and a terminal having each delivery cycle is added in order. The simulation results are the communication load at the point in time for the number of terminals indicated by the lateral axis, and the greater the number of terminals, the higher the communication load of the entirety of the system.

According to SD method, the source node directly delivers the sensor data stream, and therefore, there is no redundant transmission or reception of data, and thus, the communication load of the entirety of the system has a minimum value. As can be seen from Fig. 5.8(a), the loads of the entirety of the system respectively in LCF, LLF, LLF-H method are equal to that in SD method. This is because there is no redundant communication load, for example, the same sensor data are received from some nodes. In LCF, LLF, LLF-H method, transmission processes are executed by the nodes which relate to the delivery cycles instead of the source node. Therefore, only the transmission nodes are changed, and the total number of transmission/reception processes in the whole system is not changed. The value of the load is the same as in SD method having the minimum value, and therefore, it can be seen that the communication load in LLF-H method is also kept to a minimum.

5.2.3.3 Loads for Source Node

In the case where the transmission and reception loads per data piece are $\alpha = \beta = 1$ as in the previous section, Fig. 5.8(b) shows the communication loads for the source node. The longitudinal axis is the communication load of the source node and the lateral axis indicates the number of nodes.

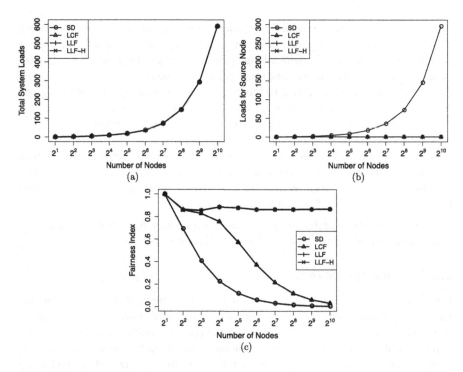

FIGURE 5.8

Results by the number of nodes. (a) Total system loads; (b) Loads for source node; (c) Load distribution.

In SD method, the source node delivers directly, and therefore, the load on the source node is increased by the number of destination nodes. Meanwhile, the destination nodes redeliver in the other methods, and therefore, the influence of the number of destination nodes is small.

5.2.3.4 Load Distribution

In the case where the transmission and reception loads per data piece are $\alpha = \beta = 1$ as in the previous section, the load Distribution to the destination nodes is shown below.

Fig. 5.8(c) shows the results where the longitudinal axis is *FI* and the lateral axis is the number of destination nodes. In Fig. 5.8(c), the greater the number of destination nodes, the smaller the Fairness Index, and the more the loads are unbalanced. This is because the greater the number of destination nodes, the longer the longest delivery cycle, and the greater the difference in the delivery cycle. In LLF and LLF-H method, the communication load can be dispersed more than in LCF method, and in particular, the difference is significant in an environment having a great number of destination nodes. This is because in LLF and LLF-H method, the destination node having the smallest load at a point in time sends the sensor data stream to another

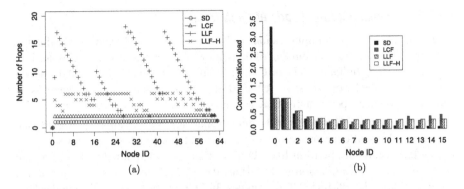

FIGURE 5.9

Results of each node. (a) The number of hops; (b) Communication loads.

destination node so that the load can be equalized. Meanwhile, in SD method, the load is concentrated in the source node. In LCF method, the load concentrates in a destination node having a long cycle when there are many destination nodes at the same time.

5.2.3.5 The Number of Hops

Fig. 5.9(a) shows the number of hops until each node receives data at the time of the lowest common multiple of delivery cycles. All nodes receive data at the time of the lowest common multiple. In this simulation, the number of nodes is $2^6 = 64$ and delivery cycle of each node is determined at random between 1 to 10. The lateral axis shows each node as ID $0 \sim 63$. The longitudinal axis is the number of hops until each node receives data at the time of the lowest common multiple. The node IDs in the lateral axis are sorted from the shorter cycle node. The node with ID $= 0$ is the source node. In this simulation, the allowable number of hops in LLF-H is 6.

As can be seen in Fig. 5.9(a), the number of hops of the source node (ID $= 0$) is always zero in all methods. The number of hops of the longest cycle node at the time of the lowest common multiple (ID $= 63$) is one in all methods. In SD method, the source node delivers to all destination nodes. The numbers of hops of the nodes without the source node are one. In LCF method, the longest cycle node (ID $= 63$) sends to all destination nodes. The numbers of hops of the nodes without the source node (ID $= 0$) and the longest cycle node (ID $= 63$) are two. In LLF method, the delivery paths are constructed without considering the number of hops and the maximum number of hops is 18. In this simulation, the number of nodes is 2^6 and the number of hops seems large comparing with $\log_2 n = \log_2 2^6 = 6$. Since the number of hops is increased by the number of nodes, the number of hops in LLF method is large and a delivery delay happens. On the other hand, in LLF-H method, another low load node sends data if the number of hops seems over the specified number 6. Thus, the number of hops of each node is less than six.

5.2.3.6 Communication Loads of Each Node

Fig. 5.9(b) shows the communication loads of the source node and the destination nodes in the case where the number of destination nodes is ten and the cycles are 1, 2, $\cdots$, 16. The longitudinal axis is the communication load of each node, and the lateral axis indicates the ID of nodes where the ID of the source node is 0. The transmission and reception loads per data piece are $\alpha = \beta = 1$ as in the previous section.

In SD method, the source node directly delivers the sensor data stream, and therefore, the load concentrates on the source node of which the ID is 0. The existing P2P techniques perform like SD method in this environment where there are no same cycles in destination nodes. Thus, the load extremely concentrates on the source node in the existing P2P techniques. In LCF method, destination nodes having a long cycle send the sensor data stream with priority, and therefore have a load greater than that of the destination nodes of which the cycle is close to the median. Meanwhile, in LLF and LLF-H method, loads can be equalized between the destination nodes having a delivery cycle at a certain level or higher as compared to LCF method.

5.3 TECHNOLOGIES ON THE CLOUD

5.3.1 ADDRESSED PROBLEMS

In the sensor data stream delivery system that we assume in this chapter, computers (nodes) to relay sensor data streams constructs P2P overlay network on the cloud. Sensors are periodically sent from their sources through the Internet or private line to a node on the cloud. The destinations also connect to a node on the cloud likewise and request sensor data streams with *sensor-ID* and *delivery cycle*. Then the sensor data is delivered to the destinations by the hops among nodes on the cloud. We assume that selectable delivery cycles for each sensor data stream are given in advance. Nodes are able to send sensor data to other nodes anytime, and sensor data are distributed for each sensor data stream and time.

Hereafter, the sensor data streams are denoted as S_i ($i = 1, \cdots, l$), destinations are denoted as D_i ($i = 1, \cdots, m$), and nodes are denoted as N_i ($i = 1, \cdots, n$). Fig. 5.10 shows a model of delivery system. The number of sensor data streams $l = 2$, the number of destinations $m = 4$ and the number of nodes $n = 3$. The 'a' represents the sensor ID of the source S_1, and the 'b' represents the sensor ID of the source S_2. In this example, the source S_2 corresponds to a camera sensor and acquires an image once every second, and D_1 does not view the image, D_2 and D_3 view the image once every second, and D_4 views the image once every three seconds, for example. Table 5.2 shows the delivery cycle of each destinations and the sensor data to be received in the example in Fig. 5.10. In this chapter, we assume that selectable delivery cycles are represented as C_i ($i = 1, 2, \cdots$). The sensor data delivery system assigns delivery cycles or times to relay sensor data streams to nodes. The nodes

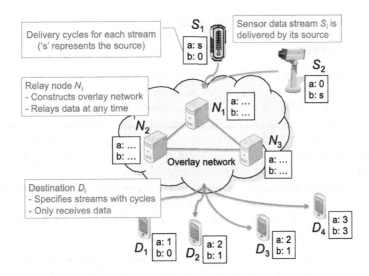

FIGURE 5.10

System model.

Table 5.2 An Example of Delivered Data									
Time		**0**	**1**	**2**	**3**	**4**	**5**	**6**	**...**
D_1	$C_{1a}=1$	o	o	o	o	o	o	o	...
	$C_{1b}=0$								...
D_2	$C_{2a}=2$	o		o		o		o	...
	$C_{2b}=1$	o	o	o	o	o	o	o	...
D_3	$C_{3a}=2$	o		o		o		o	...
	$C_{3b}=1$	o	o	o	o	o	o	o	...
D_4	$C_{4a}=3$	o			o			o	...
	$C_{4b}=3$	o			o			o	...

constructs delivery paths mutually and send various sensor data stream to each other according to the paths. By relaying sensor data among multiple nodes, it is difficult to keep the order to receive or send data sequentially. However, general streaming applications have restrictions about delivery time and allow buffering time satisfying those restrictions. Therefore, nodes or destinations can sort data correctly within the buffering time at each delivery process.

The purpose is same to the case of P2P-based technologies to disperse the communication loads. Therefore, the objective function is SL and $1-FI$, and the delivery path is determined to make these values minimum.

5.3.2 LOAD DISTRIBUTION METHOD

5.3.2.1 Overview

Each node is able to construct delivery paths autonomously by putting nodes on a hash space and determining relay nodes based on distributed hashing, which is called "consistent hashing (Karger et al., 1997)." Currently many techniques have been proposed to construct a P2P overlay network like Skip Graphs (Aspnes and Shah, 2007) and P2PR-tree (Mondal et al., 2004). Although we can apply those P2P overlay networks to construct delivery paths, in this chapter, we assume a circular hash space like Chord (Stoica et al., 2003). Chord is applied widely as a typical P2P technique because it uses one-dimensional values and has a simple structure. In addition, a technique to create "virtual nodes" is mentioned in Chord. Virtual nodes are created by physical computers and enables to reduce a load imbalance among physical computers.

In the case of distributed hashing by sensor data streams, a relay node is determined for each sensor data stream. The relay node receives sensor data from its source and sends them to destinations. We call the node that first receives sensor data from its source "root node." Fig. 5.11(a) shows an example of distributed hashing by sensor data stream in the case of sensor data stream 'a.' Same processes are executed for other sensor data streams. Sensor data stream 'a' is assigned to N_1 in Fig. 5.11(a) because the hashed value of 'a' is assigned to the N_1 on the hash space, which is shown in the upper left within Fig. 5.11(a). On the other hand, N_2 and N_3 do not relate to sensor data stream 'a.' The load of each node at the process of sensor data stream 'a' is $L_1 = 20$, $L_2 = 0$, and $L_3 = 0$. In this case, $SL = 20$ and $FI = 0.333$. The number of sensor data and destinations of relaying sensor data stream unbalance the load of each node. In addition, the nodes that have no relaying sensor data streams occur in the case where the number of sensor data stream is small.

In the case of distributed hashing by combination of sensor data stream and delivery cycle, each combination of sensor data stream and delivery cycle is assigned to a specific node. Sensor data are relayed among nodes. For example, the node that has cycle 3 sends sensor data on time 0, 6 ⋯ to the node that has cycle 2. While the communication loads occur on nodes, loads are avoided concentrating to a specific node and time. Fig. 5.11(b) shows an example of distributed hashing by combination of sensor data stream and delivery cycle in the case of sensor data stream 'a.' The root node and delivery cycle 3 of sensor data stream 'a' are assigned to N_1 in Fig. 5.11(b) because the hashed value of the combination of 'a' and cycle 3 is assigned to the N_1 on the hash space, which is shown in the upper left within Fig. 5.11(b). Similarly, delivery cycle 1 of sensor data stream 'a' is assigned to N_2, and delivery cycle 2 of sensor data stream 'a' is assigned to N_3. The load of each node at the process of sensor data stream 'a' is $L_1 = 15$, $L_2 = 12$, and $L_3 = 11$. In this case, $SL = 38$ and $FI = 0.982$. However, the load of each node is unbalanced by assigned combinations of sensor data stream and delivery cycle in the case where the number of sensor data streams is small.

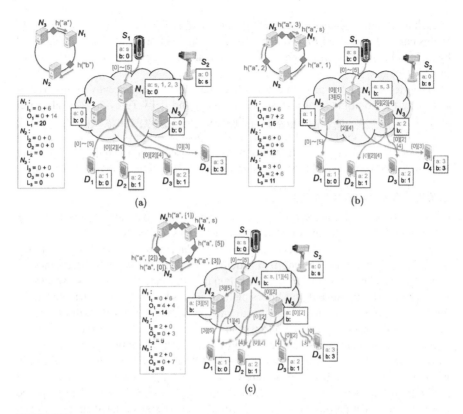

FIGURE 5.11

Examples of distributed hashing. (a) Stream hashed; (b) Cycle hashed; (c) Time hashed.

In the case of distributed hashing by combination of sensor data stream and delivery time, each combination of sensor data stream and delivery time is assigned to a specific node. The length of timetable for delivery is the least common multiple, and the length is 6 in the case where the delivery cycle that destinations can specify is 1, 2 or 3. In this case, each time from 0 to 5 is assigned to a specific node. The root node of sensor data stream sends sensor data to assigned nodes on each time. Combinations of sensor data stream and delivery time are assigned to nodes more finely compared with distributed hashing by sensor data stream or combination of sensor data stream and delivery cycle. Fig. 5.11(c) shows an example of distributed hashing by combination of sensor data stream and delivery time in the case of sensor data stream 'a.' The root node and delivery time 1, 4 of sensor data stream 'a' are assigned to N_1 in Fig. 5.11(c) because the hashed value of the combination of 'a' and time 1, 4 is assigned to the N_1 on the hash space, which is shown in the upper left within Fig. 5.11(c). Similarly, delivery time 3 and 5 of sensor data stream 'a' are assigned to N_2, and delivery time 0 and 2 of sensor data stream 'a' are assigned to N_3. The load of each node at the process of sensor data stream 'a' is $L_1 = 14$, $L_2 = 5$, and $L_3 = 9$. In this case, $SL = 28$ and $FI = 0.865$. However, the load of each node is unbalanced by assigned combinations of sensor data stream and delivery time such as the case of Fig. 5.11(c).

Therefore, we have proposed a method that divides nodes into groups represented by the combination of sensor data stream and delivery cycle (Kawakami et al., 2014a). In addition, the proposed method assigns delivery times to nodes for each group of delivery cycle. The proposed method distributes processes by assigning a node based on delivery time. In addition, the propose method avoids concentrating loads to a specific node and time by assigning a node for each group of delivery cycle.

5.3.2.2 Grouping of Nodes

On the basis of the idea in Section 5.3.2.1, we explain how to divide nodes into groups represented by the combination of sensor data streams and delivery cycles.

In sensor data stream delivery, the amount of data to send/receive varies among different delivery cycles. The shorter the delivery cycle is, the larger the data amounts and loads are. Therefore, the proposed method first generates circular hash spaces for each sensor data stream and puts nodes on hash spaces based on the distributed hashing of the combination of sensor data stream and node ID. After that, the proposed method divides each hash space into partial hash spaces as groups for each delivery cycle so that the partial hash spaces of shorter cycles have more nodes. The size of each partial hash space is determined based on its cycle. For example, in the case where the selectable delivery cycles are $C_i = i$ ($i = 1, 2, 3$), the ratio of the sizes of partial hash spaces is $1/C_1 : 1/C_2 : 1/C_3 = 1/1 : 1/2 : 1/3 = 6 : 3 : 2$. The proposed method treats each partial hash space as circular and assigns related times for each cycle to nodes on its partial hash space. In the case where there are no nodes on the partial hash space, the proposed method assigns the partial hash space to the nearest neighbor node in the next partial hash space. In addition, the proposed method determines the root node on the partial hash space of the shortest cycle based on distributed hashing such as the least common multiple of cycles. The root node first receives data from the source of sensor data stream.

Fig. 5.12 shows an example of the case where the number of nodes is $n = 8$, cycles are $C_i = i$ ($i = 1, 2, 3$), and the size of a hash space is 2^p. The beginning values of each partial hash space are $2^p \times 0/11$, $2^p \times 6/11$ and $2^p \times 9/11$.

5.3.3 NODE ASSIGNMENT AND CONSTRUCTION OF DELIVERY PATHS

On the basis of the idea in Section 5.3.2.1, we explain how to assign the times on each group to nodes.

To construct delivery paths, nodes first calculate the least common multiple of selectable delivery cycles for each sensor data stream. After that, the nodes search the sender nodes for each related time that the same time is assigned to in the other cycle groups. The nodes determine the cycle groups to search for each time based on approaches such as LCF method (Kawakami et al., 2014b), and the node that

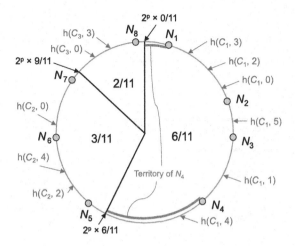

FIGURE 5.12

Assignment to a group of cycle.

```
Input:
 1: cycles: Array of delivery cycles of the nodes sorted in ascending order (cycle of source node is −1 and at index 0)
 2: ownId: An identification of own (node)
 3: assignedCycleIndex: An index of own assigned delivery cycle in cycles
Output:

 4: cycleLcm ← calculateLCM(cycles);
 5: if assignedCycleIndex ≠ 0 or searchNode(0, cycleLcm) ≠ ownId then
 6:     time ← 0;
 7:     while time < cycleLcm do
 8:         assignedNode ← searchNode(assignedCycleIndex, time);
 9:         if assignedNode = ownId then
10:             longCycleIndex ← calculateLongestCycleIndex(cycles, time);
11:             relayNode;
12:             if longCycleIndex = assignedCycleIndex then
13:                 relayNode ← searchNode(0, cycleLcm);
14:             else
15:                 relayNode ← searchNode(longCycleIndex, time);
16:             end if
17:             requestToSend(relayNode, ownId, time);
18:         end if
19:         time ← time + cycles[assignedCycleIndex];
20:     end while
21: end if
```

FIGURE 5.13

Pseudocode to construct delivery paths by nodes.

belongs to the longest cycle on each time receives data from root node and sends to the nodes that belong to the other cycle groups. The node that does not belong to the longest cycle group on each time searches for the node on the longest cycle group and requests to send data. On the other hand, the node that belongs to the longest cycle group on each time searches the root node of the sensor data stream and

```
Input:
 1: cycles: Array of delivery cycles of the nodes sorted in ascending order (cycle of source node is −1 and at index 0)
 2: ownId: An identification of own (destination)
 3: requestCycleIndex: An index of request delivery cycle in cycles
Output:

 4: cycleLcm ← calculateLCM(cycles);
 5: time ← 0;
 6: while time < cycleLcm do
 7:     targetCycleIndex ← getRandomCycleIndex(cycles, time);
 8:     relayNode ← searchNode(targetCycleIndex, time);
 9:     requestToSend(relayNode, ownId, time);
10:     time ← time + cycles[requestCycleIndex];
11: end while
```

FIGURE 5.14

Pseudocode to construct delivery paths by destinations.

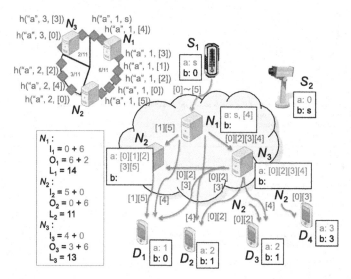

FIGURE 5.15

An example of the proposed method.

requests to send data. Fig. 5.13 shows the pseudocode to construct delivery paths by a node. Similarly, destinations first calculate the least common multiple of selectable delivery cycles for each sensor data stream. After that, the destinations determine the cycle group for each time at random among the related cycles. The destinations search the senders in the determined cycle groups for each time and request to send data. Fig. 5.14 shows the pseudocode to construct delivery paths by a destination.

Fig. 5.15 shows an example of the proposed method in the case of sensor data stream 'a.' N_1 and N_2 belong to the cycle group 1, and N_3 belongs to the cycle group 3, which is shown in the upper left within Fig. 5.15. No nodes belong to the cycle group 2. Therefore, N_3 is also assigned to delivery related to the cycle group 2 since N_3 is located next to the cycle group 2. Destinations determine the cycle group for each time at random among related cycle groups. For example, D_1 selects the cycle group 3 at random at time 0, 2, and 3 from cycle groups 1 and 3. The load of each node at the process of sensor data stream 'a' is $L_1 = 14$, $L_2 = 11$, and $L_3 = 13$. In this case, $SL = 38$ and $FI = 0.990$. Same processes for node assignment and construction of delivery paths are executed for other sensor data streams. The loads to deliver sensor data streams are integrated on each node.

5.3.4 EVALUATION

In this section, we describe the evaluation of the proposing method in Section 5.3.2 by simulation.

5.3.4.1 Simulation Environment

We compared the proposed method (Cycle-Time) with the methods based on distributed hashing of "sensor data stream (Source)," "combination of sensor stream and delivery cycle (Cycle)," and "combination of sensor data stream and delivery time (Time)." Destinations request all streams to be delivered. The delivery cycles are $C_i = i$ ($i = 1, \cdots, 10$) and determined at random from 1 to 10 for each sensor data stream. Each destination receives sensor data from the nodes whose number is equal to the number of streams at most.

Selectable delivery cycles are represented as a multiple of a basic delivery cycle. We expect requests for a delivery cycle are satisfied over a certain number of selectable patterns and ten patterns are enough in general applications. In this environment, the maximum of the least common multiple of delivery cycles is 2520, and then the timetable for delivery is from time 0 to 2519. Nodes in this simulation are placed to a circular hash space with a nearly equal distance not to occur a imbalance of territories among nodes.

As evaluated values, we calculated the load of each node, system total loads (SL) and fairness index (FI) among the time of the least common multiple of selectable delivery cycles. We executed this simulation ten times for each method and environments changing the number of nodes, streams. We assume that the number of destinations is larger than the number of nodes or streams.

5.3.4.2 Results by the Number of Nodes

Fig. 5.16(a) shows the maximum instantaneous load in an environment where the number of nodes is $n = 100i$ ($i = 1, \cdots, 5$). The transmission and reception loads per data piece as described in Section 5.2.1.4 are $\alpha = \beta = 1$ ($i = 1, \cdots, n$). The maximum instantaneous load means the maximum load for each node and time. The longitudinal axis is the maximum instantaneous load, and the lateral axis is the number of nodes. The number of sensor data streams is $l = 500$, and the number of destinations is $m = 2500$. In the future Internet of Things / M2M network, we assume that a huge number of devices collect, analyze, and utilize sensor data. Therefore, we

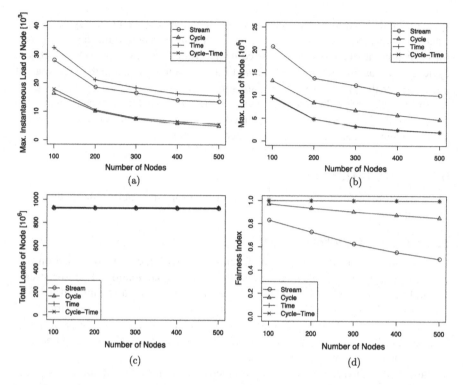

FIGURE 5.16

Results by the number of nodes. (a) Maximum instantaneous load; (b) Maximum load of node; (c) System total loads; (d) Load balance.

evaluate in the environments which have many destinations compared with nodes. In the case of many destinations, the loads to deliver sensor data must be reduced even if the number of streams is small. Therefore, we evaluate in the environments which have both few and many streams compared with nodes. The maximum instantaneous load of node in the proposed method is lower than the values in the methods of "Stream" and "Time." This result means that the proposed method can disperse such instantaneous loads to more nodes.

Fig. 5.16(b) shows the maximum load of node, and Fig. 5.16(c) shows SL. The maximum load of node is the maximum of the loads of nodes among the time of the least common multiple of cycles. The maximum load of node in the proposed method is lower than the values in the methods of "Stream" and "Cycle." This result means that the proposed method can disperse the loads continuously for a long period. On the other hand, the differences of SL among the methods are small by the number of nodes.

Fig. 5.16(d) shows FI in the same environment described above. FIs in the methods of "Stream" and "Cycle" are decreased in the case of many nodes, and the loads

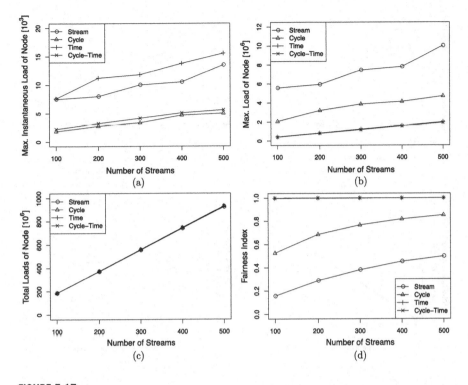

FIGURE 5.17

Results by the number of sensor data streams. (a) Maximum instantaneous load;
(b) Maximum load; (c) System total loads; (d) Load balance.

of nodes are unbalanced. On the other hand, the proposed method keeps *FI* high in
the case of many nodes, and the loads of nodes are balanced highly.

5.3.4.3 Results by the Number of Streams

Fig. 5.17(a) shows the maximum instantaneous load in an environment where the
number of sensor data streams is $l = 100i$ $(i = 1, \cdots, 5)$. The longitudinal axis is
the maximum instantaneous load, and the lateral axis is the number of sensor data
streams. The number of destinations is $m = 2500$, and the number of nodes is $n =
500$. The maximum instantaneous load is increased by the number of sensor data
streams and lowest in the method of "Cycle." The maximum instantaneous load in
the proposed method is little different from the value in the method of "Cycle" and
lower than the values in the methods of "Stream" and "Time."

Fig. 5.17(b) shows the maximum load of node, and Fig. 5.17(c) shows *SL*. The
maximum loads are lowest in the methods of "Time" and the proposal. The differences of *SL*s among the methods are small by the number of streams.

Fig. 5.17(d) shows *FI*. *FI*s in the methods of "Stream" and "Cycle" are largely
decreased in the case of a few sensor data streams, and the loads of nodes are unbal-

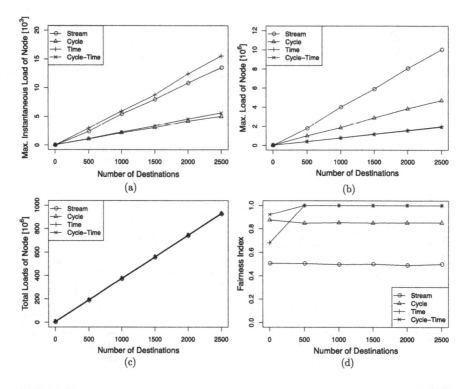

FIGURE 5.18

Results by the number of destinations. (a) Maximum instantaneous load; (b) Maximum load; (c) System total loads; (d) Load balance.

anced. On the other hand, the proposed method keeps *FI* high also in the case of a few sensor data streams, and the loads of nodes are balanced highly.

5.3.4.4 Results by the Number of Destinations

Fig. 5.18(a) shows the maximum instantaneous load in an environment where the number of destinations is $m = 500i$ $(i = 0, \cdots, 5)$. The number of sensor data streams is $l = 500$, and the number of nodes is $n = 500$. The case of $m = 0$ means that there are no destinations, and the results show the loads to relay data to the assigned nodes. The maximum instantaneous load is increased by the number of destinations and lowest in the method of "Cycle." The maximum instantaneous load in the proposed method is little different from the value in the method of "Cycle" and lower than the values in the methods of "Stream" and "Time."

Fig. 5.18(b) shows the maximum load of node, and Fig. 5.18(c) shows *SL*. The maximum load is increased by the number of destinations and lowest in the methods of "Time" and the proposal. The differences of *SL* among the methods are small by the number of destinations.

Fig. 5.18(d) shows *FI*. *FI*s in the methods of "Time" and the proposal are low in the case where the number of destinations is zero, and the loads of nodes are unbalanced. However, in the case where there are destinations, *FI* is increased by the loads to send to destinations, and the loads of nodes are more balanced.

5.4 DISCUSSION

Various techniques for dispersing the communication load in the delivery of streams have been studied.

A P2P stream delivery technique according to which a stream is delivered using a P2P technology for sending and receiving data between terminals having no servers in between has been proposed in order to disperse the communication load among the terminals (Zhang et al., 2005; Liao et al., 2006; Magharei and Rejaie, 2007; Yu et al., 2011; Sakashita et al., 2010). The P2P stream delivery technique is divided into a pull type technique and a push type technique. In a pull type technique, such as PPLive (http://www.pplive.com/), DONet (Zhang et al., 2005), and SopCast (http://www.sopcast.com/), the reception terminal that receives data requests data from another terminal and acquires it. Though communication is carried out in order for the reception terminal to find a terminal that has data that has not yet been received by the reception terminal, no such redundant communication that the data that has already been received by the reception terminal is again requested is carried out. In a push type technique, such as AnySee, data is sent from the transmission terminal that sends data to another terminal (Liao et al., 2006). Though communication is carried out in order for the transmission terminal to find a terminal that has not yet received data received by the transmission terminal, no such redundant communication is carried out that the data that has already been received by the reception terminal is distributed again. A technique where a pull type and a push type are combined, such as PRIME, has been proposed (Magharei and Rejaie, 2007).

In P2P stream delivery techniques, a case where the same data stream is delivered to a number of terminals is assumed. In the delivery of the sensor data streams, however, there are cases where a data stream of the same sensor having different delivery cycles is delivered. In this case, sensor data streams having different delivery cycles are delivered as different data streams. Thus, the communication load on the source node cannot be efficiently dispersed.

Several techniques for preventing the communication load from being concentrated on a particular terminal by constructing a data delivery path, which is referred to as a multicast tree, in advance so that a data stream is delivered have been proposed (Banerjee et al., 2002; Tran et al., 2003; Jin et al., 2007; Silawarawet and Nupairoj, 2011; Le and Nguyen, 2012). In the ZIGZAG method, a multicast tree is constructed by clusters that are collections of terminals (Tran et al., 2003). The number of clusters included in each depth of a multicast tree is made the same, and thus, the load is dispersed. Multicast trees are constructed only of information gained in the application layer, and it is not necessary to understand the physical network structure.

In the MSMT/MBST method, the communication load can be prevented from concentrating on a particular terminal as compared to ZIGZAG by taking the communication delay between terminals into consideration in the case where the physical network structure can be understood (Jin et al., 2007). The implementability of the MSMT/MBST method was poor because it is necessary to understand all the network structures between the terminals concerning stream delivery. In LAC (locality aware clustering), a load dispersion higher than that in ZIGZAG is achieved by taking into consideration the communication delay between only some terminals, though the physical network structure cannot be understood (Silawarawet and Nupairoj, 2011).

In the case where sensor data streams have different delivery cycles as in the above-described P2P stream delivery technique, sensor data streams having different delivery cycles are delivered as different data streams. Thus, the communication load on the source node cannot be efficiently dispersed. On the other hand, our proposing methods introduced in this chapter enable to disperse the loads among the nodes in the network efficiently.

5.5 CONCLUSION

In this chapter, we introduced smart sensor data stream delivery technologies. We described our proposing methods in a P2P model or cloud model and those experimental results in simulations. Our proposed method determines delivery paths for sensor data streams with different delivery cycles. The simulation results of the evaluations confirmed that the proposed method can widely disperse both instantaneous and continuous loads among the nodes.

In the future, we will study a technique to enhance robustness against the joining/leaving of nodes. As one of our ideas, we plan to employ a technique of virtual nodes in Stoica et al. (2003) to reduce the imbalance among physical computers. In addition, we currently assume that nodes send data individually even if those receivers are the same node or destination at the same time. General streaming applications have restrictions about delivery time and allow buffering times that satisfy those restrictions. By using those buffering times, sending data at once can be merged to enhance the efficiency of delivery. Moreover, actual nodes have different and often limited performances including α, β, and network environments. Therefore, we will study a technique to distribute loads considering node performance and network environment. By these future techniques, we can achieve more smart and highly scalable stream delivery services with flexible delivery cycles.

ACKNOWLEDGMENTS

This research was partly supported by the collaborative research of National Institute of Information and Communications Technology (NICT) and Osaka University (Research on high functional network platform technology for large-scale distributed computing). This research was partly supported by JSPS KAKENHI Grant-in-Aid for Scientific Research (B) numbered 15H02702.

REFERENCES

Aspnes, J., Shah, G., 2007. Skip graphs. ACM Transactions on Algorithms 3 (4), 37.

Banerjee, S., Bhattacharjee, B., Kommareddy, C., 2002. Scalable application layer multicast. In: Proceedings of the ACM Conference on Applications, Technologies, Architectures, and Protocols for Computer Communications. SIGCOMM 2002, pp. 205–217.

Jain, R., Chiu, D.-M., Hawe, W., 1984. A Quantitative Measure of Fairness and Discrimination for Resource Allocation in Shared Computer Systems. DEC Research Report TR-301.

Jin, X., Yiu, W.-P.K., Chan, S.-H.G., Wang, Y., 2007. On maximizing tree bandwidth for topology-aware peer-to-peer streaming. IEEE Transactions on Multimedia 9 (8), 1580–1592.

Karger, D., Lehman, E., Leighton, T., Panigrahy, R., Levine, M., Lewin, D., 1997. Consistent hashing and random trees: distributed caching protocols for relieving hot spots on the World Wide Web. In: Proceedings of the 29th Annual ACM Symposium on Theory of Computing. STOC '97, pp. 654–663.

Kawakami, T., Ishi, Y., Yoshihisa, T., Teranishi, Y., 2014a. A load distribution method based on distributed hashing for P2P sensor data stream delivery system. In: Proceedings of the 3rd IEEE International Workshop on Modeling and Verifying of Distributed Applications (MVDA 2014) in Conjunction with the 38th Annual International Computer, Software and Applications Conference. COMPSAC 2014, pp. 716–721.

Kawakami, T., Ishi, Y., Yoshihisa, T., Teranishi, Y., 2014b. A P2P-based sensor data stream delivery method to accommodate heterogeneous cycles. Journal of Information Processing 22 (3), 455–463.

Le, T.A., Nguyen, H., 2012. Application-aware cost function and its performance evaluation over scalable video conferencing services on heterogeneous networks. In: Proceedings of the IEEE Wireless Communications and Networking Conference: Mobile and Wireless Networks (WCNC 2012 Track 3 Mobile and Wireless), pp. 2185–2190.

Liao, X., Jin, H., Liu, Y., Ni, L.M., Deng, D., 2006. AnySee: peer-to-peer live streaming. In: Proceedings of the 25th IEEE International Conference on Computer Communications. INFOCOM 2006, pp. 1–10.

Magharei, N., Rejaie, R., 2007. PRIME: peer-to-peer receiver-driven mesh-based streaming. In: Proceedings of the 26th IEEE International Conference on Computer Communications. INFOCOM 2007, pp. 1415–1423.

Mondal, A., Lifu, Y., Kitsuregawa, M., 2004. P2PR-tree: an R-tree-based spatial index for peer-to-peer environments. In: Proceedings of the International Workshop on Peer-to-Peer Computing and Databases in Conjunction with EDBT, pp. 516–525.

Sakashita, S., Yoshihisa, T., Hara, T., Nishio, S., 2010. A data reception method to reduce interruption time in P2P streaming environments. In: Proceedings of the 13th International Conference on Network-Based Information Systems. NBiS 2010, pp. 166–172.

Silawarawet, K., Nupairoj, N., 2011. Locality-aware clustering application level multicast for live streaming services on the Internet. Journal of Information Science and Engineering 27 (1), 319–336.

Stoica, I., Morris, R., Liben-Nowell, D., Karger, D.R., Kaashoek, M.F., Dabek, F., Balakrishnan, H., 2003. Chord: a scalable peer-to-peer lookup protocol for Internet applications. IEEE/ACM Transactions on Networking 11 (1), 17–32.

Tran, D.A., Hua, K.A., Do, T., 2003. ZIGZAG: an efficient peer-to-peer scheme for media streaming. In: Proceedings of the 22nd Annual Joint Conference of the IEEE Computer and Communications Societies. INFOCOM 2003, vol. 2, pp. 1283–1292.

Yu, L., Liao, X., Jin, H., Jiang, W., 2011. Integrated buffering schemes for P2P VoD services. Peer-to-Peer Networking and Applications 4 (1), 63–74.

Zhang, X., Liu, J., Li, B., Yum, T.-S.P., 2005. CoolStreaming/DONet: a data-driven overlay network for peer-to-peer live media streaming. In: Proceedings of the 24th Annual Joint Conference of the IEEE Computer and Communications Societies. INFOCOM 2005, pp. 2102–2111.

ACRONYMS AND GLOSSARY

List of acronyms with explanation

α	a load to receive one piece of sensor data
β	a load to send one piece of sensor data
C	cycle, an array of delivery cycles
D	destination, an array of destination nodes
FI	fairness index
FPS	frames per second
L	load, an array of node loads
LCF	longest cycle first
LLF	lowest load first
LLF-H	lowest load first considering hops
M2M	machine-to-machine
N	node, an array of relay nodes
P2P	peer-to-peer
S	source, an array of source nodes
SD	source direct
SL	system load

Glossary of terms with explanation

Communication load a load of computers and it is defined as the amount of messages to send/receive.

Delivery cycle a period requested by the destination to receive a sensor data stream.

Destination a computer which expects to receive sensor data streams with specific delivery cycles.

Distributed hashing a method to provide a function similar to a hash table in distributed computing environments.

Fairness index an index to quantify the fairness among specified values.

Internet of Things (IoT) connecting to the Internet by physical devices, sensors, actuators, vehicles, buildings, and other items to realize various services/applications.

LCF a method to make the longest cycle computer at each time to send sensor data stream to other computers.

Least common multiple (lowest common multiple, LCM) the smallest positive integer that is divisible by all specified values.

LLF a method to estimate the loads of each computer and make the lowest load computer at each time to send sensor data to other computers.

LLF-H a method to estimate the loads of each computer and make the lowest load computer satisfying the specified hops at each time to send sensor data to other computers.

Maximum instantaneous load the maximum load for each node and time.

Maximum load the maximum value of the loads of nodes among the time of the least common multiple of cycles.

Node a computer or such a computing device.

Overlay network virtual links constructed over a physical network.

SD a method to make the source node to send sensor data to destinations directly.

Sensor data stream delivery to deliver the periodically gained sensor data continuously.

Source node a computer which gaines sensor data periodically and sends them as sensor data streams to other computers.

System load the sum of the loads of nodes in the whole system.

Root node a relay node which receives sensor data first from source in the cloud-based approach.

The number of hops the number of sending messages from the source node to the destination.

SCALABLE PROCESSING OF MASSIVE TRAFFIC DATASETS

6

Sergio Di Martino*, Simon Kwoczek[†], Wolfgang Nejdl[‡]

**University of Naples "Federico II", Italy [†]Group Research, Volkswagen AG, Germany
[‡]L3S Research Center, University of Hanover, Germany*

6.1 INTRODUCTION

Traffic congestions are one of the most hated phenomena by road users. A report from CEBR (2014) reveal that people in Europe and the US are wasting on average 111 hours annually in gridlocks. This is not only a key source of pollution and stress, but it costs billions of dollars to the society every year. Without any significant introduction of Intelligent Transportation Systems (ITS), the combined cost of traffic delays in UE and USA is expected to rise up to $293.1 billion per year by 2030, almost a 50% increase from 2013, according to CEBR (2014). On the other hand, a report by Manyika et al. (2011) from the McKinsey Global Institute estimates that the use of smarter technologies able to help vehicles in avoiding congestions and reducing idling at red lights could lead to savings up to "about $600 billion annually by 2020."

Given these numbers, clearly many research efforts are currently being aimed world-wide at improving mobility predictions. Latest developments within the ITS community, based on smart sensor networks including sophisticated sensing and communication technologies has led to a sudden availability of massive mobility datasets with high spatial and temporal resolution. The information contained in these datasets is usually generated from a combination of various different sources. These include Floating Car Data (originated from millions of vehicles equipped with GPS sensors), GSM probe data, and data from stationary sensor (e.g. loop detectors, camera sensors) obtained from local traffic management centers. More information about the different sources and their aggregation can be found in International (2009). The total amount of available data is becoming impressive, since sensor networks providing real-time traffic information have reached a significant spatio-temporal coverage of the road network, world wide, as described in Kargupta et al. (2010).

These datasets are enabling a new approach to mobility predictions, where the future state of the road infrastructure is no more estimated using simulations based on a-priori modeling of the Origin-Destination matrices (e.g.: Yang et al., 1992; Krajzewicz et al., 2002), but rather it is predicted using data-driven solutions able to extract new knowledge from historical data, like in Kwoczek et al. (2014a),

Smart Sensors Networks. DOI: 10.1016/B978-0-12-809859-2.00008-5

123

Pan et al. (2013) or in Kwoczek et al. (2015). One representative example is the service of vehicular traffic prediction. This services requires to:

1. collect data from real-time traffic providers,
2. create an historical dataset for each geographic region of interest,
3. analyze this dataset to learn mobility models, and
4. use these models to predict future traffic situations.

The resulting volume of available spatial data (both historical and real time) for a geographic region like a nation, where the situation about hundreds of thousands of road segments has to be constantly monitored, exceeds by far the capability of traditional storing systems, being in the range of Terabytes per Week per Nation. Moreover, since data is collected continuously, preprocessing, aggregating, and storing the received data streams on time in a scalable manner leads to great challenges in terms of computational resources required to provide predictive results on time.

In this chapter, we describe our experiences and our approaches in providing a solution to store and analyze massive amounts of traffic data from an industrial provider in order to enable the Knowledge Discovery process able to scale up to any geographic level, thanks to an intrinsic parallelizability of the solution. More in details, the contributions to the Body of Knowledge can be summarized as follows:

1. we propose an architecture to store, analyze, and visualize massive amounts of traffic data from a commercial provider;
2. we propose and compare two different data models, based on relational and NoSQL Database Management Systems;
3. we provide preliminary benchmarks on the reference implementation of the architecture.

The remainder of the chapter is structured as follows: In Section 6.2 we describe current state of the art in the field of scalable solutions for ITS. In Section 6.3 we describe the problem at the hand, also in terms of data sources and potential use cases. In Section 6.4 we describe the parallel architecture we propose to tackle the requirements posed by the massive datasets and the use cases. In Section 6.5 we provide a detailed description of two potential data models, using a relational database and a NoSQL one. Then, in Section 6.6 we present some preliminary results of the performance comparison between the two data models. Finally, in Section 6.7 we describe the lessons we learned, while some concluding remarks and future research directions are pointed out.

6.2 BACKGROUND AND STATE OF THE ART

Traffic congestion predictions have been widely studied within the research communities of ITS, Smart Cities, and others, leading to a rich body of knowledge. Indeed, by knowing in advance the traffic patterns, route calculation engines can compute more efficient routes, saving also time for the drivers. Due to the lack of real mobility data, early approaches for traffic predictions were mainly based on simulations and theoretical modeling (e.g. Chrobok et al., 2004; Clark, 2003). Nowadays thanks to the spreading of GPS and high-speed cellular networks, new massive datasets on traffic are available. Consequently, several statistic and data driven approaches for traffic predictions have been presented to the community. Examples include generalized linear regression (Zhang and Rice, 2003), non-linear time series (Ishak and Al-Deek, 2002), Kalman filters (Van Lint, 2008), support vector regression (Wu et al., 2004), and various neural network models (Park et al., 1999; Van Lint, 2008; Vanajakshi and Rilett, 2004). A combination of these approaches is also used by current commercial navigation solutions, able to predict recurring congestions by learning characteristic traffic flow patterns for street segments from historical data. On top of that, these commercial systems can also optimize the route planning based on the real-time traffic situation International (2009). Nevertheless, many of these solutions based their experiments on so called "small world scenarios" whereby their investigations were focused on either limited geographic regions (e.g.: Kwoczek et al., 2014b; Pan et al., 2012) or on data for limited time spans (e.g.: Dunne and Ghosh, 2012; Zhu et al., 2014). While these constraints are acceptable for research scenarios, solutions in an industrial context need to be based on architectures and algorithms able to easily scale to handle broader spatial and temporal coverages.

These challenges are currently being targeted, in a general manner, within the Big Data research community, especially from those research activities about so called *Spatial Big Data* (SBD). In Shekhar et al. (2012), authors highlight that

> *[...] the size, variety, and update rate of [mobility] datasets exceed the capacity of commonly used spatial computing and spatial database technologies to learn, manage, and process the data with reasonable effort.*

Moreover, they point out that new Cloud computing paradigms, tentatively called *spatial cloud computing*, could be a viable solution for these problems. Some generic technological solutions to deal with SBD are presented in Di Martino et al. (2011a, 2011b), Amato et al. (2014), Evans et al. (2014), Aji et al. (2013), Vatsavai et al. (2012), but, to the best of our knowledge, to date there is no specifically tailored solution or architecture being proposed, able to tackle the spatio-temporal complexity of these massive mobility datasets.

6.3 THE PROBLEM DESCRIPTION

In order to understand the motivations underlying the proposed scalable architecture, as a first step we describe the problem and the typical use cases to be faced in order to provide traffic predictions. Then we will present the type and the size of the data we are dealing with.

The problem we are facing is to provide traffic predictions on a world-wide scale, based on mobility patterns that are automatically learned from real traffic data. For this class of problems, the typical use case is to process a dataset of historical traffic information in order to learn some recurring patterns over some road segments. To this aim, we start by describing the employed data sources, then we exemplify two possible use cases.

6.3.1 THE DATA SOURCES

Learning traffic models in order to perform predictions of the state of the road infrastructure requires two main types of data sources:

1. A map of the area to be investigated (potentially world-wide), and
2. A continuous stream of real time traffic information (potentially world-wide), which has to be stored in order to generate a historical dataset.

Actually it is possible to find on the market many providers and many formats for this kind of data. In the following we describe the data types we used in our experiments.

6.3.1.1 The Digital Map

The map data contains a detailed description of the road network for the region of interest. In our case, we adopted a commercial automotive solution based on the Navigation Data Standard (NDS) (http://www.nds-association.org/), which is a global physical storage standard employed by many players in the automotive domain. Within this map format, each road is divided into several segments, also named *links*, whose length varies from a few meters (mostly at intersections and in urban scenarios) up to a couple of kilometers. The road segments are connected through nodes, called *intersections*, to form a graph. Usually each road segment has information about its geographic position, shape, type of road (highway, urban, etc.), allowed driving directions and further more. The ecosystem of features that can be contained on a NDS map is shown in Fig. 6.1.

For our experiments we employed the map of the entire Germany, which is about 2 GB and contains millions of links.

6.3.1.2 The Traffic Dataset

As for the real traffic information, we employed two different data sources, with a different level of spatial and temporal granularity:

- The Traffic Flow (TF) dataset covers traffic information for major road segments. The data received uses a table-based referencing system, called Traffic Message Channel (TMC), where there is a fixed number of predefined road segments identified by an ID, for each nation. Each traffic message from TMC, which can be also broadcasted via FM radio, includes the IDs of the involved segments plus some information about the current speed on that stretch (in km/h). Since there is a totally different level of spatial granularity between TMC segments (coarse grained) and NDS links (fine grained), a decoding step is required to match the TMC message onto an actual position on the digital map. As an example, the TMC table for entire Germany contains only 56,000 segments and is updated twice per minute.

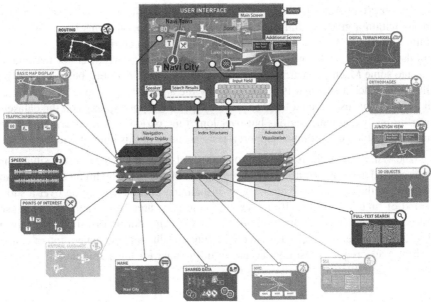

Navigation Data Standard (NDS) e.V. (www.nds-association.org), EMDE GRAFIK (www.emde-grafik.de)

FIGURE 6.1

The scope of NDS features.

- The Traffic Flow Detailed (TFD) dataset comes from a commercial traffic provider and contains the same kind of information as the TF dataset, but with a significantly better spatial precision. It is based on an dynamic referencing system called OpenLR (http://www.openlr.org/), which covers major and side roads of the street network. Again, as an example, the TFD dataset for Germany contains about 750,000 locations and is updated once a minute. To match these data onto a road network, an open source software to decode OpenLR binary codes can be used (http://www.openlr.org/decoder.html). This dataset leads to a continuous massive stream of traffic data, whose storage can pose severe challenges to standard database solutions.

6.3.2 DESCRIPTION OF THE USE CASES

To define an architecture able to leverage the vast amounts of traffic data, both in terms of spatial and temporal dimensions, we need to have in mind what could be the possible use cases. Two example of potential applications exploiting on our architecture, are the *Analyzer* and the *Visualizer* components, intended as external software modules interacting with our system. We selected these use cases because they are quite at the extremes of a spectrum of functionality required to the underlying architecture, having contrasting requirements.

The *Analyzer* Actor is meant to run data analytic methods for selected road segments within a given spatial region on rather long temporal spans (i.e. extract the traffic model for Wednesdays for one selected road segment using 52 weeks as input data). Indeed, to learn the historical traffic patterns, clearly it is not feasible to execute some Machine Learning algorithms on the whole datasets, but data must be filtered to reach a size that is manageable by the algorithms, as done in Shekhar et al. (2011). This use case requires to perform spatio-temporal queries on the datasets over limited geographic regions but over long time periods. As a consequence, the underlying storage layer must support a fast execution of this kind of queries.

The *Visualizer* Actor is intended to show or also replay historic traffic situations in a selected geographic area and time span. It is mainly used by human operators to understand the dynamics of mobility by visualizing traffic information over a short period of time in a potentially rather big geographic area. An example of this use case could be to visualize the traffic in Berlin, Germany during rush-hour on a given day. The use case, especially for the replay functionality and the generation of charts or heat maps, requires fast data access to traffic data for huge amounts of selected locations over relatively small time spans.

6.4 THE PROPOSED ARCHITECTURE

In this section we provide a description of the architecture we propose by introducing the overall system design, describe the data preparation process and introduce the system interaction. The system is composed of three main components, named *Import*, *Access*, and *Database Representation*, as shown in Fig. 6.2. In the following we provide a detailed description for each of these components.

6.4.1 THE IMPORT COMPONENT

The *Import* component is responsible for receiving, preprocessing, and storing the massive amount of incoming data. Data packets (in the following called "messages") are received by an online connection using the Protocol Buffer format (https://developers.google.com/protocol-buffers). Each of these messages contains information about the traffic situation for a specific stretch of road (encoded in a map-independent referencing system) at a given timestamp t. After decoding the message, the *Import* Component starts the preprocessing pipeline that mainly consists of two sub-steps:

1. *Map Matching.* It is required due to the map-independent referencing format of the different traffic data types (as introduced in Section 6.3). Each map-independent reference code received is decoded using different open source software components and mapped onto the underlying NDS digital map. With the resulting NDS references, a standard geo-line object is calculated that serves as

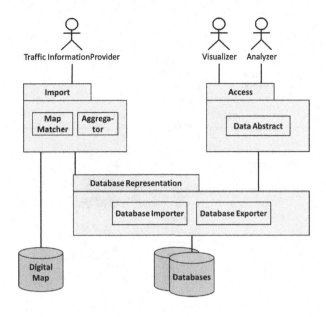

FIGURE 6.2

Component view of the system architecture.

spatial reference in our architecture and then it is indexed in the database to facilitate spatial queries.

2. *Spatial Aggregation*. It describes the process of transforming data into a different aggregation format. As described in Section 6.3, traffic data is updated once or twice per minute. Each update thereby contains all traffic information for a vast geographic region (mostly nation wide) for the time of the update. However, moving from this temporal aggregation (i.e. traffic information for one minute for entire Germany) to a more spatial oriented (i.e. traffic information for several months of selected street segments) is a key requirement for many use cases.

Fig. 6.3 illustrates the preprocessing described above. The data received from the Traffic Information Provider is decoded from the Protocol Buffer format, the location references are map matched, and the resulting data packets are stored in a temporary database. After having collected data in this database for a predefined time period (in our setup we picked 24 hours as time granularity), the spatial aggregation task begins. Thereby, the entire corpus of data is split up into separate data blocks for each location. Each block holds the traffic information for a specific location for the last 24 hours. These data elements are exported from the temporary database and imported to the final storage solution using the database connector elements within the Database Representation component.

The main focus of the way we designed the preprocessing pipeline is its applicability on distributed or cloud environments. With the data granularity provided, the

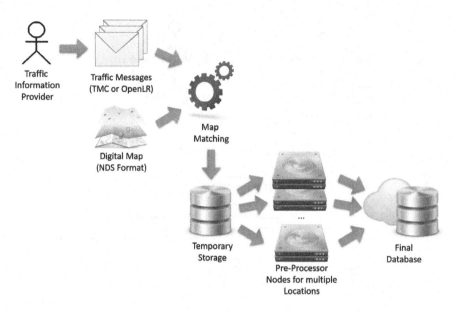

FIGURE 6.3

The pipeline of the parallelized preprocessing.

mentioned preprocessing process can easily be parallelized, also on highly scalable environments, according to some temporal or spatial criteria of the received data.

6.4.2 THE ACCESS COMPONENT

The *Data Access* acts as an access layer between the Database Representation and the external Actors of the system. It receives information from the underlying data structures and provides a unified interface to access traffic data for the use cases. The data is always transformed to a uniform representation used by the Visualizer and Analyzer, so it is totally abstracted from the actual data model used in the database.

6.4.3 THE DATABASE REPRESENTATION COMPONENT

The *Database Representation* component handles a variety of connectors for different DBMS and data models, which will be introduced in Section 6.5, that can be used for both the temporary and the final databases. It offers wrapper objects based on different mapping-frameworks for each database setup, including also different storage paradigms, to allow the other layers to easily and transparently handle the underlying data storage.

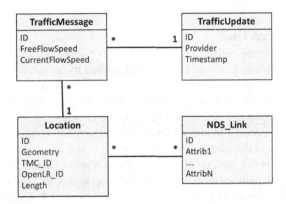

FIGURE 6.4

The designed relational schema for storing traffic information.

6.5 THE DATABASES

While there are multiple options for designing an architecture suitable to process traffic data, there are mainly only two solutions for the data storage: those based on Relational Database Management Systems (RDBMS) and those based on NoSQL Data Management Systems. In our proposal, we investigated both solutions, trying to understand and measure their pro and cons. In this section we present how we modeled the data in both the options.

6.5.1 THE RELATIONAL DATABASE MODEL

The first implementation is done using a RBDMS, namely the H2 DBMS (http://www.h2database.com). H2 is an open source RDBMS, written in Java. It has a transactional support and it is partly ACID (Atomicity, Consistency, Isolation, Durability) compliant. With H2 it is possible to create both in-memory tables, as well as disk-based tables. In our case, given the amount of the managed data, we always used disk-based tables.

6.5.1.1 The H2 Schema

The schema we designed to store traffic data in H2 is shown in Fig. 6.4, and consists of four different entities: *TrafficUpdate*, *TrafficMessage*, *Location*, and *NDS_Link*.

The *TrafficUpdate* is a high level entity that models the concept of an update on the traffic situation received by a traffic provider. It contains an *ID*, the name of the *Provider*, and the *Timestamp*. As described in Section 6.3, each Traffic Update we got at predefined time intervals contains a massive amount (usually nation-wide) of traffic incidents, which are modeled by the following entity. This entity is also used to monitor the import process to recognize duplicate import attempts or fill gaps of missing data at a later time.

The *TrafficMessage* entity is the key table, containing the information about a specific traffic incident. Such an incident is described by an *ID*, the *FreeFlowSpeed* attribute, that represents the drive speed (in km/h) under non-congested conditions and the *CurrentFlowSpeed* attribute, that represents the speed driven (in km/h) at the specific time of the traffic situation. The temporal coverage of the message can be retrieved by the reference to the related tuple in the table *TrafficUpdate*, while its spatial coverage is in the related table *Location*.

The *Location* is the spatial reference of the traffic messages. This table can be either pre-filled or built incrementally. Indeed, it contains the geometries for each of the spatial descriptor (TMC for the TF dataset or OpenLR for the TFD one) of a traffic message. Since TMC and OpenLR locations are fixed and known a priori, this table could be pre-compiled at the first run of the system. Of course in this case there is the risk to have a huge amount of useless information, in case there are some stretches of road where traffic never happens. The other solution is to add a tuple to this table every time we get a message for a Location which is not already in the DB. We opted for this second solution. As for the attributes of this table, we have an *ID*, the *Geometry* (as a LineString), and the *Length* of the stretch in meters.

Finally, the *NDS_Link* table comes from the proprietary NDS format, which allows data analysis of links contained in different Traffic Messages. For sake of confidentiality we cannot describe the content of this table. This entity is essential in order to use generated traffic predictions for applicable navigation solutions, since the predicted traversal time of each link is the key information used by Route Calculation algorithms.

The described relational database model allows easy access of traffic information, both in terms of spatial and temporal queries. For example, using the *geometry* attribute of the *Location* table, we can use spatial queries within our database to filter out locations that traffic information exist for in a geographic area of interest. Similarly, a query on the *timestamp* attribute of the *TrafficUpdate* table allows us to filter the dataset only for a give time frame.

6.5.2 THE NoSQL DATA MANAGEMENT MODEL

The second instance of our architecture is more Cloud-oriented, being based on a *NoSQL* distributed database, namely *Apache Cassandra* (http://cassandra.apache. org/).

Cassandra is a free and open-source distributed database management system, designed to handle large amounts of data across many computational nodes. It was initially developed at Facebook to power their Inbox Search feature. In March 2009, it became an Apache Incubator project, and the version 1.0 was released on October 2011. It falls in the category of *NoSQL* (also known as "non-SQL" or "non-relational") database management systems, which uses a different data model to traditional Relational DBMS, operating without putting the data into a fixed schema involving tables and references. This relaxation of the constraints dramatically simplifies the distribution of data, with the key advantage of being therefore easily executed on distributed environments, like the Cloud, leading to higher scalability and faster access speed.

On the other hand, there are also some pitfalls in this new model. Indeed, NoSQL DBMS mostly lack true ACID transactions, but instead provide a concept called "eventual consistency" whereby a lack between changes of data and its propagation might exists. More in details, on Cassandra databases no foreign key references can be defined among tables. Thus a much higher effort is required by the application developers to ensure data integrity and consistency. Additionally, data selection via SQL WHERE clauses is restricted in many ways. It is only possible to filter on columns which are either part of the primary key or which are indexed. Furthermore only comparison with the $=$ (equals) operator are supported and no $<$, $<=$, $>$ or $>=$ comparison operators can be used under the most of the circumstances. Another difference with relational databases is the write behavior of the Cassandra database. There is no primary key (or other constraint) checking prior to write operations. This means that data is directly persisted in the database, irrespectively whether the given key is already existing. To avoid unintended overriding of data, before any write operation the existence of data for the key must be checked before manually.

6.5.2.1 The Cassandra Schema

Cassandra is essentially a hybrid between a key-value and a tabular DBMS, where rows are organized into denormalized tables. The schema we designed to store traffic data in Cassandra is shown in Fig. 6.5, and is built upon three different tables within the Cassandra database, called *Day*, *Location*, and *Flow*. This simplified data model is possible because NoSQL databases typically support higher level data structures like lists (or arrays), maps, and sub-documents, even with the possibility to index them. For example the list of NDS references in the *Location* can be indexed and queried efficiently.

The table *day* contains the header information for a specific local date in a specific country. The table contains the following fields:

- **start_utc** is the time stamp in UTC for the start of the day.
- **country_code** is the ISO 3166 ALPHA 2 country code for the region this data set is belonging to.
- **resolution** is the measurement resolution used for the representation of the imported traffic data.
- **flow_counter** is the map of time stamp (these time stamps are day locale and thus starting to count from 0 at the start of the day) in milliseconds to counter values in which for each measurement period (with respect to the measurement resolution) the number of measurements is stored. This information can later be used in the Traffic Visualizer to calculate a confidence for the quality of measurements for the specific day.

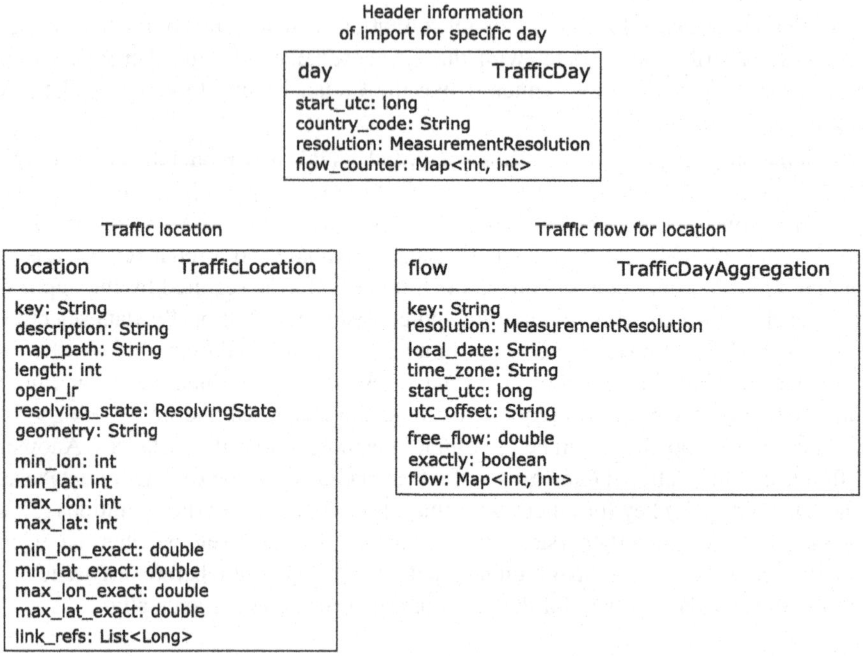

Header information
of import for specific day

day	TrafficDay
start_utc: long
country_code: String
resolution: MeasurementResolution
flow_counter: Map<int, int>	

Traffic location

location	TrafficLocation
key: String
description: String
map_path: String
length: int
open_lr
resolving_state: ResolvingState
geometry: String
min_lon: int
min_lat: int
max_lon: int
max_lat: int
min_lon_exact: double
min_lat_exact: double
max_lon_exact: double
max_lat_exact: double
link_refs: List<Long>	

Traffic flow for location

flow	TrafficDayAggregation
key: String
resolution: MeasurementResolution
local_date: String
time_zone: String
start_utc: long
utc_offset: String
free_flow: double
exactly: boolean
flow: Map<int, int>	

FIGURE 6.5

Datamodel of the Cassandra database.

start_utc and **country_code** are building the primary key. Therefore in the Traffic Visualizer, for each country and each start of day timestamp, there will be only one data set at maximum.

The table *Location* holds all information about the spatial units where there is available traffic data. It is structured in a similar way as the *Location* entity described in Section 6.5.1 but it also contains the references to the underlying map references. The table contains the following fields:

- **key** is the distinct key which is built up from the openLR code or the TCM information.
- **description** gives some additional information for the location, if it is present in the imported data.
- **map_path** is the path to the map used for location resolving.
- **length** is the length, in meters, of the location where a traffic accident has happened.
- **open_lr** the openLR string stored in an encoded BASE 64 manner.
- **resolving_state** the state of the traffic location concerning the resolving process.

- **geometry** is the JSON encoded geometry which was resolved. This information is used in the Traffic Viewer to render the location in the map and to calculate the bounding box of the location.
- **link_refs** the link references for the location in list manner.

Additionally to the geometry, the bounding box parameters for the location are stored in the table.

- **min_lon**, **min_lat**, **max_lon**, and **max_lat** stores the minimal longitude in integer manner in a floored representation of the exact longitude and latitude values. This is required to implement a way to filter the location via the SQL WHERE clause and to only use = operator for this.
- **min_lon_exact**, **min_lat_exact**, **max_lon_exact**, and **max_lat_exact** are the exact longitude and latitude values which are not rounded. They come directly form the envelope of the NDS map during resolving process.

As primary key fields only the **key** is used. When selecting location data from the table, due to restrictions in SQL WHERE clauses for $<$, $<=$, $>$ or $>=$ comparison operators, the selection is performed iteratively for all those longitude and latitude values being requested. This approach clusters the location data into a grid of geographical cells. To enhance performance for the application, location clusters loaded once are cached for later usage.

The table *Flow* holds the actual traffic information for an entire day, together with a timestamp and the id of the location that the data is referencing to. The *freeflow* attribute holds the information about the normal travel speed on the referenced location on that particular day. This information comes directly from the traffic data provider. The data is stored in a list of timestamps (minute of day) and the current_speed (in km/h) during that timestamp. The table contains the following fields:

- **key** is the key of the location the flow data is belonging to. Because Cassandra databases do not contain foreign key constraints, data integrity and consistence must be ensured by the developer.
- **resolution** the measurement resolution of traffic data. This field should always have the same value of the resolution defined in the day data set belonging to the same import date.
- **local_date** is an ISO-string representation of the local import date.
- **time_zone** stands for the time zone id for the zone the traffic location is belonging to. With respect to this value all general (UTC) time stamps are converted into local data and time information.
- **start_utc** is the UTC time stamp for the start of the day.
- **utc_offset** holds the offset from the local time to UTC time in milliseconds.
- **free_flow** is the free flow speed for the import day. For detailed TTF data, this value is calculated.
- **exactly** boolean flag indicating if the free flow was calculated exactly.

- **flow_map** the map of time stamps to actual free flow over every time slot for the day. The actual free flow is given as integer in percent of the free flow speed for that date. It is assumed, that the free flow for the location is not changing in between the day.

The primary key for this table is consisting of the **key** and the **start_utc** for the start of the day time stamp. When traffic data is imported for a specific traffic day, a check that no data for that day was imported before is performed.

6.6 PRELIMINARY EXPERIMENTAL RESULTS

In order to understand if there are differences in the scalability of the proposed technological solutions for the data models, we conducted a set of preliminary experiments. In particular, we analyzed the performance for the two main tasks that all use cases have in common: *Data Insertion* of a significant amount of data, with different loading scenarios for the Database, and *Data Extraction* of a different number of locations. For that benchmark, all experiments were conducted on the same hardware set-up to ensure the comparability of the results, which was a cluster based on the Intel i7 architecture, 16 GB RAM, and a set of Hard Disk Drives at 15K RPM, in configuration RAID $0 + 1$.

6.6.1 PERFORMANCE OF DATA INSERTION

As described in Section 6.3, to predict traffic situation, there is the need to constantly store, for further processing, the stream of real-time traffic data, which in our case is updated each minute. As a consequence, in order to process this vast amount of mobility data in reasonable time, efficient insertion of the received data packets is a key aspect. We have measured the performance of the two DBMS with respect to the insert functionality, to assess the manageable workload of a node of a cluster, handling the data of a certain number of vehicles. In particular, we aimed at simulating the workload that could be imposed by 1,000 connected cars sending basic XFCD messages to a remote data center. To this aim, we executed the insertion of 1000 records, logging the running time in milliseconds, with different amounts of data already stored in the database, ranging from 1 to more than 850,000 records in the involved tables. The upper bound is around the number of messages that could be collected from 1,000 vehicles driving for 14 hours, which could be a possible workload for a futuristic scenario of car sharing. Results are plotted in Fig. 6.6, where on the X-Axis there is the number of records already in the database, and on the Y-Axis there is the execution time in milliseconds.

As for the Relational Database, we have some B-Tree indexes on part of the attributes used for the queries, that must be updated for each data entry. As a con-

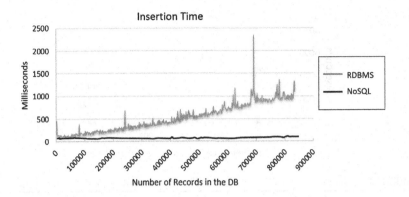

FIGURE 6.6

Experimental time for record insert in the RDBMS.

sequence, for each insert there is a quite linear relationship between the required time and the number of traffic messages already in the database.

As for the NoSQL database, since the amount of entities is highly decreased due to simplified data model, the insertion time is always almost constant, even for very large data sets. Indeed, it takes between 20 and 30 milliseconds to execute the insert of 1000 messages.

The general conclusion we can get from these results is that one node of a cluster, with a significant hardware equipment, is able to deal a very limited number of vehicles, if equipped with a RDBMS. Indeed, it could barely deal with one week of basic XFCD collected from 1,000 cars. Let us observe that the top car manufacturers are selling more than 10 Million vehicles per year, and that, in order to properly learn seasonal phenomena, it would be necessary to store and process some months, or even years of XFCD. Thus, in this scenario, the NoSQL model is clearly winning against the RBDMS, which is showing many difficulties in scaling up to this amount of data.

6.6.2 DATA EXTRACTION PERFORMANCE

After the data is preprocessed and stored in the underlying storage system, efficient query times are crucial for most use cases. A good example is the described *Visualizer* use case. To replay a historic traffic situation, queries for those traffic situations need to be very fast. The query we used for the evaluation requires traffic data for an entire day, for a variable number of locations. The filter fields are indexed in the relational data model. Fig. 6.7 shows the results of our benchmark analysis between the query times of the H2 and of the Cassandra databases. The X-Axis shows the amount of locations that we queried, while the Y-Axis shows the resulting running time of the query in milliseconds.

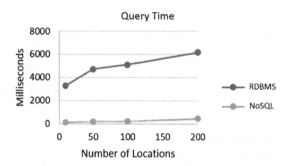

FIGURE 6.7

Experimental time for record extraction both databases.

The results show that the H2 RDBMS is quite slow in providing the first results, probably due to the fact that is Java-based and thus the communication pipeline is longer than with Cassandra, which is written in C++. Nevertheless, from Fig. 6.7 we can see that for both the solutions there is an interesting stability and scalability, since the performance are only slightly degrading, even in presence of significant changes in the amount of processed data.

6.7 CONCLUSIONS AND LESSON LEARNED

One of the greatest challenges mobility providers are facing today is to alleviate the social end economic negative effects of traffic congestions. In times of the big data hype, lots of research is currently being devoted, within the Intelligent Transportation Systems, to develop data-driven solutions. However, to the best of our knowledge, only few research attention has been dedicated to a world-wide scalability of these proposals, leveraging the findings and the proposals of the Spatial Big Data community.

To investigate this research area, we have proposed a parallelizable architecture characterized by a high degree of scalability. Indeed, all the needed preprocessing steps can be performed in a distributed environment, such as a cluster or the Cloud, and thus being able to process the massive amount of data that characterize the problem. Moreover, we have implemented and tested two different storage solutions, based respectively on a relational and on a NoSQL data model. In a preliminary experimental assessment of performances, we have shown that both the data models are able to tackle the query of the data, scaling in a comparable way. However, given the massive amount of incoming traffic data updates, the RDBMS show a lack of performance in managing these continuous updates, not being able to scale as easily as the NoSQL solution, which in the end seems to be preferable for this specific problem we are addressing. As future research direction, other than providing a broader investigation of the performance of the architecture under different use cases, it will be interesting also to test new emerging technologies. Especially the newly emerged solutions from the Spatial Big Data community, like for instance Spatial Hadoop (Eldawy and Mokbel, 2015), should be evaluated for the considered use cases.

ACKNOWLEDGMENTS

This research has been partly funded by the European Community Seventh Framework Programme (FP7/2007–2013) under grant agreement no. 610990 – Project COMPANION.

REFERENCES

Aji, A., Wang, F., Vo, H., Lee, R., Liu, Q., Zhang, X., Saltz, J., 2013. Hadoop GIS: a high performance spatial data warehousing system over mapreduce. Proceedings of the VLDB Endowment 6 (11), 1009–1020.

Amato, F., Mazzeo, A., Moscato, V., Picariello, A., 2014. Exploiting cloud technologies and context information for recommending touristic paths. In: Intelligent Distributed Computing VII. Springer, pp. 281–287.

CEBR, 2014. The Future Economic and Environmental Costs of Gridlock in 2030. Tech. Rep. cucs-012-011, Inrix.

Chrobok, R., Kaumann, O., Wahle, J., Schreckenberg, M., 2004. Different methods of traffic forecast based on real data. European Journal of Operational Research 155 (3), 558–568.

Clark, S., 2003. Traffic prediction using multivariate nonparametric regression. Journal of Transportation Engineering 129 (2), 161–168.

Di Martino, S., Giorio, C., Galiero, R., Ferrucci, F., Sarro, F., 2011a. A matching-algorithm based on the cloud and positioning systems to improve carpooling. In: Proceedings of DMS 2011 – 17th International Conference on Distributed Multimedia Systems, pp. 90–95.

Di Martino, S., Giorio, C., Galiero, R., 2011b. A rich cloud application to improve sustainable mobility. In: Web and Wireless Geographical Information Systems. In: Lecture Notes in Computer Science, vol. 6574. Springer, Berlin, Heidelberg, pp. 109–123.

Dunne, S., Ghosh, B., 2012. Regime-based short-term multivariate traffic condition forecasting algorithm. Journal of Transportation Engineering 138 (4), 455–466.

Eldawy, A., Mokbel, M.F., 2015. Spatialhadoop: a mapreduce framework for spatial data. In: Proceedings of the IEEE International Conference on Data Engineering (ICDE'15). IEEE.

Evans, M.R., Oliver, D., Yang, K., Zhou, X., Shekhar, S., 2014. Enabling spatial big data via cyberGIS: challenges and opportunities. In: Wang, S., Goodchild, M. (Eds.), CyberGIS: Fostering a New Wave of Geospatial Innovation and Discovery.

International, T., 2009. White paper – how TomTom's HD traffic and iq routes data provides the very best routing. Tech. rep., TomTom International. http://www.tomtom.com/lib/doc/download/HDT_White_Paper.pdf.

Ishak, S., Al-Deek, H., 2002. Performance evaluation of short-term time-series traffic prediction model. Journal of Transportation Engineering 128 (6), 490–498.

Kargupta, H., Gama, J., Fan, W., 2010. The next generation of transportation systems, greenhouse emissions, and data mining. In: Proceedings of the 16th ACM SIGKDD International Conference on Knowledge Discovery and Data Mining. ACM, pp. 1209–1212.

Krajzewicz, D., Hertkorn, G., Rössel, C., Wagner, P., 2002. Sumo (simulation of urban mobility)—an open-source traffic simulation. In: Proceedings of the 4th Middle East Symposium on Simulation and Modelling (MESM20002), pp. 183–187.

Kwoczek, S., Di Martino, S., Nejdl, W., 2014a. Predicting and visualizing traffic congestion in the presence of planned special events. Journal of Visual Languages and Computing 25 (6), 973–980.

Kwoczek, S., Di Martino, S., Nejdl, W., 2014b. Predicting traffic congestion in presence of planned special events. In: Proceedings of the Twentieth International Conference on Distributed Multimedia Systems. DMS, pp. 357–364.

Kwoczek, S., Di Martino, S., Nejdl, W., 2015. Stuck around the stadium? An approach to identify road segments affected by planned special events. In: 2015 IEEE 18th International Conference on Intelligent Transportation Systems (ITSC).

Manyika, J., Chui, M., Brown, B., Bughin, J., Dobbs, R., Roxburgh, C., Byers, A.H., 2011. Big Data: the Next Frontier for Innovation, Competition, and Productivity. Tech. rep., McKinsey Global Institute.

Pan, B., Demiryurek, U., Shahabi, C., 2012. Utilizing real-world transportation data for accurate traffic prediction. In: 2012 IEEE 12th International Conference on Data Mining (ICDM). IEEE, pp. 595–604.

Pan, B., Demiryurek, U., Shahabi, C., Gupta, C., 2013. Forecasting spatiotemporal impact of traffic incidents on road networks. In: 2013 IEEE 13th International Conference on Data Mining (ICDM), pp. 587–596.

Park, D., Rilett, L., Han, G., 1999. Spectral basis neural networks for real-time travel time forecasting. Journal of Transportation Engineering 125 (6), 515–523. http://dx.doi.org/10.1061/(ASCE)0733-947X(1999)125:6(515).

Shekhar, S., Evans, M.R., Kang, J.M., Mohan, P., 2011. Identifying patterns in spatial information: a survey of methods. Wiley Interdisciplinary Reviews: Data Mining and Knowledge Discovery 1 (3), 193–214.

Shekhar, S., Gunturi, V., Evans, M.R., Yang, K., 2012. Spatial big-data challenges intersecting mobility and cloud computing. In: Proceedings of the Eleventh ACM International Workshop on Data Engineering for Wireless and Mobile Access. ACM, pp. 1–6.

Van Lint, J.W.C., 2008. Online learning solutions for freeway travel time prediction. IEEE Transactions on Intelligent Transportation Systems 9 (1), 38–47. http://dx.doi.org/10.1109/TITS.2008.915649.

Vanajakshi, L., Rilett, L., 2004. A comparison of the performance of artificial neural networks and support vector machines for the prediction of traffic speed. In: 2004 IEEE Intelligent Vehicles Symposium, pp. 194–199.

Vatsavai, R.R., Ganguly, A., Chandola, V., Stefanidis, A., Klasky, S., Shekhar, S., 2012. Spatiotemporal data mining in the era of big spatial data: algorithms and applications. In: Proceedings of the 1st ACM SIGSPATIAL International Workshop on Analytics for Big Geospatial Data. ACM, pp. 1–10.

Wu, C.-H., Ho, J.-M., Lee, D., 2004. Travel-time prediction with support vector regression. IEEE Transactions on Intelligent Transportation Systems 5 (4), 276–281. http://dx.doi.org/10.1109/TITS.2004.837813.

Yang, H., Sasaki, T., Iida, Y., Asakura, Y., 1992. Estimation of origin-destination matrices from link traffic counts on congested networks. Transportation Research Part B: Methodological 26 (6).

Zhang, X., Rice, J.A., 2003. Short-term travel time prediction. Transportation Research Part C: Emerging Technologies 11 (3–4), 187–210. http://dx.doi.org/10.1016/S0968-090X(03)00026-3.

Zhu, J.Z., Cao, J.X., Zhu, Y., 2014. Traffic volume forecasting based on radial basis function neural network with the consideration of traffic flows at the adjacent intersections. Transportation Research Part C: Emerging Technologies 47, 139–154.

ACRONYMS AND GLOSSARY

List of acronyms with explanation

ACID Atomicity, Consistency, Isolation, Durability. Desirable properties of transactions in a DBMS.

DBMS Data Base Management Systems. A collection of programs running on one or more computers, providing users with a systematic way to create, retrieve, update, and manage data.

FCD Floating Car Data. The collection of information about position, speed, direction of travel, and time from driving vehicles.

GPS Global Positioning System. A global navigation satellite system providing geolocation to receivers.

GSM Global System for Mobile communications. A set of protocols for second-generation (2G) digital cellular networks.

ISO 3166 ALPHA 2 A part of the ISO 3166 standard, published by the International Organization for Standardization (ISO), to represent countries and other areas of geographical interest.

ITS Intelligent Transportation Systems. Various Information and Communication Technologies synergically applied to improve various modes of passenger and freight transport.

JSON JavaScript Object Notation. An open-standard format to exchange data objects, based on attribute–value pairs.

NDS Navigation Data Standard. A standard for automotive maps.

NoSQL A model to store and retrieve data, operating without putting the data into a fixed schema involving tables and references.

OpenLR Open Location Referencing. A royalty-free location referencing method for digital maps of different vendors and versions.

RAID Redundant Array of Inexpensive Disks. A data storage technique combining two or more physical disk drives into a single logical unit, to improve redundancy, performance, or both.

RDBMS Relational Data Base Management Systems. A DBMS based on the relational model, where data is represented by means ore potentially correlated tables.

RPM Rounds per Minute. Unit of measure of the speed of a mechanical hard drive.

SQL Structured Query Language. A language to interact with RDBMS.

SBD Spatial Big Data. Datasets exceeding the capacity of commonly used spatial computing and spatial database technologies.

TF Traffic Flow. Traffic information represented by means of TMC messages.

TFD Traffic Flow Detailed. Traffic information represented by means of OpenLR messages.

TMC Traffic Message Channel. A technology to deliver traffic information via conventional FM radio broadcasts.

UTC Coordinated Universal Time. The primary standard to represent time information.

XFCD eXtended Floating Car Data. The collection of advanced information collected from the sensors installed in the vehicles.

Glossary of terms with explanation

Cassandra A free and open-source distributed DBMS, designed to handle large amounts of data across many computational nodes.

Cloud Computing A paradigm of computing where processing and storage resources are shared and allocated to clients on demand.

Connected Vehicle A vehicle able to interact with a back-end using some wireless communication technology.

Database Schema The structure of the information contained within a Database.

Data Model A model to specify the organization of the data and the handling of the inherent relationships.

Digital Map A digital representation of a map, with additional information useful for routing.

Eventual consistency A situation in NoSQL databases, where a lack between changes of data and its propagation might exists.

Hadoop An open source software stack developed by Apache, that can be installed on a commodity Linux cluster to support large scale distributed data analysis.

H2 An open source RDBMS, written in Java.

Knowledge Discovery Process A process composed of multiple steps, to discover useful knowledge from data.

Map Matching A process of matching raw GPS points presenting some noise on the road network of a digital map.

Origin-Destination Matrix A key component of the transportation planning model, intended as matrix where rows and columns represent origins and destinations of a road network, and values in the cells are the number of trips going from each origin to each destination.

Primary Key One or more attributes of a table in a RDBMS with the property that there are no two distinct tuples having the same values. It is used to uniquely identify each tuple of the table.

Protocol Buffer format A method to serialize structured data, developed by Google, and suitable for exchanging data among programs.

Smart Cities A vision for urban scenarios where the smart use of ICT will improve the quality of life by improving the efficiency of services, like for instance the usage of the road infrastructure.

Traffic Information Provider A company of a public institution providing updated information about the traffic on a part of the road network.

Traffic Message A message sent by a Traffic Information Provider to broadcast a change in the state of the traffic.

Traffic Model A model suitable to represent and predict the dynamics of the traffic on a given road segment, over the time.

Wrapper Object An intermediate object that allows the interface of an existing class to be used as another interface. It is usually used to make existing classes work with others expecting a different signature of methods, without modifying their source code.

BOUNDED ERROR DATA COMPRESSION AND AGGREGATION IN WIRELESS SENSOR NETWORKS

7

Ray-I Chang*, Meng-Han Li*, Polon Chuang†, Jeng-Wei Lin‡

*National Taiwan University, Taiwan †Institute for Information Industry, Taiwan
‡Tunghai University, Taiwan

7.1 INTRODUCTION

Wireless sensor networks (WSNs) (Dasgupta et al., 2003) have been widely used in environment monitoring, target tracking, and so on. As sensor nodes usually work with restrictive power supply, such as batteries, energy efficiency has become an important issue to increase the system lifetime. In order to reduce the power consumption in data transmission, compression mechanisms have been applied to WSNs to reduce the total size of transmission data. However, the errors, i.e., differences between the compressed and original data, are usually unbounded. These lossy data compression algorithms are not suitable for many real world applications that require bounded errors between the original and received data. Furthermore, if the errors are bounded, many bounded aggregation operations can be done much faster, such as Min(), Max(), and Sum().

Under the manufacturing process, sensor itself is inevitable to have error and this error has bounded value. On the other hand, the demand output for real-world applications may allow limited error. We can take advantage of this bounded error concept to improve the processing results. In this article, given an error bound, an efficient and effective data compression and aggregation algorithm, called BEDCA (Bounded Error Data Compression and Aggregation), is proposed. BEDCA uses data, temporal, and spatial correlations to compress sensed data within a given error bound. In the first step, data correlation of the monitored environment is taken into consideration. BEDCA uses observed data to construct a codebook for reference. Then, given an error bound, BEDCA determines whether a new sensed data should be compressed or not by referring to data such as the previous sensed data (for temporal correlation), the neighboring sensed data (for spatial correlation), and the codebook data (for data correlation).

Smart Sensors Networks. DOI: 10.1016/B978-0-12-809859-2.00009-7

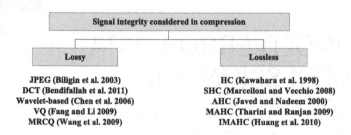

FIGURE 7.1

The classification of different signal integrities considered in data compression mechanisms.

The data compression in WSN can use the bounded error concept to improve its ratio. Computer algorithm to calculate the processing results may be simplified under the bounded error. Nowadays, data fusion appears to be important in WSN to combine sensory data derived from disparate sources such that the resulting information has less uncertainty. BEDCA can compress sensory data derived from disparate sources by referring their previous sensed data, their neighboring sensed data, and their code-book data. It is very suitable for data fusion applications of WSN where sensory data derived from disparate sources are temporal, spatial, and data correlated. Even under the bounded error, data fusion can result information with less uncertainty by combining sensory data derived from disparate sources.

The remainder of this paper is organized as follows. In Section 7.2, we survey the literature for bounded and unbounded error data compression and aggregation with signal integrity and correlations. In Section 7.3, we present BEDCA and then describe aggregation query processing without decompression. In Section 7.4, the performance evaluation results of BEDCA are shown. Finally, conclusions and intended future works are given in Section 7.5.

7.2 BACKGROUND AND LITERATURE REVIEW

Signal integrity considered in data compression mechanisms of WSNs can be divided into two categories, lossless and lossy, as shown in Fig. 7.1.

Lossy data compression mechanisms usually give much better compression ratios based on signal fidelity lost. Examples of mechanisms in this category include JPEG (Biligin et al., 2003; Higgins et al., 2010), DCT (Bendifallah et al., 2011), Wavelet-based (Chen et al., 2006), Vector Quantization (VQ) (Fang and Li, 2009), and Multiresolution Spatial and Temporal Coding (MRCQ) (Wang et al., 2009). However, the errors between the compressed and original data are unbounded when these mechanisms are applied. They may cause serious accidents because the errors cannot be controlled in decision making processes.

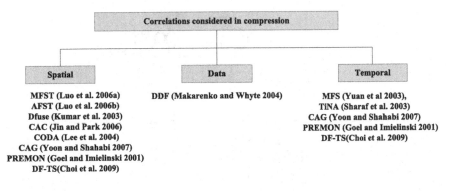

FIGURE 7.2

The classification of different signal correlations considered in data compression mechanisms.

Lossless data compression mechanisms maintain complete signal integrity but have worse compression ratios than lossy data compression mechanisms. Lossless data compression mechanisms such as Huffman Coding (HC) (Kawahara et al., 1998), Static Huffman Coding (SHC) (Marcelloni and Vecchio, 2008), Adaptive Huffman Coding (AHC) (Javed and Nadeem, 2000), Modified Adaptive Huffman Coding (MAHC) (Tharini and Ranjan, 2009), and Improved MAHC (IMAHC) (Lin et al., 2010) are introduced in this category. HC refers to the use of a variable-length codebook where symbols with higher probabilities to appear are encoded with fewer bits. Given statistics information of source symbols, SHC is proposed to construct a codebook with an optimal prefix code. Adaptive Huffman Coding separates encoded data to prefix and suffix based on weight of data. Prefix can be redefined adaptively to have better compression ratio. MAHC merges the codebook with an optimal prefix code of SHC and the codebook with dynamic suffix of AHC to have better compression ratio. IMAHC applies dynamic codebook to adjust to rapid environment changes for WSNs.

Correlations considered in compression mechanisms of WSNs can be divided into three categories (spatial correlation, data correlation, and temporal correlation) as shown in Fig. 7.2.

Spatial correlation suggests that the sensor nodes located closely in an intensive area usually have similar sensed data. Thus, we can apply the spatial correlation to compress transmission data. Example mechanisms of this category include MTST (Luo et al., 2006a), AFST (Luo et al., 2006b), DFuse (Kumar et al., 2003), CAC (Jin and Park, 2006), CODA (Lee et al., 2004), CAG (Yoon and Shahabi, 2007), PREMON (Goel and Imielinski, 2001), and DF-TS (Choi et al., 2009).

Different from spatial correlation, data correlation uses information theory to encode high probability data with short length output. An example mechanism of this category is DDF (Makarenko and Whyte, 2004). Temporal correlation is based on the idea that sensor nodes usually have similar sensed data in short period of

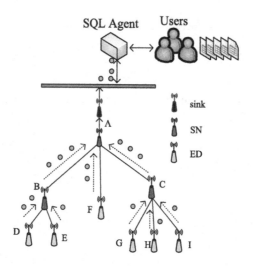

FIGURE 7.3

System architecture of a WSN.

time. Example mechanisms of this category include MFS (Yuan et al., 2003), TiNA (Sharaf et al., 2003), CAG (Yoon and Shahabi, 2007), PREMON (Goel and Imielinski, 2001), and DF-TS (Choi et al., 2009).

7.3 THE PROPOSED APPROACH

Fig. 7.3 shows a simple WSN architecture, where many sensor nodes are deployed. Sensor nodes are divided into two categories: end device (ED) and super node (SN). SNs are usually more powerful and faster than EDs, and also much more expensive than EDs. In this article, without loss of generality, we simply assume all EDs and SNs are homogeneous. They are deployed in the monitored environment randomly.

In this article, we propose an efficient and effective data compression algorithm, called Bounded Error Data Compression and Aggregation (BEDCA) for WSNs. Given a bounded error between the compressed and original data, BEDCA tries to minimize the size of transmission data.

The processing of BEDCA can be divided into two phases: offline phase and online phase. In the offline phase, BEDCA evaluates the data correlation of sensed data to construct a codebook in the first step. At the same time, BEDCA assigns bounded errors to all sensor nodes in the WSN.

In the online phase, the types of data processing required by EDs and SNs are different. Under the given bounded error constraint, an ED decides how to compress a new sensed data by comparing it with the previous sensed data (for temporal correlation) and codebook encoded data (for data correlation); meanwhile, a SN decides how

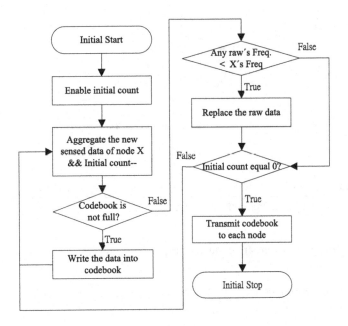

FIGURE 7.4

Flowchart of the codebook generation.

to compress a new received data by comparing it with the neighboring sensed data (for spatial correlation) and codebook encoded data (for data correlation). Finally, the compressed data are decompressed and results are reported to users by sink.

BEDCA needs a period of time to cluster sensed data to the sink. The sink is going to record these data in the buffer until the quantity of sensed data reaches a threshold which is designed by different environment. Then, the sink analyzes the correlations by the differences between sensed data to build a codebook. Finally, the sink sends the trained codebook to all EDs and SNs. The flowchart of codebook generation is shown as Fig. 7.4. We must note that codebook generation runs just once in the offline. We can ignore the energy cost of the codebook generation.

Before a WSN system starts, the bounded error (τMAX) must be set according to the application requirement. When the codebook is under construction, the sink also learns the range of nodes in this system and then it must assign the bounded error to each node. For example, the bounded error assignment for "SUM" aggregation is shown in Fig. 7.5. In this case, there are night nodes in this system. The max bounded error is set with one. Thus, the bounded error (τ) of each node is set with one ninth.

BEBDCA designs a compression method with three types of correlation considered under bounded error constraint: temporal, spatial, and data correlations. Tempo-

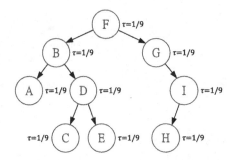

FIGURE 7.5

Bounded error assignment for nodes in a WSN.

ral and spatial correlations are defined as Eqs. (7.1) and (7.2).

$$\text{Temporal correlation} = \frac{1}{n} \times \sum_{p=3}^{n} (\Delta_{p,p-1} - \Delta_{p-1,p-2}) \tag{7.1}$$

$$\text{Spatial correlation} = \frac{1}{n} \times \sum_{p=3}^{n} \sqrt{\frac{1}{k} \times \sum_{i=1}^{k} (\Delta_{p,p-1}^{i} - M(\Delta_{p,p-1}))} \tag{7.2}$$

n is the number of samples in a time sequence, $\Delta_{p,p-1}$ is the difference between two successive samples sensed at time p and $p - 1$, k is the number of EDs, and $M(\Delta_{p,p-1})$ is the average of $\Delta_{p,p-1}$ of all EDs.

An ED first determines whether the new sensed data (V_{new}) would be compressed or not based on the difference between V_{new} and the previous sensed data (V_T). If the difference is within the bounded error (τ), V_{new} is compressed. The ED replaces V_{new} with a _T_NoOpt, which is encoded as a zero bit (0). When a SN receives a _T_NoOpt, it does not apply other compression mechanism to compress it. Otherwise, V_{new} is not compressed. Data correlation compression is applied.

For example, as shown in Fig. 7.6 with $\tau = 0.5\,°C$, EDs B, C, and D obtain sensed data in serial. Node B compresses V_{new} because $|V_{new} - V_T| = |25.0 - 24.5| = 0.5 \le \tau$. It sends one bit 0 (where _T_NoOpt = 0) to SN A. On the other hand, for node C, since $|V_{new} - V_T| = |24.5 - 23.5| = 1.0 > \tau$, it applied data correlated compression. It looks up the codebook and sends three bits 1|10 (where not(_T_NoOpt) = 1 and codebook(−1.0) = 10) to SN A. Similarly, for node D, since $|V_{new} - V_T| = |22.8 - 24.0| = 1.2 > \tau$, it sends three bits 1|11 (where not(_T_NoOpt) = 1 and codebook(−1.2) = 11) to SN A.

Since an ED has overhead to check data of its neighboring nodes, the spatial correlation data compression is used only in SNs. When a SN receives a _T_NoOpt, it sends two bits 0|0. If the difference between the received compressed data (V_{new}) and the data of the neighboring ED (V_s) is within τ, V_{new} is compressed furthermore.

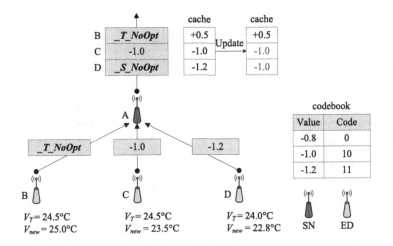

FIGURE 7.6

Bounded error data compression with temporal and spatial correlation.

The SN replaces V_{new} with a _S_NoOpt_ and sends two bits 0|1. Otherwise, data correlated compression is used.

As shown in Fig. 7.6, three bit streams 0, 1|10, and 1|11 are sent to SN A by EDs B, C, and D, respectively. When SN A receives _T_NoOpt_ from ED B, it sends two bits 0|0, to the sink. From the cache, it knows the data from ED B is +0.5 °C. Later, SN A receives −1.0 °C from ED C. Since the difference, $|{-1.0 - 0.5}| = 1.5$, is larger than τ, data correlation data compression is used. SN A looks up the codebook and sends three bits 1|10. Later, SN A receives −1.2 °C from ED D. Since the difference, $|{-1.2 - (-1.0)}| = 0.2$, is smaller than τ, it sends two bits 0|1. In addition, the changes are updated in the cache.

Data correlation data compression is used in EDs and SN. Among all codes in the codebook that encode values within the bounded error, the one that has a shortest length and smallest error is chosen. To encode a value V_{new}, V_{DC} is first identified, which is a set of code candidates in the codebook, such that the differences between V_{new} and values encoded by these candidates are smaller than or equal to τ. V_{CC} is then reduced from V_{DC}, which is the set of the shortest candidates in V_{DC}. V_D is finally determined, which is the most similar reference to V_{new} in V_{CC}.

Fig. 7.7 shows how BEDCA works when further data correlation is used in bounded error compression. τ is 0.5 °C. When SN A receives 1|10 (it decoded as $V_{new} = -1.0$ °C) from ED C, data correlation data compression is used. $V_{DC} = \{-0.8\,°C, -1.0\,°C, -1.2\,°C\}$ is identified. The difference between any member in V_{DC} and V_{new} is smaller than τ. In the codebook, −0.8 °C requires one bit, while −1.0 °C and −1.2 °C require two bits. Thus, $V_{CC} = \{-0.8\,°C\}$ is reduced from V_{DC}. Finally, $V_D = -0.8$ °C is chosen. SN A sends two bits 1|0 and updates the cache.

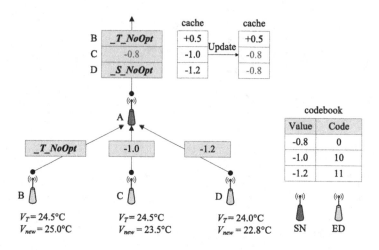

FIGURE 7.7

Bounded error data compression with data correlation.

Later, SN A receives $-1.2\,°C$ from ED D. Since the difference, $|-1.2-(-0.8)| = 0.4$, is smaller than τ, it sends two bits 0|1 and updates the cache.

Given an error bound, BEDCA considers temporal, spatial, and data correlations to improve the performance of compression and thus save energy of a WSN. EDs use temporal and data correlations, while SNs use spatial and data correlations. To save more energy, EDs gather more sensed data until buffer is full. Thus, overhead for packet header is reduced. Lossless data compression method such as SFALIC (Starosolski, 2007) can be integrated to further reduce the number of required bits and thus the consumed energy. Figs. 7.8 and 7.9 show algorithms EDC and SNC with the pseudocodes of BEDCA used for ED compression and SN compression, respectively.

7.4 PERFORMANCE EVALUATION

In our experiments, there is a WSN with a specified error bound. EDs monitor the target environment, compress the sensed data by BEDCA with temporal and data correlations, and send the resultant bit streams to SNs; meanwhile, SNs receive the bit streams from EDs, compress the received data by BEDCA with spatial and data correlations, and send the final bit streams to the sink.

We evaluate BEDCA by four real world datasets: forest temperature from Chequamegon Ecosystem Atmosphere Study (ChEAS), seismic wave from Earthquake Hazards Program (U.S. Geological Survey), electrocardiogram (ECG), and

---

Algorithm EDC ED (End Device) Compression.

Input: The new sensed data (V_{new})
Output: The compressed data
1:   **if** $|V_{new} - V_T| \leq \tau$ **then**
2:       **return** _T_NoOpt
3:   **end if**
4:   V_{DC} = FindDataCorellatedCandidates(V_{new})
5:   **if** V_{DC} is empty **then**
6:       **return** V_{new}
7:   **end if**
8:   V_{CC} = FindSmallestCandidates(V_{DC})
9:   V_D = FindMostSimilarCode(V_{CC})
10: **return** Encoded(V_D)

FIGURE 7.8

BEDCA pseudocode for end devices.

Algorithm SNC SN (Super Node) Compression.

Input: The new sensed data (V_{new})
Output: The compressed data
1:   **if** V_{new} is _T_NoOpt **then**
2:       **return** _T_NoOpt
3:   **end if**
4:   **if** $|V_{new} - V_S| \leq \tau$ **then**
5:       **return** _S_NoOpt
6:   **end if**
7:   V_{DC} = FindDataCorellatedCandidates(V_{new})
8:   **if** V_{DC} is empty **then**
9:       **return** V_{new}
10: **end if**
11: V_{CC} = FindSmallestCandidates(V_{DC})
12: V_D = FindMostSimilarCode(V_{CC})
13: **return** Encoded(V_D)

FIGURE 7.9

BEDCA pseudocode for super nodes.

electroencephalogram (EEG) from PhysioNet (PhysioNet). They are with different temporal and spatial correlations. Table 7.1 shows some properties of the four datasets.

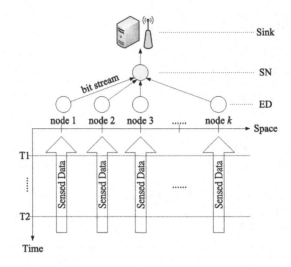

FIGURE 7.10

A hierarchical WSN.

Table 7.1 Data Source			
Source	**Temporal Correlation**	**Spatial Correlation**	**Categories**
Forest Temperature	11.7	488.2	High temporal correlation High spatial correlation
Seismic Wave	41.6	146.9	Low temporal correlation High spatial correlation
Electrocardiogram	10.8	2080.2	High temporal correlation Low spatial correlation
Electroencephalogram	86.4	1312.0	Low temporal correlation Low spatial correlation

Given a bounded error, we simulate the WSN with the four datasets by BEDCA and other lossy data compression methods. Without loss of generality, a simplified hierarchical WSN as shown in Fig. 7.10 is simulated. In each dataset, 5 near nodes are selected. The scale of WSN is $k = 5$. It is not difficult to extend the proposed method to WSN with a higher scale.

In the codebook generation of BEDCA, two HC-based methods, SHE (Marcelloni and Vecchio, 2008) and IMAHC (Lin et al., 2010), are adopted. They result in BE(SHE) (bounded error SHE) and BE(IMHAC) (bounded error IMHAC). We increase the received error bound at the sink from 0 to 2% to evaluate the performance of BE(SHE) and BE(IMHAC). At the same time, three lossy data compression methods (JPEG, DCT, VQ), which do not guarantee error bound, are also examined. With unbounded error (UBE), they are referred to as UBE(JPEG) (Biligin et al., 2003;

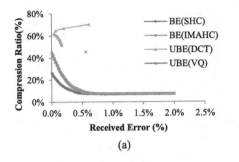 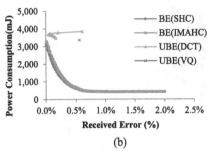

FIGURE 7.11

Compression ratio and power consumption for forest temperature dataset.

Table 7.2 Parameters of Energy Consumption

Parameter		Value	Parameter		Value
Current per bit	I_{TX}	17.4 mA	Current	I_{cpu}	31 mA
	I_{RX}	19.7 mA	Frequency	F_{cpu}	48 MHz
Time per bit	T_{TX}	3.2×10^{-5} s			E_{add}	2.13 nJ
	T_{RX}	3.2×10^{-5}s	Energy per	E_{mul}	6.39 nJ
Voltage per bit	V_{TX}	3.3 V	instruction	E_{cmp}	2.13 nJ
	V_{RX}	3.3 V		E_{sht}	4.26 nJ

Higgins et al., 2010), UBE(DCT) (Bendifallah et al., 2011), and UBE(VQ) (Fang and Li, 2009) in this study. We vary their different proprietary parameters to depict the compression ratio and resultant errors when different datasets are simulated. However, when UBE(JPEG) is used, the received error at the sink always excesses 2%. As a result, UBE(JPEG) is not furthermore considered in the following discussion.

To evaluate the energy consumption, parameters shown in Table 7.2 are adopted (Liang and Peng, 2010). According to these parameters, energy consumed in data compression is calculated according to the number of different instructions executed in BEDCA, including ADD (addition), MUL (multiply), CMP (comparison), and SHT (shift). As well, energy consumed in data transmission (send and receive) is calculated according to the number of bits compressed by BEDCA.

When the first dataset, forest temperature, is examined, Fig. 7.11(a) and (b) show the compression ratio and power consumption of the proposed BEDCA methods, BE(SHC) and BE(IMAHC), and two lossy data compression methods, UBE(DCT) and UBE(VQ). Because UBE(JPEG), UBE(DCT), and UBE(VQ) cannot bound errors, different parameters are used to depict the following figures. The errors received by using UBE(JPEG) are always larger than 2%, the curve of UBE(JPEG) is not shown. UBE(DCT) may compress data at a same compression ratio, but with different errors. It is also possible that UBE(DCT) may compress data with a same error level, but at different compression ratios. The curve of UBE(DCT) is not a function

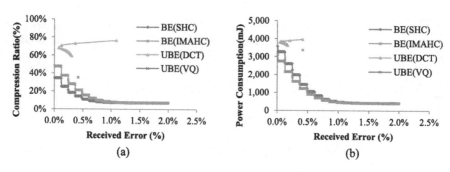

FIGURE 7.12

Compression ratio and power consumption for seismic wave dataset.

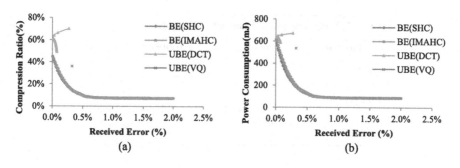

FIGURE 7.13

Compression ratio and power consumption for ECG dataset.

of received error. It is clear that BE(SHC) and BE(IMACE) are much more effective than UBE(DCT) and UBE(VQ).

Since BEDCA can reduce the data size of the dataset, the power consumption in data transmission (send and receive) are reduced. BE(SHC) and BE(IMACE) consumed less energy than UBE(DCT) and UBE(VQ). For the other three datasets, the experiment results are very similar, as shown in Figs. 7.12, 7.13 and 7.14.

Although the four datasets have different temporal and spatial correlations, given a bounded error, BEDCA outperforms UBE(DCT) and UBE(VQ) significantly. Even the bounded error is set to a small value (under 0.5%), the proposed method can reduce a lot of (over 70%) from the uncompressed data.

7.5 CONCLUSION AND FUTURE WORKS

In this study, we introduce BEDCA, a data compression and aggregation mechanism whose error can be bounded. The data compression in WSN can use BEDCA to im-

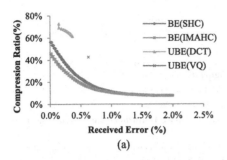 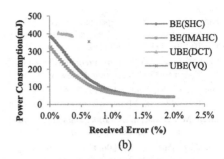

FIGURE 7.14

Compression ratio and power consumption for EEG dataset.

prove its ratio. We use four real world datasets with different temporal and spatial correlations to evaluate the proposed algorithm. Experiment results show that even if the error bound is set to a small value (under 0.5%), BEDCA can reduce a lot of transmission energy (over 70%). Lessons learned that, using conventional JPEG mechanism, the worst errors received are all larger than 2% even applying the best compression parameters. Without the bounded error concept, conventional lossy data compression mechanisms are difficult to guarantee their errors. Our insight into the compression and aggregation mechanism is that, BEDCA can compress sensory data derived from disparate sources by referring their previous sensed data, their neighboring sensed data, and their codebook data. BEDCA can make WSNs more power efficient. It is very suitable for data fusion applications of WSN where sensory data derived from disparate sources are temporal, spatial, and data correlated. There are many possible future research issues. For example, in BEDCA, it is difficult for an ordinary user to properly set the bounded error in unknown situation. A possible future research issue to study dynamic error bound assignment to all EDs.

REFERENCES

Bendifallah, A., Benzid, R., Boulemden, M., 2011. Improved ECG compression method using discrete cosine transform. Electronics Letters 47 (2), 1–2.

Biligin, A., Marcellin, M.W., Altbach, M.I., 2003. Compression of electrocardiogram signals using JPEG2000. IEEE Transactions on Consumer Electronics 49 (4), 833–840.

ChEAS. Chequamegon ecosystem atmosphere study. Available at http://cheas.psu.edu/data.

Chen, J., Ma, J., Zhang, Y., Shi, X., 2006. A wavelet-based ECG compression algorithm using Golomb codes. In: 2006 International Conference on Communications, Circuits, and Systems. IEEE, pp. 130–133.

Choi, K., Kim, M.-H., Chae, K.-J., Park, J.-J., Joo, S.-S., 2009. An efficient data fusion and assurance mechanism using temporal and spatial correlations for home automation networks. IEEE Transactions on Consumer Electronics 55 (3), 1330–1336. http://dx.doi.org/10.1109/TCE.2009.5277996.

Dasgupta, K., Kalpakis, K., Namjoshi, P., 2003. An efficient clustering-based heuristic for data gathering and aggregation in sensor networks. In: IEEE Wireless Communications and Networking Conference. IEEE, pp. 1948–1953.

Fang, J., Li, H., 2009. Hyperplane-based vector quantization for distributed estimation in wireless sensor networks. IEEE Transactions on Information Theory 55 (12), 5682–5699. http://dx.doi.org/10.1109/TIT.2009.2032856.

Goel, S., Imielinski, T., 2001. Prediction-based monitoring in sensor network: taking lessons from MPEG. Computer Communication Review 31 (5), 82–98. http://dx.doi.org/10.1145/1037107.1037117.

Higgins, G., Mc Ginley, B., Glavin, M., Jones, E., 2010. Low power compression of EEG signals using JPEG2000. In: 4th International Conference on Pervasive Computing Technologies for Healthcare.

Javed, M.Y., Nadeem, A., 2000. Data compression through adaptive Huffman coding schemes. In: Proceedings of TENCON 2000. IEEE, pp. 187–190.

Jin, G.-Y., Park, M.-S., 2006. CAC: context adaptive clustering for efficient data aggregation in wireless sensor networks. Lecture Notes in Computer Science 3976, 1132–1137.

Kawahara, M., Chiu, Y.-J., Berger, T., 1998. High-speed software implementation of Huffman coding. In: Proceedings of '98 Data Compression Conference, p. 553.

Kumar, R., Wolenetz, M., Agarwalla, B., Shin, J., Hutto, P., Paul, A., Ramachandran, U., 2003. DFuse: a framework for distributed data fusion. In: 1st International Conference on Embedded Networked Sensor Systems. ACM, pp. 114–125.

Lee, S., Yoo, J., Chung, T., 2004. Distance-based energy efficient clustering for wireless sensor networks. In: 29th Annual IEEE International Conference on Local Computer Networks. IEEE, pp. 567–568.

Liang, Y., Peng, W., 2010. Minimizing energy consumptions in wireless sensor networks via two-model transmission. Computer Communication Review 40 (1), 12–18.

Lin, C.-C., Chuang, C.-C., Chiang, C.-W., Chang, R.-I., 2010. A novel data compression method using improved JPEG-LS in wireless sensor networks. In: 12th International Conference on Advanced Communication Technology, pp. 346–351.

Luo, H., Liu, Y., Das, S.K., 2006a. Routing correlation data with fusion cost in wireless sensor networks. IEEE Transactions on Mobile Computing 5 (11), 1620–1632. http://dx.doi.org/10.1109/TMC.2006.171.

Luo, H., Luo, J., Liu, Y., Das, S.K., 2006b. Adaptive data fusion for energy efficient routing in wireless sensor networks. IEEE Transactions on Computers 55 (10), 1286–1299. http://dx.doi.org/10.1109/TC.2006.157.

Makarenko, A., Whyte, H.D., 2004. Decentralized data fusion and control in active sensor networks. In: 7th International Conference on Information Fusion, pp. 479–486.

Marcelloni, F., Vecchio, M., 2008. A simple algorithm for data compression in wireless sensor networks. IEEE Communications Letters 12 (6), 411–413. http://dx.doi.org/10.1109/LCOMM.2008.080300.

Physionet. Available at http://www.physionet.org/cgi-bin/ATM.

Sharaf, M.A., Beaver, J., Labrinidis, A., Chrysanthis, P.K., 2003. TiNA: a scheme for temporal coherency-aware in-network aggregation. In: 3rd ACM International Workshop on Data Engineering for Wireless and Mobile Access. ACM, pp. 69–76.

Starosolski, R., 2007. Simple fast and adaptive lossless image compression algorithm. Software, Practice & Experience 37 (1), 65–91. http://dx.doi.org/10.1002/spe.746.

Tharini, C., Ranjan, P.V., 2009. Design of modified adaptive Huffman data compression algorithm for wireless sensor network. Journal of Computer Science 5 (6), 466–470.

U.S. Geological Survey. Earthquake Hazards program. Available at http://earthquake.usgs.gov/.

Wang, Y.-C., Hsieh, Y.-Y., Tseng, Y.-C., 2009. Multiresolution spatial and temporal coding in a wireless sensor network for long-term monitoring applications. IEEE Transactions on Computers 58 (6), 827–838. http://dx.doi.org/10.1109/TC.2009.20.

Yoon, S., Shahabi, C., 2007. The clustered aggregation (CAG) technique leveraging spatial and temporal correlations in wireless sensor networks. ACM Transactions on Sensor Networks 3 (1). http://dx.doi.org/10.1145/1210669.1210672.

Yuan, W., Krishnamurthy, S.V., Tripathi, S.K., 2003. Synchronization of multiple levels of data fusion in wireless sensor networks. In: Global Telecommunications Conference. IEEE, pp. 221–225.

ACRONYMS AND GLOSSARY

List of acronyms with explanation

$\_S\_NoOpt$ a replacement to represent a V_{new} when $|V_{new} - V_T|$ is within the bounded error

$\_T\_NoOpt$ a replacement to represent a V_{new} when $|V_{new} - V_S|$ is within the bounded error

AFST	Adaptive Fusion Steiner Tree
AHC	Adaptive Huffman Coding
BEDCA	Bounded Error Data Compression and Aggregation
CAC	Context Adaptive Clustering
CAG	Clustered AGgregation
CODA	Cluster-based self-Organizing Data Aggregation
DCT	Discrete Cosine Transform
DDF	Decentralized Data Fusion
DF-TS	Data Fusion using Temporal and Spatial correlations
DFuse	Data Fusion
ECG	electrocardiogram
ED	End Device
EEG	electroencephalogram
HC	Huffman Coding
IMAHC	Improved MAHC
JPEG	Joint Photographic Experts Group
MAHC	Modified Adaptive Huffman Coding
MFS	Multi-level Fusion Synchronization
MRCQ	Multiresolution Spatial and Temporal Coding
MTST	Minimum Fusion Steiner Tree
PREMON	PREdiction based MONitoring
SFALIC	Simple Fast and Adaptive Lossless Image Compression
SHC	Static Huffman Coding
SN	Super Node
TiNA	Temporal coherency-aware in-Network Aggregation
V_{CC}	subset of V_{DC} that includes the shortest candidates
V_D	the most similar reference candidate to V_{new} in V_{CC}
V_{DC}	a set of code candidates in the codebook, such that the differences between V_{new} and values encoded by these candidates are smaller than or equal to the bounded error
V_{new}	the new sensed data in a ED or received data in a SN from EDs
VQ	Vector Quantization
V_S	the data of the neighboring ED
V_T	the previous sensed data in a ED
WSN	Wireless Sensor Network
τ**MAX**	the bounded error set in the target WSN application

Glossary of terms with explanation

Bounded error the differences between the compressed and original data are bounded.

APPLICATION OF DATA ANALYSIS IN WELLNESS AND HEALTH SENSOR NETWORK ENVIRONMENT

8

Jian-hua Yeh, Wei-ting Chen
Aletheia University, Taiwan

8.1 INTRODUCTION

8.1.1 BACKGROUND

In 2010, the elderly population aged 65 and above in Taiwan reached 10.7%, which has exceeded the level of aging population countries defined by World Health Organization (WHO). The elderly population of Taiwan reached 12.51% in 2015, which means that Taiwan will face an aging population society. When the population of a country begin to age, it will bring some problems. Along with the growth of the elderly population, chronic diseases and dysfunction of the proportion of people will rise. According to the Ministry of Health and Welfare of Taiwan, the disability ratio of the number of the total population exceeds 3% (about 0.76 million), and over 65 years old accounted for 63.2% (approximately 0.48 million disability population) in 2015, as shown in Fig. 8.1.

The disability population usually means those who has no ability for daily living, including the incapacitated person, physical aging, patients of specific diseases, and infants. These are the people who needs long-term care from others because they have no self-care ability. According to the statistics from Ministry of Health and Welfare of Taiwan, the population which need long-term care grow year by year: 293,466 persons in 2013, 307,676 persons in 2014. The funding plan for long-term care also grows fast by year: 3.229 billion NTD in 2013, 4.163 billion NTD in 2014, and 4.8 billion NTD in 2015, as shown in Table 8.1.

According to the statistics of the Ministry of the Interior of Taiwan, the newborn population is 196,486 in 2008, 191,310 in 2009, and drop down sharply to 166,886 in 2010. The phenomenon of declining birthrate of the population also occurred in Taiwan, the elderly people are more and more, the younger population are fewer and fewer, resulting in the gradually needs for more long-term care services, as shown in Table 8.2.

Smart Sensors Networks. DOI: 10.1016/B978-0-12-809859-2.00010-3

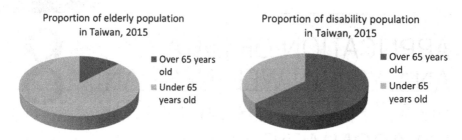

FIGURE 8.1

The proportion of the chronic diseases in Taiwan.

Table 8.1 The Funding Plan for Long-Term Care Grows Fast by Year

Year	2013	2014	2015
People of long-term care services (people)	293,466	307,676	–
People of long-term care services (billion NTD)	3.229	4.163	4.80

Table 8.2 The Newborn Population Drops down in Taiwan

Year	New Born Population (people)
2007	204,414
2008	196,486
2009	191,310
2010	166,886

In addition to be used in the procurement of equipment and facilities for the preparation of the fund, the largest consumption of the funding for long-term care services are human resources. Currently, the long-term care services are in the traditional way of doing care, which will cause more funding needs of human resources. According to the trend of elderly population in the last section, the needs of human resources will grow under traditional way of long-term care services.

8.1.2 PROBLEM STATEMENT

The increasing proportion of people who need long-term care services in Taiwan mentioned in the last section reveals that the need of technological way to resolve the shortage problem on human resources of long-term care services exists. New technologies are able to improve the quality of life for the public, they are certainly possible to transform the way of long-term care services for the elderly population. Sensor technologies used to deploy in the environment are able to sense the environmental conditions of the users. Since the adoption of cameras and other vision-based sensing systems are less acceptable by the public because people are not obliged to expose their privacy, the research in this study will not reply on vision-based systems, which is more acceptable by the public.

In this research, we propose the use of sensor technologies with information analysis and prediction techniques to reduce the needs of human resources in long-term care environment, which may cause the Government's financial burden occur in the future long-term care environment.

8.1.3 RESEARCH CONTRIBUTIONS

In this research, several aspects of problems in the long-term care environment are discussed, and a system architecture to reduce the human resource problem is proposed. The research significance of this research is listed as below:

1. A system architecture for health analysis and prediction in the long-term care environment called WHSNS is defined.
2. An application for health analysis and prediction under long-term care environment, that is, an application of WHSNS structure is developed.
3. A real world case study for the application of WHSNS has been elaborated in this article to demonstrate the possibility of health analysis and prediction with the proposed WHSNS structure.

8.1.4 ARTICLE OUTLINE

This article is broadly divided into five sections. Section 8.1 begins with the introduction of the aging society problem and discusses the motivation and objectives of this research. Section 8.2 deliberates the literature review related to the elderly population, long-term care environment for monitoring the elderly and disability people, and the current technologies. Section 8.3 describe the designed and developed computing techniques for sensor data analysis for health analysis and prediction in the long-term care environment. Section 8.4 describes one of the applications for health analysis and prediction under long-term care environment, with a real world case study to demonstrate the possibility of health analysis and prediction with the proposed WHSNS structure. Section 8.5 presents the conclusion of the present research study and suggestions for the future works.

8.2 LITERATURE REVIEW

8.2.1 INTRODUCTION

This section is about the studies on the wellness and health sensor environment and the technologies supporting long-term care (Health Canada, 2012). From the perspective of literature review, it discusses how to design and develop the methods of establishing Wellness and Health Sensor Network Systems (WHSNS).

The so-called "care environment" refers to the place where the elderly and those suffering life-long trauma caused by accidents can enjoy health service and medical facilities (Patient Safety and Quality, 2008). In Taiwan, an aging society, the most frequently-seen care environment is long-term care center. As a care environment, a long-term care center is home to those who need long-term care support; hence, it is equipped with complete health care, providing food, and a space for relaxation. In the long-term care environment, professional medical care workers are deployed to offer medical care services.

The current long-term care environment features a traditional operation model – those in need of long-term care (care recipients) are cared through manual management. The so-called manual management refers to the management free from information technology (Braddock, 2011). For instance, caregivers check if care recipients need assistance on a regular basis; instead of using computer and information technology, they offer care services entirely according to their own experience. Therefore, this study aims to assist caregivers with computer and information technology.

8.2.2 DEFINITION OF HEALTH CARE

In the past decade, the proportion of the elderly (those aged over 65) has become larger; it is estimated that it will have accounted for 19.3% of the global population by 2050 (Gavrilov and Heuveline, 2003). In the 21st century, the lifespan of human has increased from 46–89 to 66–93 of age. The proportion of the elderly will grow from 24% to 32% (Global Health and Aging, 2011). In the coming three decades, it is estimated that the proportion of the elderly aged over 75 will hit 17%, up from 8.5%.

According to the prediction of the United Nations, the drastic decline in the birth rate will result in a large proportion of the elderly and children (UN Documents, 2012). It is estimated that the ratio of children to the elderly will have declined from 9:1 to 4:1 by 2050. Additionally, industrialized countries will encounter the following problem: the proportion of the disabled and the handicapped will become larger, and the annual cost for the treatment of Alzheimer's disease will increase from USD 33 billion to USD 61 billion. Such a growth shows that the elderly will be confronted with more serious health problems. The fast growth in the demand for medical care and the incorporation of most public expenses into the scope of medical care have consumed 13% of the gross national income. These phenomena have demonstrated an increase in the demand for the long-term care environment. Hence, the long-term care environment needs to be improved with more computer and information technologies.

As the awareness of health intensifies and the quality of food and medicine improves, the life of human has been lengthened. Nevertheless, people are vulnerable to various injuries and accidents when they are old; therefore, the elderly need more care facilities. Once the expense for the health care of the elderly increases, it would impose a heavier financial burden on care service providers in the long-term care environment; hence, it is necessary to achieve the support of long-term care monitoring with new methods and systems. According to this study, the sensor analysis and prediction of health care can help care recipients with the evaluation of personal health status and reduce the cost of health care. Moreover, it can reduce the workload on caregivers. Hence, it has become increasingly critical to study how to assess the physical health, living conditions and behaviors of the elderly, and the disabled with an intelligent system. In a health care system, a sensor is supposed to help care-related professionals collect the data about health or gather the information about the lifestyle of care recipients.

8.2.3 WELLNESS AND HEALTH SENSOR NETWORK SYSTEMS

The so-called Wellness and Health Sensor Network Systems (WHSNS) is a systematic architecture which deploys a number of sensors in a care environment. Featuring various sensors, it offers continuous environment monitoring and thus provides care recipients with a more comfortable residence, and lightens the burden of caregivers. With the advance of the sensor technology, the embedded processor, and the communication system, the WHSNS-related devices can be deployed in a simpler way and make it easier for caregivers to manage those who need long-term care.

The WHSNS architecture comprises three essential components: first, the physical sensing components (the sensor and the embedded control board); second, the communication system components (wired or wireless networks), usually connected with the physical sensing components; third, the data analysis components, which achieve monitoring and management through information processing in the machine learning procedure.

The WHSNS architecture enables the procedure to analyze and predict the environmental data detected by the sensor deployed in the long-term care environment, so that caregivers will be able to monitor the situations of care recipients without being around them. The sensor plays an important role in the WHSNS architecture – collecting sensor data. Only with these relevant sensor data can the procedure achieve the analysis and prediction. In the WHSNS architecture, the information is usually delivered in a wireless manner, so that WHSNS can be installed in a place unsuitable for the wired networks.

The functions of WHSNS can be divided into three types – the care sensor, the server, and the caregiver client. The care sensor is a sensing device in the long-term care environment. Normally, it is a physical sensing component designed to collect and analyze sensor data, and decide if notifications should be sent. The server stores the raw data and analytic results uploaded by the care sensor; its another task is to provide a data analysis model for the care sensor. Meanwhile, it delivers necessary notifications to caregivers. The caregiver can receive the notifications from the server, browse sensor data through the server, and manage the server.

8.2.4 COMPONENTS OF WHSNS

In the WHSNS, the fundamental operation is to sense various kinds of environment data for latter analysis and prediction. The WHSNS consists of several modules, in-

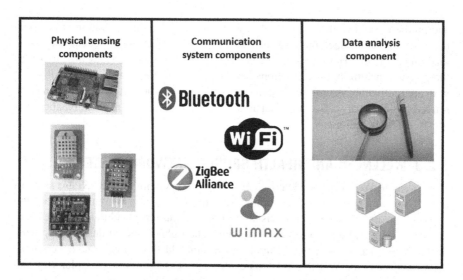

FIGURE 8.2

The basic components of WHSNS.

cluding the physical sensing components for monitoring, the communication system components for wireless transmission, and the data analysis components for analysis and prediction. The basic components of WHSNS are shown in Fig. 8.2.

The physical sensing components (sensors) in the WHSNS can obtain the sensor data in the external environment and transmit the data to the server through wireless networks to process care recipients-related analysis and prediction. Fig. 8.3 depicts the relations among the basic WHSNS components.

8.2.4.1 Physical Sensing Components

In the WHSNS, the physical sensing components represent the deployed sensor-related hardware, which is extremely important in the WHSNS. The physical sensing components can be divided into two types – the sensor and the embedded control board. The sensor is supposed to sense the data about the current long-term care environment. For instance, the temperature and humidity sensor can measure the current temperature and humidity, while the pressure sensor can detect if there is any pressure on objects. The embedded control board serves as the hardware that connects the sensors. It obtains the data about the actual environment through the sensors and then uploads the data into the server for analysis through wireless communication.

8.2.4.2 Communication System Components

The communication system components offer the communication between the physical sensing components and the server. In real operation scenario, the WHSNS architecture can be equipped with both wired and wireless communication. However,

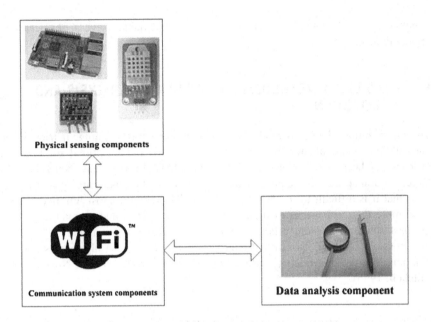

FIGURE 8.3

The relations among the basic WHSNS components.

there may be problems in the wired network, which would affect the flexibility of the installation of the WHSNS architecture. In contrast, the wireless network provides greater flexibility in installation of the WHSNS and avoids the location impact in the installation. To decrease the inconvenience of frequent replacement of batteries, it is possible to use household electricity. The wireless technologies frequently used in the WHSNS architecture include Bluetooth (Milestones in the Bluetooth, 2004), Wi-Fi (Authorization of Spread Spectrum, 1985), WiMAX (Marks, 2006), and ZigBee (ZigBee, 2012).

8.2.4.3 Data Analysis Component

The data analysis component in the WHSNS refers to the server mentioned above. The functions of data analysis component are as follows:

- Store the raw sensor data
- Analyze data
- Manage the physical sensing components

The server stores all detected environment data and takes them as the basis for future information analysis and inquiry. The information analysis targets at the analysis and prediction of historical information; the management over the physical sensing components means that a single server can connect and manage several physical sensing components.

8.2.5 REVIEW OF METHODOLOGIES ON HEALTH ANALYSIS AND PREDICTION

The optimization algorithm is also a part of the machine learning. It tries different solutions and evaluates their cost so as to decide if the optimal solution to a problem has been found. The algorithm is suitable for the problems consisting of a large solution space. The feature of these problems is that there are so many solutions that it is difficult to try them one by one. Usually, the most intuitive method is no more than trying tens of thousands of solutions in a random manner to seek the best solution. A more efficient method is to make a systematic approaching of a possible solution. The most-frequently-seen optimization algorithms are hill-climbing, simulated annealing, and the genetic algorithm (Russel and Norvig, 2009; Mitchell, 1996).

8.2.6 ADVANTAGES AND CURRENT LIMITATIONS OF THE WHSNS

One of the advantages of the WHSNS is that it is easy to install, causing no dramatic change to the original environment. Thanks to the existing electronic technologies, sensors can be made into small ones with complete functions, and some can even generate over three kinds of sensor data. The embedded control board, though small, can still handle basic calculation. The above physical sensing components are of great help for the installation of the WHSNS. The physical sensing components are small, so they can be easily incorporated into the original long-term care environment, which would have positive effects on the care recipients and caregivers in the long-term care environment. The devices of the WHSNS would not affect the daily life and activities of concerned parties, which ensures a regular level of environmental comfort. Moreover, the cost of these devices is rather low.

The data collected and analyzed by the sensors and the embedded control board can also assist caregivers, for the WHSNS would automatically send necessary notifications to caregivers according to the result of data analysis, so that they can take appropriate processing before the appearance of problems. Moreover, the WHSNS architecture needs to collect environment data and even the physiological information like electronic cardiogram of care recipients through the sensors. But there would be some difficulties in the data collection in the society dominated by the law of Personal Information Protection Act (Privacy Act, 2015; Human Rights Act, 1998; Personal Information Protection Act, 2012). Meanwhile, the data collected in the WHSNS should not be used without any constraint because all these data are personal. Therefore, there may be some legal limitations on the collection of environment data for the WHSNS.

8.3 DEPLOYMENT OF THE WELLNESS AND HEALTH SENSOR NETWORK SYSTEMS

8.3.1 INTRODUCTION

With the fast advance of smart sensors and communication technologies, the systems based on the wireless sensor network have become increasingly popular; moreover, since the wireless sensor network systems usually target at environment, therefore, it is easy to understand that these systems are designed to record the information in the physical world. As the processor technology and the wireless communication system develop, the sensor network has become smaller, cheaper, and lower power-consuming in recent years. Additionally, such wireless sensor networks have made it faster for users to observe and explore the phenomena in the real world with an unprecedented accuracy. The earliest wireless sensor network came from military and scientific research projects, but its application scope has been expanded with the reduction in the cost of sensors. In recent years, the wireless sensor network has attracted the attention of those networks and databases research communities. For example, environment researchers are interested in temperature-sensing reading, while ecologists want the information about the sensed humidity of soil. Similarly, this study processes the sensed information about the elderly and patients (commonly known as care recipients) in the long-term care environment. And the sensor system of the long-term care environment is called the Wellness and Health Sensor Network Systems (WHSNS). One of the objectives of this study is to use the sensor data collected by the heterogeneous sensors installed of the WHSNS in the long-term care environment to recognize the physical conditions of care recipients. Hence, the WHSNS can cover various sensors, such as the current sensor, the stress sensor, the contact sensor, the temperature and humidity sensor, and even the passive infrared sensor. Then, the information processing and transmission of the micro-controller modules like Arduino (Kushner, 2011), ZigBee (ZigBee, 2012), and even Raspberry Pi (Bush, 2011), which read the data from the sensing components, is done to form a complete sensing and processing procedure. Normally, the development and installation of the WHSNS is a non-invasive one featuring flexibility and security in use. In terms of information integration, transmitting the signals of wireless sensors to the central processing system is the most widely-used method at present. Meanwhile, recent studies have switched attention to how to lengthen the operation of sensing nodes and satisfy specific application demands. The two main concepts of the studies on information integration are as follows: 1) the relationship among sensor data, such as the relationship between time and space; 2) there may be change to the target data in many deterministic studies. The above concepts have influenced the deployment of most WHSNS, including the consideration of temporal precision and energy conservation. This study will give a detailed description of how to collect, store, and process near real-time sensor data.

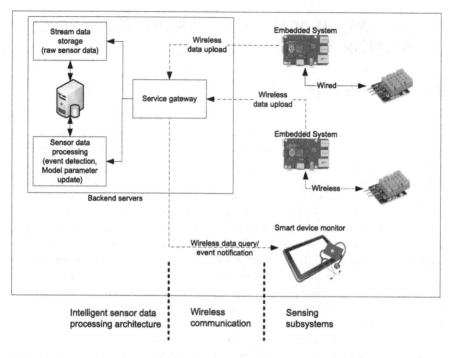

FIGURE 8.4

A typical WHSNS architecture.

8.3.2 BRIEF DESCRIPTION OF WHSNS

The wireless sensing system is a module furnished with one or more sensing devices, combined wireless transmitting modules, and modules with limited calculation capability. Such a module can sense the physical environment in temperature and humidity and the use of other objects. A WHSNS has a group of backend servers which are supposed to receive the heterogeneous data sensed from subsystems and make specific calculation of the data to finish predefined tasks. Fig. 8.4 shows a typical WHSNS architecture which includes sensing subsystems and a backend server group.

The information integration architecture of an entire WHSNS has two important modules: 1) the wireless sensing subsystems; 2) the intelligent sensor data processing architecture, which collects, analyzes, and predicts heterogeneous sensor data to observe specific changes to the conditions of care recipients in the long-term care environment and decide if notifications or alarms should be sent according to the changes. In general, many objects in the long-term care environment can be equipped with sensors according to the objectives of application, and the sensors and the sensing modules can be connected and transmitted through wired and wireless manner. The reason for equipping the objects with sensors is that the care recipients in the

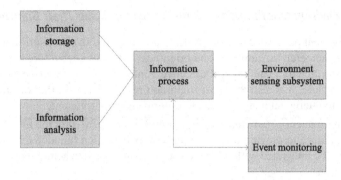

FIGURE 8.5

The function blocks of the WHSNS system.

long-term care environment would frequently interact with these objects. Fig. 8.5 describes the function blocks of the system.

Such a WHSNS will be of great use in checking the conditions of care recipients in the long-term care environment. Aside from the parts where sensing is used in the entire system, there is a notification mechanism of emergent assistance which can help caregivers handle problems according to the processing of the real-time senor data about care recipients. Hence, this study attempts to define a systematic architecture which can check if a care recipient is "healthy" or not. Meanwhile, a small number of sensing components are used to develop the demonstration of the systematic architecture according to the general WHSNS architecture. The subsystems consisting of these sensing components have been applied in other studies and occasions; hence, this study is not the first to use them nor the only testing case at present. Moreover, these subsystems are compatible with the operation architecture of the Internet of Things (IoT) today. The entire systematic operation can be done through distant operation; hence, it features great flexibility in operation.

8.3.3 APPLICATION OF WHSNS IN WELLNESS AND HEALTH SENSING

The selection of WHSNS sensing subsystems is based on the living conditions of the care recipients in the long-term care environment. According to the demand of the long-term care environment, the sensing subsystems can be classified as the following groups. These sensing subsystems operate in cooperation with the objects in the environment. The description of these sensing subsystems is as follows:

Group 1: Environment parameter sensing systems and their control
Group 2: Contact sensing systems and their control
Group 3: Passive infrared sensing systems and their control
Group 4: Electrical apparatuses and their sensing control
Group 5: Non-electrical apparatuses and their sensing control.

8.3.3.1 Environment Parameter Sensing Systems and Their Control

The environment parameter sensing systems are supposed to sense surroundings, including the temperature, humidity, and brightness in the long-term care environment. Such sensing function can suggest effective control over the environmental conditions; meanwhile, it is also the most important research field in the transition from traditional to intelligent control in the long-term care environment.

The environment parameter sensing systems are usually combined with a graphical user interface to provide users with necessary operation functions. More importantly, it can be integrated with machine learning algorithms to achieve smart sensing and prediction.

8.3.3.2 Contact Sensing Systems and Their Control

Such objects include office desks, cabinets, and the doors of refrigerators. The contact sensing systems can be connected with these objects to check if the drawers and the refrigerator doors are open or not, so as to check if the desks, cabinets, and refrigerators are in use. The results of the sensing are presented in the form of ON/OFF status.

8.3.3.3 Passive Infrared Sensing Systems and Their Control

The passive infrared sensing systems are designed to monitor all possible actions in the scope they cover. Their components are small, cheap, energy-saving, and flexible, with a long term of service. Normally, they are called IR motion sensors, which usually offer simple ON/OFF output to show if any action is detected.

8.3.3.4 Electrical Apparatuses and Their Sensing Control

Such sensing system aims at monitoring the electrical equipment in the long-term care environment. From the perspective of user, monitoring electrical equipment is to sense if they are still in use. Such a sensing module checks if electrical equipment are in use by testing the current that flows through their wires.

8.3.3.5 Non-Electrical Apparatuses and Their Sensing Control

The frequently-seen objects connected to the sensing system of non-electrical equipment include bed, chair, close stool, and sofa. Such equipment are usually deployed with an extremely thin and flexible non-invasive pressure sensor. According to the analog pressure readings in such a system, the WHSNS can judge if these equipment are still in use or not. The measurement method is to compare the pressure reading before the use with that in the use. The WHSNS can invoke corresponding processing according to the status of the equipment, such as recording the activities of care recipients.

8.3.4 NETWORK COMMUNICATION OF WHSNS

Normally, the sensing systems mentioned in the previous section do not have complicated demands on processing; hence, the composition of sensing modules (or sensing nodes or sensing subsystems) is simple. In this study, the embedded system development board and the general purpose input/output ports (GPIO) are used to connect the sensing components; meanwhile, they are integrated with the simple sensor data reading logic to form into standard sensing subsystems. In respect of communication protocol, the simplest 802.11 series protocol is adopted for the transmission, which is believed to be an intuitive and most accessible transmission model in a WI-FI environment.

8.3.5 TOPOLOGICAL ARCHITECTURE OF WHSNS

Apart from the group of backend servers which analyzes and store data, the entire WHSNS comprises of several sensing subsystems. In the WHSNS planning, an adequate number of Wi-Fi access points are installed in the long-term care environment. In other words, Wi-Fi signals would cover the whole long-term care environment. Therefore, compared with the naive wireless sensing system in the environment, the sensing subsystem has better signal transmission and quality of data. For that reason, the sensing subsystems are taken as the data generator in the topological architecture of WHSNS, while the group of backend servers are regarded as the data sink which would develop into a star-shaped architecture. Fig. 8.6 depicts the star-shaped topological transmission structure in the long-term care environment.

8.3.6 INTEGRATION AND ANALYSIS OF REAL-TIME HETEROGENEOUS SENSOR DATA

WHSNS is designed to obtain sensor data from surroundings. These data are acquired at a different time and flow into the group of processing backend servers in the form of streaming. In the current WHSNS architecture, there are two main models of inquiring sensors data in the sensing subsystems – the PUSH-based model and the PULL-based model. In the PUSH-based model (Zanikolas and Sakellariou, 2005) sensor data are inquired in the group of backend servers; during the same period, sensor data would constantly flow into the group of servers to satisfy the demand for the real-time inquiry into sensor data. Such an inquiry operation model is the most frequently-adopted and practical method today. A typical inquiry into WHSNS sensor data includes the following information: 1) sensing frequency, which is used to describe the frequency of sensors' sampling in the environment; 2) sensing features, the features used to form the results of inquiry, such as temperature and humidity; 3) limitation on the inquiry return value, which is used to filter the unexpected return value data.

In the PULL-based model, the results of inquiry into sensor data constitute the snapshots of sensor data. More specially, such results of inquiry would circulate in

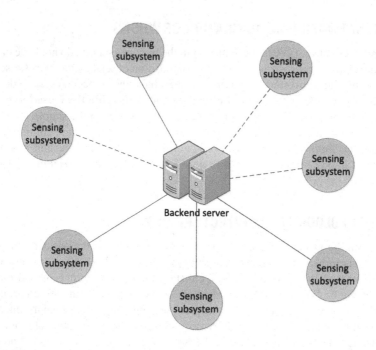

FIGURE 8.6

The star-shaped topological transmission structure.

the WHSNS network environment on a regular basis. After receiving the request on the inquiry into sensor data, the sensing subsystems would return the current sensor data; after collecting all sensor responses, the group of backend servers would calculate the ultimate results to be shown to users. The greatest difference between the two inquiries into sensor data is as follows: in the PUSH-based model, the sensing subsystems would reply with a data stream as the result of return; in the PULL-based model, the sensing subsystems would merely reply with the data about sensing status. In this study, the PUSH-based model is adopted to process the sensor data. In other words, the sensor data would be returned to and processed in the group of backend servers in the form of streaming.

The quantity of the sensor data collected in the WHSNS is huge. Therefore, the quantity of the information necessary for the control over and processing of the real environment is more important than the quantity of data generated by the sensing subsystems. In this study, the WHSNS offers a real-time user interface; hence, the information about the long-term care environment and the conditions of care recipients detected in the WHSNS can be immediately obtained.

8.3.7 THE SOFTWARE SYSTEM USED TO OBTAIN SENSOR DATA

The information processing of the WHSNS begins with the IoT sensing subsystems. The three reasons for using the IoT architecture to obtain sensor data are as follows: 1) low cost; 2) long-term operation; 3) distributed and flexible information transmission. The features of the IoT sensing architecture are as follows: 1) a real-time or a near real-time acquisition of sensor data in the database of the sensing subsystems; 2) a logic for simple decision making on the raw sensor data; 3) the ability to detect abnormal status and send signals to the group of backend servers.

In this study, a typical WHSNS consists of the following software components:

(i) A real-time heterogeneous sensor data processing program based on Python (van Rossum, 2009).
(ii) A group of SQL (Chamberlin and Boyce, 1974) statements which are used to store the raw sensor data.
(iii) A group of Java Server Pages (JSPs) (Bergsten, 2003) programs which are used to process and present real-time or near real-time sensor data.
(iv) A group of Java (Gosling and McGilton, 1996) programs which are used to classify and recognize care recipients' conditions represented by sensor data.
(v) A distributed calculation and storage platform which can process a huge amount of sensor data in a distributed manner. The Hadoop (White, 2009) platform is adopted to meet the demand in this study.

8.3.8 OTHER DEMANDS

According to the requirement of efficient analysis, the following system parameters must be taken into account in the installation of a WHSNS:

(i) The sampling rate that must be considered in the acquisition and processing of data
(ii) The quantity of transmitted data in the system environment
(iii) The parameters of prediction model which are used to classify the conditions of care recipients.

The sampling rate parameter involves the frequency and total quantity of the data which is transmitted from the sensing subsystems to the group of backend servers. The first, fourth, and fifth sensing subsystem patterns are all equipped with the processing components of analog digital converter (ADC) (Lathi, 1998) which are used to convert sensor data into digital form to be sent. Therefore, the ADC sampling rate has direct effects on the quantity of transmitted data. Additionally, the sampling rate also directly influences the temporal reasoning of information.

8.3.9 SUMMARY

This section is about the systematic architecture of processing real-time heterogeneous sensor data of this study, including the design and installation of the sensing subsystems, the description of data input/output, and the information processing in the group of backend servers. In the next section, a real-world experiment will be demonstrated, and the processing of sensor data and the prediction of the conditions of care recipients will be discussed in detail.

8.4 CASE STUDY: APPLICATION OF WHSNS IN HEALTH ANALYSIS AND PREDICTION

8.4.1 INTRODUCTION

In the previous section, a WHSNS systematic architecture is explained. In this section, an intelligent medical bed is taken as a case to illustrate how the devices most frequently used by care recipients are combined with the processing of sensor data to provide health monitoring services for care recipients in the long-term care environment. In general, in a long-term care center, the proportion of the care recipients relying on medical beds is larger than that of the average lying on medical beds. In other words, care recipients have to lie on medical beds for a longer time than the average because of their poor health. An observation on the operation of the long-term care center shows that caregivers would help long-term care recipients turn over. There are two reasons for turning over in bed: 1) maintain a good lying position, reduce the stress on physical tissues, and decrease the incidence of pressure ulcers; 2) maintain the limb functions and positions of care recipients and prevent joint contracture or the complication caused by long-term lying. In practice, there is a principle that bodies must be turned over at least once every two hours. As different care recipients have different physical health conditions, some of them would perspire within the two hours of lying on bed while some would perspire after two hours. The reason why perspiration is mentioned is that it is a major factor that results in the incidence and deterioration of pressure ulcers harassing long-term care recipients. Additionally, the number of the caregivers in the long-term care center is limited, so making full use of the limited labor resources in helping care recipients turn over would have positive effects in saving the resources. In this study, an actual perspiration prediction system in the WHSNS architecture is proposed as a case. The intelligent medical bed for perspiration prediction is used to optimize the laborer use in the long-term care center. Most of the previous studies on intelligent medical bed merely focus on the monitoring of notification of leaving medical bed (Travis, 1994; Scott, 2000), without any kind of prediction function. In this study, a perspiration prediction model is proposed and combined with the operation of the entire WHSNS to develop a WHSNS which can predict and alarm perspiration. The effective perspiration alarm will remind caregivers of immediately helping care recipients turn over, so as to save the laborer resources and reduce the incidence of pressure ulcers.

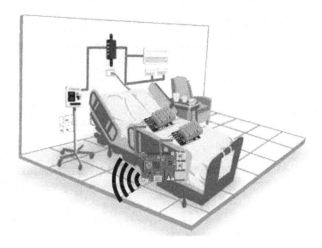

FIGURE 8.7

The intelligent medical bed designed in this study.

8.4.2 HEALTH ANALYSIS AND PREDICTION-ORIENTED SYSTEM DESIGN

The intelligent medical bed designed in this study is shown in Fig. 8.7, where two DHT22 (DHT22 data sheet, n.d.) sensors for sensing temperature and humidity are placed under the bed sheet which are located at the center of the back and waist of a care recipient respectively. Since the size of DHT22 is small enough that the care recipient will not able to detect them, therefore the arrangement of these sensors are non-invasive. The DHT22 sensor features a high precision of sensing and can measure temperature and humidity at the same time; thus, it is suitable for this experiment. Additionally, the reason for adopting DHT22 sensor for the measurement was that the nearly closed space between the back and the bed would become an area under the effects of the thermal dissipation of the back when a care recipient lies on the bed. Fig. 8.8 shows the results measured by the DHT22-sensor in a lying-on-bed test. Fig. 8.8(a) is the curve consists of temperature and humidity measured on the center of the back of the subject, while Fig. 8.8(b) is a curve comprising temperature and humidity measured on the center of the waist of the subject. In this case, the subject perspired around 75 minutes after the beginning of the experiment. In a more official sense, the so-called perspiration refers to "perceived perspiration" (Lamke et al., 1977). In the perspiration, the temperature curve was in saturation, and the sensor data about temperature declined in the perceived perspiration, which indicated the thermal dissipation of the body surface of the respondent. The above observation on the temperature and humidity curves shows the perspiration of the subject. Meanwhile, Fig. 8.8 reveals that the sensor data in the center of the back were not as

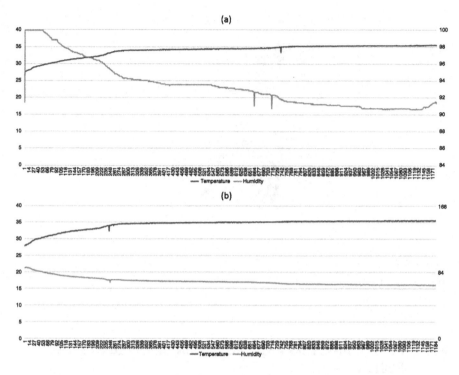

FIGURE 8.8

The results measured by the DHT22-sensor in a lying-on-bed test (X axis: time, Y axis: temperature and humidity readings). (a) Sensor: Back; (b) Sensor: Waist.

sensitive as the one in the center of the waist; therefore, the latter was taken as the target of processing in this case.

As the observation on temperature and humidity can confirm perspiration, this study proposed a model of predicting perceived perspiration. In comparison with multiple curves, a single curve is simpler in prediction; hence, this study combined temperature and humidity into a single-scale value. Currently, there are two common single-scale temperature and humidity indexes – Humidex (Meteorological Service of Canada, 2016) and Heat Index (Steadman, 1979). The description of the two indexes is as follows:

(i) Humidex: Humidex is an index to describe how hot the weather feels to the average person, by combining the effect of heat and humidity. The Humidex is a dimensionless quantity based on the dew point. The Humidex formula is as follows:

$$\text{Humidex} = T_{\text{air}} + 0.5555 \left[6.11 e^{5417.7530 \left(\frac{1}{273.16} - \frac{1}{T_{\text{dew}}} \right)} - 10 \right] \qquad (8.1)$$

where T_{air} is the air temperature in °C and T_{dew} is the dew point in K.

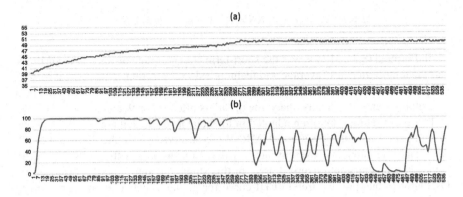

FIGURE 8.9

(a) Heated Index (HI) curve combining temperature and humidity measurement (b) Stochastic Oscillator curve (MMRSV curve) derived from Heat Index data (X axis: time, Y axis: (a) Heat index value, (b) Stochastic Oscillator value).

(ii) Heat Index (HI): Heat Index is an index that combines air temperature and relative humidity in an attempt to determine the human-perceived equivalent temperature. The formula below approximates the Heat Index in degrees Fahrenheit:

$$HI = c_1 + c_2T + c_3R + c_4TR + c_5T^2 + c_6R^2 + c_7T^2R$$
$$+ c_8TR^2 + c_9T^2R^2 \tag{8.2}$$

where HI is the heat index (in degrees Fahrenheit), T is the ambient dry-bulb temperature (in degrees Fahrenheit), R is the relative humidity (percentage value between 0 and 100), $c_1 = -42.379$, $c_2 = 2.04901523$, $c_3 = 10.14333127$, $c_4 = -0.22475541$, $c_5 = -6.83783 \times 10^{-3}$, $c_6 = -5.481717 \times 10^{-2}$, $c_7 = 1.22874 \times 10^{-3}$, $c_8 = 8.5282 \times 10^{-4}$, $c_9 = -1.99 \times 10^{-6}$.

According to the discussion of Engber (2005), Humidex is defective in the calculation of human body's reaction to a high temperature; therefore, this study adopted Heated Index (HI) to combine temperature and humidity into a single-scale numerical value. Fig. 8.9(a) describes the HI curve derived from the combination of the measure of temperature and humidity. The observation on the curve shows that there was still drop in the HI curve at the time of perspiration.

In the HI curve-based prediction of the time of perspiration, the Stochastic Oscillator was proposed to use in the observation. The Stochastic Oscillator refers to an evaluation index (Murphy, 1999) for combined calculation according to the ascending and descending trend of the curve. The formula is shown as follows:

$$RSV_n = \frac{C_n - L_n}{H_n - L_n} \times 100,$$
$$MRSV_n = \alpha \times RSV_n + (1 - \alpha) \times MRSV_{n-1}, \tag{8.3}$$

$$MMRSV_n = \alpha \times MRSV_n + (1 - \alpha) \times MMRSV_{n-1}.$$

In Formula (8.3), MRSV was the short-term (fast) average of RSV, and MMRSV was the long-term (slow) average and features a slower reaction. Therefore, if MRSV > MMRSV, then there would be an ascending trend in the curve, or there would be a descending trend. Practically, the value of "n" is often set as 9, and the value of "α" as 1/3, these settings are often adapted in financial analysis domain. With the above system parameters, if MRSV and MMRSV are observed above the threshold of 80, it indicates an entry into overbuy area; if they are lower than the threshold of 20, it indicates an entry into oversell area. In this study, the author used the Stochastic Oscillator to interpret the trend of HI curve for giving effective alarm of perceived perspiration. Fig. 8.9(b) describes the MMRSV curve derived by the Stochastic Oscillator formula from the Heat Index data.

8.4.3 HOW TO USE HEALTH ANALYSIS AND PREDICTION TO CHECK THE PHYSICAL HEALTH OF CARE RECIPIENTS

In this study, perspiration prediction was taken as the demonstration case of the WHSNS; hence, how to train the perspiration prediction model according to sensor data would be the definition of "Wellness" in our scenario. According to the previous section, the temperature and humidity data obtained from sensors are combined to generate the sensing curve based on the definition of HI. The Stochastic Oscillator obtained from the HI curve would become the basis for the observation on the ascending or descending trend of the curve. According to the description of the Stochastic Oscillator above, there are two system parameters which influence the curve of the Stochastic Oscillator: one is the length of the window of historical information "n," also called the length of window, which refers to the quantity of the historical information to be calculated by the Stochastic Oscillator; the other is the weighted smoothing parameter "α" or the degree to which the Stochastic Oscillator is smoothed. After the values of "n" and "α" are set, it is possible to decide the curve of the Stochastic Oscillator. Finally, the thresholds necessary for overbuy and oversell will be proposed. In this study, there is no need to define the threshold of overbuy; only the signals of the descending trend of the HI curve is needed. Therefore, the oversell threshold "θ" is the target value of this case. In short, the information processing procedure was divided into the steps shown in Fig. 8.10.

In Fig. 8.10, the threshold defined for MMRSV determined the time of issuing an alarm of perspiration. Therefore, it was possible to change the alarming prediction into a problem of optimization – seeking the best parameter combination (n, α, θ) and obtaining the time where the MMRSV of the Stochastic Oscillator was below the threshold "θ." The time was set as the time when caregivers could come and handle the problem before the time of perspiration:

$$T_{predict} = T_{sweat} - T_{preparation}. \tag{8.4}$$

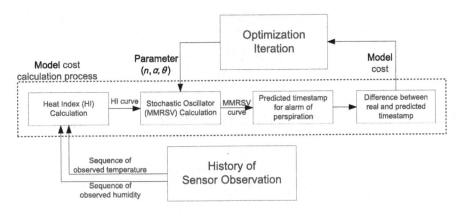

FIGURE 8.10

The information processing procedure in this study.

In the formula, "$T_{predict}$" is the target time of prediction; "T_{sweat}" refers to the time of perspiration detected in the experiment; "$T_{preparation}$" is the time by which caregivers will arrive, and it is usually set as a constant value.

8.4.4 OPTIMIZATION PROCEDURE

In terms of optimization procedure, multiple experimental data about lying on bed and perspiration were used to achieve optimal prediction. Each experiment included two kinds of temperature and humidity sensor data – the center of the back and the center of the waist. According to the observation in Section 8.4.2, only the sensor data about the center of waist were adopted for the prediction. In this experiment, three groups of sensor data about lying on bed and perspiration of a subject are collected, and the HI curves and Stochastic Oscillator curves are calculated for optimization. In respect of optimization procedure, the genetic algorithm (GA) (Mitchell, 1996) was employed to seek the optimal parameter solution (n, α, θ). The reason for using the genetic algorithm was that the solution space consisting of the possible parameter combinations based on solutions (n, α, θ) was too large. In this study, the possible value scopes of "n," "α," and "θ" are as follows:

n: Practically, the lower bound is "9" and the upper bound is the size of the data.
α: The smooth proportion is a real number ranging from 0 to 1. Its resolution (the digits under the decimal) defines the possible number of possible values.
θ: In general, threshold is calculated as an integer ranging from 0 to 100. If it is calculated as a real number, the number of possible values is infinite.

In this case, the following constraints were made according to the requirements on practical operation:

n: $9 \leq n \leq 120$; n is a positive integer
α: $0 \leq \alpha < 1$; α's resolution is as low as 10^{-5}
θ: $0 < \theta < 100$; θ is a positive integer.

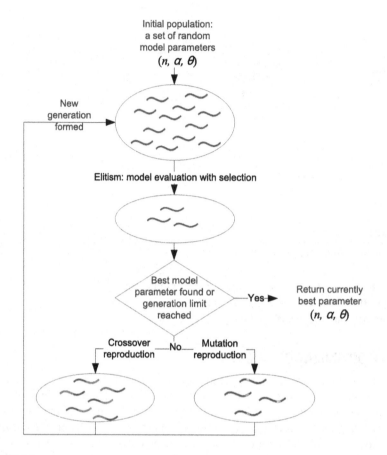

FIGURE 8.11

The genetic algorithm for optimizing system parameters.

The genetic algorithm was executed according to the above constraints, and the processing procedure is shown in Fig. 8.11.

In Fig. 8.11, the optimization process of the genetic algorithm was to achieve generation reproduction of the subsets of the solution space and repeatedly seek the possible optimal solution. In the reproduction, new-generation members, or new solution sets, were created through crossover and mutation process.

With the time (defined in Formula (8.4)) of perspiration alarm as the target of prediction, the time from lying-on-bed to perspiration in each group could form an optimization problem to seek the optimal solution. Table 8.3 shows the solutions obtained from the three lying-on-bed sensing experiments with the genetic algorithm.

According to Table 8.3, the next step was to generalize the optimal solution. The generalization of optimal solution means that the system parameters found in different experiments go through the intersection calculation to seek the most general

Table 8.3 The Solutions Obtained from the Three Lying-on-Bed Sensing Experiments

Experiment	n	α	θ
I	9	0.00223	22
II	9	0.00206	19
III	9	0.00153	17

Table 8.4 The Experiment Results

Experiment Data Set	Parameter (n, α, θ)	Alarm Time (advanced) in ms	SUCC/FAIL
I (test1.csv)	Generalized	2118	SUCC
II (test2.csv)	parameter:	405945	SUCC
III (test3.csv)	(9, 0.00223, 22)	147394	SUCC

single solution which is applicable to all the experiments of the same subject. In this case, the principles of intersection were as follows:

n: the lower the better, because weaker dependence on historical sensor data would lead to better results.

α: the higher the better, because the greater effects of the latest sensor date on the Stochastic Oscillator would lead to better results (or greater sensitivity).

θ: the higher the better, in practical, the more it was approximate to the practically frequently-used value 20, the better.

According to these principles, the three experiments were processed in the 3-fold cross validation (Geisser, 1993). The experiment results are shown in Table 8.4.

According to Table 8.4, the optimal solution of the generalization was indeed useful for the prediction of perspiration. In this experiment, the accuracy of perspiration alarming was 100%.

8.4.5 SYSTEM INSTALLATION

According to the experiment results of the previous case, it is confirmed that the perspiration prediction model can effectively issue an alarm before the deadline of alarming. Such a mechanism can be installed in the WHSNS according to Fig. 8.12.

In Fig. 8.12, the embedded system on the intelligent medical bed sends the perspiration alarm, and the system parameters of the perspiration alarm is calculated or updated for the optimal parameter by the model construction servers in the backend server group. As for a new care recipient, the initial system parameters come from a care recipient who shares the most similar physiological features with him/her. In terms of the renewal time of the parameters of the perspiration prediction model, if the alarm time is too late to issue and a care recipient has perspired, the caregivers will inform the backend servers through a smart device to re-calculate the parameters. The entire system procedure is shown in Fig. 8.13.

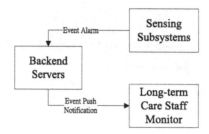

FIGURE 8.12

An alarm notification mechanism in WHSNS.

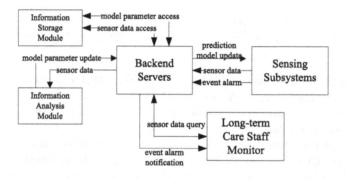

FIGURE 8.13

The entire system procedure for health analysis and prediction.

Through the above procedure, a WHSNS architecture-based intelligent medical bed perspiration prediction system can be established.

8.5 CONCLUSION AND FUTURE WORKS

In this research, we proposed a health analysis and prediction system structure called WHSNS, which is able to collect sensing data in long-term care environment and analyze and predict specific events through these data. With WHSNS, it is possible to reduce the needs of human resources in the traditional long-term care environment, which is able to relieve the burden of national budget on long-term care services, especially for the aging society like Taiwan.

With the design of WHSNS, this research also conducted a case study to demonstrate the function of health analysis and prediction. The preliminary experiment results shown in this paper reveal the possibility of perspiration prediction in the long term care environment. With this prediction, the limited number of the caregivers in the long-term care center can be fully utilized in helping care recipients turn over, which would have positive effects in saving the resources.

The next steps of this research include the enhancements of the WHSNS functions including the extension of the embedded system with pluggable prediction service module, implementation of secure communication protocol for sensor data communication, and a large-scale test of the proposed system.

REFERENCES

Authorization of Spread Spectrum Systems, 1985. Parts 15 and 90 of the FCC Rules and Regulations. Federal Communications Commission of the USA.

Bergsten, H., 2003. JavaServer Pages, 3rd ed. O'Reilly Media. ISBN 978-0-596-00563-4.

Braddock, D., 2011. Long term care spending for disability. Disability Spending in States: 1997–2008. University of Colorado, Department of Psychiatry, Boulder, CO.

Bush, S., 2011. Dongle computer lets kids discover programming on a TV. Electronics Weekly.

Chamberlin, D.D., Boyce, R.F., 1974. SEQUEL: a structured English query language. In: Proceedings of the 1974 ACM SIGFIDET Workshop on Data Description, Access and Control. Association for Computing Machinery, pp. 249–264.

DHT22 data sheet. Retrieved from https://cdn-shop.adafruit.com/datasheets/DHT22.pdf.

Engber, D., 2005. How do they figure the heat index? Slate Magazine. http://www.slate.com/articles/news_and_politics/explainer/2005/07/how_does_the_heat_index_work.html.

Gavrilov, L.A., Heuveline, P., 2003. Aging of population. Retrieved from http://health-studies.org/Population_Aging.htm.

Geisser, S., 1993. Predictive Inference. Chapman and Hall, New York, NY.

Gosling, J., McGilton, H., 1996. The Java Language Environment.

Health Canada, 2012. Long-Term Facilities-Based Care.

Human Rights Act of United Kingdoms, 1998. Retrieved from http://www.legislation.gov.uk/ukpga/1998/42/contents.

Kushner, D., 2011. The making of Arduino. IEEE Spectrum. http://spectrum.ieee.org/geek-life/hands-on/the-making-of-arduino.

Lamke, L.-O., Nilsson, G.E., Reithner, H.L., 1977. Insensible perspiration from the skin under standardized environmental conditions. Scandinavian Journal of Clinical & Laboratory Investigation 37, 325–331.

Lathi, B.P., 1998. Modern Digital and Analog Communication Systems, 3rd edn. Oxford University Press.

Marks, R., 2006. IEEE 802.16 WirelessMAN standard: myths and facts. In: 2006 Wireless Communications Conference. Washington, DC.

Meteorological Service of Canada, 2016. Humidex. Spring and Summer Weather Hazards. Environment Canada. Retrieved June 20, 2016.

Milestones in the Bluetooth Advance, 2004. Ericsson Technology Licensing.

Mitchell, M., 1996. An Introduction to Genetic Algorithms. MIT Press, Cambridge, MA.

Murphy, J.J., 1999. Technical Analysis of the Financial Markets: A Comprehensive Guide to Trading Methods and Applications. New York Institute of Finance, New York, p. 247.

National Institute on Aging, National Institutes of Health, World Health Organization, 2011. Global Health and Aging. Retrieved from http://www.nia.nih.gov/sites/default/files/global_health_and_aging.pdf.

Patient Safety and Quality, 2008. An Evidence-based Handbook for Nurses.

Personal Information Protection Act of Taiwan, 2012. Retrieved from http://law.moj.gov.tw/Eng/LawClass/LawAll.aspx?PCode=I0050021.

Privacy Act of 1974 of United States of America, 2015. Retrieved from https://www.justice.gov/opcl/overview-privacy-act-1974-2015-edition.

van Rossum, G., 2009. A brief timeline of Python. In: The History of Python. Google.

Russel, S., Norvig, P., 2009. Artificial Intelligence: A Modern Approach. Prentice Hall Press, Upper Saddle River, NJ.

Scott, T.E., 2000. Bed exit detection apparatus. US Patent 6067019.

Steadman, R.G., 1979. The assessment of sultriness. Part I: a temperature-humidity index based on human physiology and clothing science. Journal of Applied Meteorology 18 (7), 861–873.

Travis, S.C., 1994. Patient exit detection mechanism for hospital bed. US Patent 5276432.

UN Documents, 2012. Gathering a body of global agreements, Chapter 4: Population and human resources. In: Our Common Future. Retrieved from http://www.un-documents.net/ocf-04.htm.

White, T., 2009. Hadoop: The Definitive Guide, 1st ed. O'Reilly Media, p. 524.

Zanikolas, S., Sakellariou, R., 2005. A taxonomy of grid monitoring systems. Future Generations Computer Systems 21 (1), 163–188.

ZigBee, 2012. The ZigBee Alliance. Retrieved 2012 October 18.

ACRONYMS AND GLOSSARY

List of acronyms with explanation

WHSNS	Wellness and Health Sensor Network Systems
WHO	World Health Organization
NTD	New Taiwan dollar
USD	United States dollar
IoT	Internet of Things
GPIO	general purpose input/output ports
SQL	Structure Query Language
JSP	Java Server Page
ADC	analog digital converter
HI	Heat Index
RSV	Raw Stochastic Value
MRSV	Moving average of Raw Stochastic Value
MMRSV	Second order moving average of Raw Stochastic Value
GA	Genetic algorithm

Glossary of terms with explanation

Alzheimer's disease the most common form of dementia, a general term for memory loss and other intellectual abilities serious enough to interfere with daily life.

Cross validation a model validation technique for assessing how the results of a statistical analysis will generalize to an independent data set.

Disability population those who has no ability for daily living, including the incapacitated person, physical aging, patients of specific diseases, and infants.

Embedded system a computer system with a dedicated function within a larger mechanical or electrical system, often with real-time computing constraints.

Genetic algorithm a method for solving both constrained and unconstrained optimization problems that is based on natural selection.

Hadoop an open-source software framework for storing data and running applications on clusters of commodity hardware.

Health care the act of taking preventative or necessary medical procedures to improve a person's well-being.

Heat Index an index that combines air temperature and relative humidity, as an attempt to determine the human-perceived equivalent temperature.

Humidex an index number used by Canadian meteorologists to describe how hot the weather feels to the average person, by combining the effect of heat and humidity.

Internet of Things the concept of basically connecting any device with an on and off switch to the Internet.

Long-term care a variety of medical and non-medical services and supports you may need to meet your personal care needs for long periods.

Machine learning a method of data analysis that automates analytical model building without being explicitly programmed.

Personal Information Protection Act an Act to govern how private sector organizations collect, use, and disclose personal information in the course of commercial business.

Perspiration prediction the act of trying to determine the status of perceived perspiration of human beings.

PULL-based model places where data or information can be retrieved from at the audience's decision.

PUSH-based model a model form of broadcasting, where the information or data is directly to the audience.

Sensing system a system comprises one or several sensors to provide basic sensing functions.

Sensor network a network comprises a group of tiny devices that monitor conditions of environments.

Stochastic Oscillator a momentum indicator comparing the latest reading of a time series data to the range of its readings over a certain period of time.

Wellness and Health Sensor Network a systematic architecture which deploys a number of sensors in a care environment to provide care services.

PART 3

HEALTHCARE APPLICATIONS

ELECTRONIC HEALTH SYSTEM: SENSORS EMERGING AND INTELLIGENT TECHNOLOGY APPROACH

Heru Susanto*,†

*The Indonesian Institute of Sciences, Indonesia †Department of Information Management, Tunghai University, Taiwan

9.1 INTRODUCTION

SET through ICT enables organization to collect, make their work organize, and analyze data that helps them achieve their objectives. In electronic health issues, ICT is generally used for automated data collection, statistical study of data, Internet accessible shared databases, modeling and simulation, imaging and visualization of data, and investigation, Internet-based communication among researchers, and electronic dissemination of research results. For instance, ICT being used for automated gene sequencers, which use robotics to process models and computers to manage, stores, and retrieve data, has made potential the rapid sequencing of the human genome, which in turn has resulted in first time expansion of genomic databases. Shared Internet accessible databases are important in paleontology; models and databases are significantly used in population biology and ecology; and genomics are influencing many fields.

In other science fields, SET-ICT also helps in analyzing subsurface creations, mapping, and modeling complex systems. For example, seismic data used to measure earthquakes were traditionally recorded on paper or film, but today they are recorded digitally, making it possible for the researchers to analyze the data quickly. Furthermore, Internet-connected allow many researchers to obtain and contribute data to large problem. In several areas of sciences, imaging and visualization become important because they can give clear modeling that helps the researchers understand biological system such as tissues, organisms, and cells.

The American Recovery and Reinvestment Act of 2009 (ARRA) made a legislation that focus on health information technology (HIT) (The Brookings Institution, 2010; Hersh, 2009; Blumenthal, 2009). This law helps one to increase findings for health care facilities and services as well as for health investigation and makes it easier for unemployed people to buy health insurance (Blumenthal, 2009). There are three unique types of order individuals who worked in medicinal services focusing on the organization flow system. The principal type is medicinal specialists and restorative bolster staff, for example, specialists, attendants, rescue vehicle staff, and phlebotomists. The next type is individuals who bolster the primary classification, for example, chairmen, cleaners, watchmen, and personal computer specialists, and the third class is patients (Paul et al., 2012).

This study highlighted health information technology presented in many forms. Some of the well-known forms are electronic health records (EHR), electronic medical record (EMR), clinical decision support (CDSS), health information management (HIM), personal health record (PHR), and health information exchange (HIE). All these forms interrelate between information technology and health system, which therefore produces Health Information Technology (HIT). All these forms will be explained in more details about information technology that supports health system under literature review and discussion. The details on information technology and health system will be explained in literature review and will be more elaborated under the discussion. In addition, aside from explaining or proving the relationship between information technology and health system, this study also provides the usage of ICT in health systems and health facilities, and also the advantages of ICT in health system usage through sensors emerging technology.

9.2 LITERATURE REVIEW

The health care system is made available to the public and invested by government or private companies. On the other hand, ICT uses computers, networking, and other physical devices such as mobile phones, infrastructure, and procedures to create, process, store, secure, and exchange all forms of electronic data. Health information technology (HIT) is an area that involves creation, design, development, and maintenance of the information system for the healthcare industry. In addition, health information systems are predicted to improve efficiency, reduce errors, and lower cost, while also providing better consumer care and service, for instance, the management of patient care through secure use and sharing of health information.

9.2.1 THE ROLE OF ICT FOR INTELLIGENT APPS OF HEALTH SYSTEM

HIT improves the quality and patient safety and reduces the cost of healthcare. HIT includes Electronic Health Record (EHR), Electronic Medical Record (EMR), and Clinical Decision Support System (CDSS). In addition, Hersh (2009) has mentioned the relation of Health information management with ICT in health system. Health information management (HIM) is the main focus on managing medical records, and since it became electronic, the overlaps between informatics increased (Hersh, 2009).

Moreover, the most frequently used HIT is the electronic medical record (EMR), but it has been replaced by electronic health record (EHR), which shows more extra information about the patient. In addition, PHR became more appeal as it kept patient health record private. Apart from that, health information exchange (HIE) also gained interest as the health information of the patient can be exchanged with other HIE within a region.

Millery and Kukafka (2010) believes that changes in health information technology play a major role in enhancing the quality of health system especially in terms of health care. There are many aspects where information technology can improve, for example, documentation, ability in accessing any crucial information, and increase the communication availability. With improvements in information technology, decision making can be easily made of each part. Sittig and Singh (2010) introduced an eight-dimensional model that is successfully applied in real world and includes hardware (including sensors emerging technology) and software computing infrastructure, clinical contents, human–computer interface, the people, workflow communication, internal organizational features, external rules and regulations, and measurement and monitoring, which was created specifically to show the sociotechnical challenges in designing, developing, implementing, and the usage and its evaluation associated with health information technology (HIT) with the complex adaptation of healthcare systems. It was adapted to improve and understand the applications of HIT during the development and implementation processes.

Here, HIT can be implemented for such a sample of health problem monitoring. Disobedience is one of the problems among serious mental disorders patients, thus creating challenges for the mental health professionals. To overcome this problem, ICT has increased in order to prompt the patients, for example, in using text message and E-mail by stating the purpose of it being sent (Kauppi et al., 2014). The usage of ICT has increased in the health care organizations, which is similar to what happened in other companies that rely in a well-developed ICT infrastructure. ICT infrastructures may include the use of web, databases, and network infrastructures. Additionally, Health Information System (HIS) should ensure efficiency and security of information flows, and efficiency and proximity of health system. With available ICT in the health system, there is strong opposition toward the adoption of e-health systems called Electronic Medical Records (EMR). Electronic Medical Records (EMR) allows the patient data information system to save paper records in electronic format in the form of files, which is easier and more effective to manage (Jardim, 2013).

Electronic Medical Record (EMR) is one type of the Health Information Technologies (HITs). The functions of EMR are to keep and gather the patient's past medical records to inform any medical care and to publish any results they conduct in diagnostics testing (Chen et al., 2016). EMR helps to reduce cost, improve healthcare service quality, and increase productivity among physicians (Jerald et al., 2012). Electronic Health Record (EHR) provides their patients information about a summary of their recent visit, medications, drug allergies, appointments, payments, and some medical forms through the Internet. All of the patients' information is kept secure in their data (Karoly et al., 2015).

Electroencephalograph is new health technology that monitors individual brain waves from home to aid diagnosis of a disease through telemedicine. This technology hopefully is the next generation device for communication in the developments of brain science and medical area since this is useful in investigating human mental condition and health diagnosis (Motomura et al., 2015).

A clinical document consists of patient's records, notes, discharge summaries, and doctor's referral letters. Natural Language Processing (NLP) is a successful device that helps to reduce cost in medical records, improve quality of the patient's health, and the accuracy of the documents. This software is one of successful programs to keep information, provide a solution space for annotating and organizing the documents into the database for the ease of professional health analysis. The first unsupervised approach, known as PrefixSpan, was produced for medical concept extraction. Secondly, C-Value and its extraction NC-Value were produced for statistics and linguistics information (Wei et al., 2015).

The technology for dental implanting is CAD/CAM technology in the process, as it is automatic and precise. Dental implanting helps one to replace a damaged tooth with Titanium (Vedpal et al., 2015). Healthcare professionals consume extra time and resources when making decision for their patients as the decision made by Healthcare professionals will be put into HIT systems. Thus, with these capabilities, it helps doctors and other medical experts to manage their patients with ease. These systems are called Clinical Decision Support Systems (CDSS). CDSS are defined as "software applications that integrate patient data with a knowledge-base and an inference mechanism to produce patient-specific output in the form of care recommendations, assessments, alerts, and reminders to actively support practitioners in clinical decision making." Hence, CDSS can make decision based on the health situation. Additionally, a Decision Support Systems (DSS) will help surgeons and doctors to schedule their patients. The main components of DSS are the database, user interface, and DSS software system. To integrate DSS and HIS, an update service calls a web service in AIDA for a request to integrate data warehouse with data in HIS and shares the database to update the DSS database (Ahlan and Ahmad, 2014).

9.2.2 ICT IMPACTS FOR INTELLIGENT APPS OF HEALTH SYSTEM

Health care organization requires doctors or nurses or hospital staffs to comprehend that their future in health information system, the need to construct supportable well-being frameworks, is irrefutably attached to great execution administration (Lega and Vendramini, 2008).

According to Ciciriello et al. (2013), detailed and useful information for medications need to be easily understood to enable consumers to use their medications safely and effectively since some studies show that multimedia education can be more helpful than the usual care. In addition, the programs use different kinds of style to issue the information such as words, diagrams, pictures, together with audio, animation, and video. Then, the finalized information can be accessed through DVD, CD-ROM, or the Internet, and there no sufficient proof to replace written education or education by the health professionals with multimedia education. Therefore, it must be used together with usual care provided by health providers. However, there are also obstacles of health information technology such as Electronic Health Record (EHR), where elder patients find it difficult to use this technology, especially in the Consumer Health Information Technology (CHIT) since this makes them uncomfortable, less efficient, and uncontrollable because they have to go online to use this system. In future, the population of elderly will increase since the standard of living is increasing (Karoly et al., 2015). Physicians found it hard to use EMR since it makes them lose control of their workflow and there is too much irrelevant information about the patient, distraction when the physicians interact with the computer, which will lower their productivity, and this does not show any efficiency that has been proven (Jerald et al., 2012).

In addition, Paul et al. (2012) mentioned that data must be recovered from paper-based records, which frequently do not have the desire of electronic frameworks to have an inseparable tie to the patient's wellbeing. Exchanging the substance of a paper record to an EPR framework will not suffice later for increasing the framework if the framework is to be utilized for patient-driven purposes since a doctor-driven framework does not gather all the information expected to look at patients. However, information technology in health system could be risky with the patient's information due to scammers in online world (Millery and Kukafka, 2010).

9.3 DISCUSSION

9.3.1 ICT EMERGING TECHNOLOGY

The health care organization require doctors, nurses, and hospital staffs to comprehend that their future in health information system and the need to construct supportable wellbeing frameworks is irrefutably attached to great execution administration. The health care organization required scholastics and specialists to quit offering the most recent administration designs and to help the framework to contextualize its decisions over details and controllers to enhance their insight into such details, to take educated choices, and to contribute in the advancement of those territories where there is an aggregate deficit in execution administration, for example, group administrations. Part of the expansions in general wellbeing consumption has been coordinated for the utilization of steady innovations, planning to upgrade social insurance procurement. The utilization of medicinal services has advanced with the fast improvement of information technology and advances in social insurance innovation in parallel with current concerns emerging over patients' security and obviously how to cure patients effectively.

The important thing in any information system is the accuracy of the data provided. The data input is derived from the patients and other available sources that contain all the patients' medical records. The information is then processed with the output that comprises all the patients' medical records. The eight-dimensional model and the first dimension are the hardware and software computing infrastructure, which refers to the items and software that are used to control clinical appliances (Sittig and Singh, 2010). The next is the clinical content, which includes alphabetical and numerical data and graphics that represent the "language" of clinical appliances. Human–computer interface is another dimension, which comprises all characteristics of the computer that the users are able to interact easily with. When mentioning people that are under the eight-dimensional model, it refers to everybody including the developer, users, and patients communicating through the system.

The Brookings Institution (2010) stated that American Recovery and Reinvestment Act of 2009 (ARRA) believed that if HIT is implemented and used effectively, then it has the intense possibility in the improvement of patients' healthcare. The usage of electronic HIT is to store health information such as EHRs, claim data, registries, and payment system. Primary care providers in North Carolina have used health information to help better asthma care, look for their performance on a number of key metrics, and decrease hospital admission rates and emergency room admissions. One of the evidence developments of HIT is that it can be used to organize investigation including comparison of a success in observable studies. In addition, Cancer Care Outcomes Research and Surveillance Consortium (CanCORS) project uses much information that requires IT such as demographic, contact, and medical information to investigate information on lung and colorectal cancers in America. Another evidence development of HIT is the use of claim data showing that it enables to look for nearly 17,000 patients over a year to discover relative risk of heart attack among patients taking both drugs and started to warn physicians through the results.

9.3.2 THE INTELLIGENT SENSORS APPS

HIT is the combination of Information Systems, Computer Science, and Healthcare. As a result of advancement in technology, it is gaining global attention. With the available HIT systems, patients can monitor either in clinical setting or from outside, especially at home. Patients monitoring systems applied sensor network technology for collecting physiological data of a patient that are suffering from different diseases such as diabetes. For instance, Jog Falls, a diabetes management system using sensor devices for collecting physiological and activity data that monitors patients' physical activities such as food intake. Furthermore, a system of monitoring such as Type 1 diabetic patient using mobile phone for a diet management system and web-based medical diagnosis is used to predict patient's condition. Additionally, Health Information Technology is beneficial to the patients and providers since it helps to improve the patient's health and cost saving as they can easily access the information. However, despite the benefits, the use of this system as well as of Electronic Health Record (EHR) is still low (Buntin et al., 2011).

EVIDENT Program is the one of the theory based intervention programs. This program is useful for giving information specifically under healthy nutrition and exercise activities about the patients that they record daily. They can access their information through their own mobile online applications in the smart phones, and so some comparison on their everyday activities and food intake, which can make an ease to the patients to take note on their progress. Secondly, Social-Cognitive Theory is produced to monitor obesity treatment by the research group in online or traditional face-to-face methods. This program monitors the nutrition and exercise activities as they will be provided counseling. Lastly, this mobile intervention programs through sensor intelligent apps help to prevent an unhealthy behavior among the patients, such as smoking. The software provides information that can aid to develop social strategies and provide useful activities for adolescents.

Patients with severe mental health problems are more likely to disobey their medical treatment such as forgetting their appointment and when to take the medicines. This can cause them not getting the medicines at the right time, which can lead to poor health condition and hospitalization. Several strategies have been introduced to help the patients with their medications. One of the strategies is prompting, which helps to remind the patient to follow their treatments by using telephone calls, sending letters, and personal visits by the hospital staff.

Laboratory Information Systems (LIS) are complicated machines functioning in the information systems, integrated clinical information systems, and Electronic Laboratory Records (EMRs). LIS is creating a link between analyzing in the laboratory, medical technologies, and clinical providers. This can help to monitor and improve the quality of healthcare; therefore, this reduces any human errors made. In addition, LIS should provide computerization as it can support a high-performance laboratory and automatic laboratory results. However, designing a LIS is a challenge because LIS must have the ability to communicate across the technology platforms. This means that new laboratory IT systems will be incompatible with the present laboratory hardware. Furthermore, poor performance can happen because of high administration cost and limitation of information technology.

Electronic Health Record (EHR) is accepted due to chronic conditions since before it was rejected by most of the patients. This facility was introduced in other countries, as this facility was proved to improve patients' health and to be effective. With this proof, there is an increase in terms of investment. EHR is different from Electronic Medical Record or known as EMR and is the source of EHRs, where EMR provides each patient's information such as drug allergies, drug-to-drug interactions, and past treatment of each patient in one hospital. EMR also consists of sensitive personal information such as sexually transmitted disease, abortions, emotional problems, and physical abuses. EMR is only used within the hospital. However, EHRs provide a wide view of the patients' healthcare. PHRs help EMR to keep the patient information data and to help the patients by sharing the information which other hospitals may have a better cure for the patients. Furthermore, EHR elaborated the usage of these systems by taking care and giving more safety to the patients' health.

The patients' information privacy in EMRs still remains a concern to them and the hospital staff. Nurses have an important role for collecting the patients' information and keep them private, and they can secure the record-keeping efficiently. However, some of the nurses are only familiar with paper-based medical records and have no intentions to keep the information private and secure. These issues can be improved by giving an adequate training, which includes ethics, information security procedures, and IT skills of nurses. Nurses also have to know the importance of using EMRs and should know what information should be kept private (Chen et al., 2016).

In the physician case, they have difficulty in using EMR, as most physicians have practice in paper-based medical information, and now they have to change into EMR environment. This difficulty occurs among the elder physicians, who have been using paper-based medical records for many years, and who have no any technical skills. In addition, EMR gives them a hard time to all the patient medical needs, and they hope that EMR will help them to work efficiently; however this often does not happen. NVivo9 was introduced to improve the ability to access information, and they can make a better and accurate decision for the patients' care within short time. This will improve the patients' health and increase healthy population. Lastly, physicians can improve time with other staff through the EMR messaging capability (Jerald et al., 2012).

As for EMR and EHR issues, Paul et al. (2012) discussed that any composed note structure of numerous EPRs ought to bolster fantastic patient outlines attention for shared clinical consideration. Additionally, the point-by-point social insurance data in EPRs ought to make them critical wellsprings of data for clinical studies, examination, and approach. The potential advantages of EPR (Electronic Patient Record) frameworks are guaranteed to be hierarchical issues, for example, enhancing the trading of data between social insurance offices and, furthermore, the backing of institutionalized techniques that can build consistency between various administration suppliers. EPR ought to have the capacity to guarantee least principles over the direction of consideration when patients move between various specializations; these advantages have not been conveyed yet, but rather significant issues have emerged with reference to how and whether these advantages will be conceivable.

Electroencephalograph is cheap, light with approximately 100 g, easy to setup, and has a rechargeable battery. However, the battery only gives 1–2 hours of consecutive measurements, and the electrodes are too small due to the electric potential. Other problems occur when it is easy to setup because only a few electrodes used in the central part of the head, which can cause the life, and the right side is more advantageous. The second problem occurs when they have to use a simplified electroencephalograph. This simplified electroencephalograph is used for medical and brain science area research. It will give a person caution about his health conditions. One example of simplified electroencephalograph is the 3B Band produced by the B-Bridge International, and this type of headband is known in the world. This is a headband that contains NeuroSky chips. 3B Band uses Bluetooth to connect to a computer; therefore the usage of wire is not needed, and a person using this headband does not feel bound. However, these headbands have an implication that the shape of earlobe and sweat status can lead to a wrong measurement of the brain wave (Motomura et al., 2015).

NLP is an unsupervised approach to manage a clinical document, and there are three types introduced. TextRank was the first unsupervised approach familiarized by Mihalcea and Tarau (2004), It aids staff to summarize text. The second type of unsupervised approach is the C-Value and NC-Value introduced by Frantzi et al. (2000). The C-value helps to produce a unit-hood score depending on the length of the phrase entered. The NC-Value helps to enhance the accuracy and quality of the term data entered. Lastly, frequent sequence mining, known as PrefixSpan, helps to cooperate long text in little time needed, and to get use this data, only a minimal training is needed (Wei et al., 2015).

CAD/CAM technology helps the production of dental implants, and it can be applied in-house if the resources are available to a collaborating partner. This technology needs a massive investment for Research and Development (R&D), as this step needs a huge amount of money for the special kits and accessories to make successful dental implants; however, this gives long-term benefits. The benefits are saving time and money and meets the international standards in terms of technology of producing dental implants.

Whereas health consumers demand for exact and evidence-based information to be delivered in a way that is easy to understand about their health and proper treatments, health professionals want to save consultation time and improve the medication compliance. Consultation is usually presented verbally alongside with written materials. Consequently, there are high chances that the patients will forget the information delivered to them. Therefore, multimedia educational programs give more benefits as it is convenient that they can be accessed anywhere by the individuals and their families, which is cost saving in comparison with consultation of a doctor. Additionally, multimedia programs also allow individuals to alter the information according to their need.

In ICU, healthcare system is needed to give a relevant accurate information in time. However, this high technology gives some limitations such as high cost of maintaining the machines, delayed sending the products, and wrong delivery date. RFID is an enforcement technology, which monitors a manufacturing in a critical healthcare such as stents (Vedpal et al., 2015). The act of passing the duty of care for a patient to another nurses is called nursing handover. During this time, there is a possibility of getting error, especially when the important medical information is not shared efficiently, accurately, and timely and thus may result in adverse events (AEs). Therefore, Information Technology is used to support the process in order to decrease the potential risk such as miscommunication, misunderstanding, and omission of crucial information. Furthermore, poor handover may cause delays for the patients' treatments. Consequently, an accurate handover of important information is crucial to continuity and safety of care for hospitalized patients (Smeulers et al., 2014).

In any field, particularly in the well-being framework field, one of the numerous focal points of data innovations is to offer the specialists some assistance with nursing, healing facility staff, and patients to pay in contact with one another inside and outside the association structure. A standout among the most critical points of interest of data advances is the formation of one exhaustive asset, which is parallelly upgraded and utilized by specialists, medical attendants, or doctor's facility staff (Harutyunyan et al., 2015). One of the challenges for retrieving the health information is the mismatch between consumer's terms and professional vocabularies used in medical literature. To overcome this, a system called MeshMed has been introduced, where combined different functional searches are combined into a single one. It has two new search components known as term browser and tree browser, which provide unique information about the search topic. In addition, Mesh vocabulary is downloaded from the National Library of Medicine (NLM) in Extensible Markup Language (XML) format to support both browsers. Both browsers provide quick access to the information needed, which makes it more efficient and beneficial in finding the definitions and synonyms for medical terms.

9.3.3 USABILITY OF INTELLIGENT SENSORS APPS

The perceptions over the system in Health System can only be known when the users, such as doctors, surgeons, and patients, accept and use it. Therefore, there are studies that facilitate the adoption of use of HIT system. Technology Acceptance Model (TAM) is one of the popular theories for studying the perception and factors to the acceptance of a new technology, in this case, the acceptance of technology in healthcare system. TAM is designed for modeling user acceptance of Information Systems. Moreover, by promoting its acceptance, it will increase the use of IT. TAM focuses on the users' behavioral intentions toward accepting a new technology, specifically, a self-diagnosis system for reducing cost and improving the quality in healthcare system. TAM was not only reengineer to TAM2 and TAM3. TAM2 focused on identifying sources of usefulness and moderating variables, and TAM3 centered on interventions that can affect the acceptance and use of IT in a healthcare system. A new model was developed from TAM2, called Information and Communication Technology Acceptance Model (ICTAM). ICTAM is for predicting and showing consumers' health information and services usage behavior on the Internet.

An intervention program has been introduced into the online applications, and this has been beneficial to the patients. Moreover, interventions have the intention to provide a prevention or treatment to the patients who need them, which depends on the theory and model. These intervention programs have been introduced into the online applications, and this has been beneficial to the patients.

The Brookings Institution (2010) shared an example when researchers such as the clinicians use electronic health information. They can easily look for patients' health information anytime and anywhere; in other words, electronic health information is very convenient, and they can observe their quality in terms of their health care services, which may vary with the other healthcare centers. Different policies may give different outcomes like in designing formula and payment procedure. Therefore, health information technology such as Electronic Health Record (EHR) or Electronic

Medical Record (EMR) enables efficient early investigations, and this may also be one of the advantages of health information technology.

Health care has invested a large amount of money for Information Technology (IT). The utilization of IT and Information System (IS) has taken different directions due to the demands and needs in various sectors of government and public organizations. The technological advancement in the form of electronic patients' records, clinical applications, and health management information system (HMIS) have encouraged many Health Care Organizations to use it. However, this system can be hard to implement in both public and health care organizations. Guo (2008) discussed and believed overseen care focuses on decreasing conveyance costs and enhancing medicinal services financing through strict use administration, money related impetuses to doctors, and restricted access to suppliers. Right now, oversaw care exists as the overwhelming financing and conveyance framework; as anyone might expect, access, expense, and quality predicaments are imperative; procedures and arrangements, in this manner, must be received to address these issues. A methodical examination of these systems is valuable for administrators and experts endeavoring to enhance care quality under health care organization.

To add on, according to Karsh et al. (2010), researchers have argued that the acceptance of using HIT is low as some believe that it is not very useful and has less benefit. Health information technology needs to focus on changing the care services, improve patient results, created to support the needs the hospitals. In order for HIT to be successful, its focal point must be on the usage or assumption rather that the effect of the people's health. Therefore, if HIT is applied in a right way, it may be successful in terms of supporting and expand clinician and patient efforts to intensify the population's health and welfare.

A new generation in information technology helps in improving the quality of health care services, for example, in handling the documentations especially the patients' information and their confidential prescription. Decision making in each organizations also plays a major role in having the information technology, where it can fasten the decision making among the end users not only by the administration management, but also by including medical experts. The functions in IT involves interchange of any content of information, which leads continuation of health care management between departments in the organizations, helps the end user to decode any scientific documentation into practice, treatment procedures, and health care system's security.

Administration controls frameworks as an apparatus for top supervisors to control doctors and to distribute top-down assets. Few associations have grown more cooperative methodologies, in which the administration control framework gives a chance to encourage dialog between administration and doctors. Targets and assets are not just allotted top-down, but rather will be fairly arranged from the base up and take into account a common "sharing of psyches," a typical comprehension of issues and a mutual need setting where showing clinics were and still are late movers, as colleges opposed the presentation of any execution estimation framework.

9.4 CONCLUSION

Patients monitoring systems or electronic health records utilized sensor network technology for collecting physiological data of a patient suffering from different diseases such as diabetes, cholesterol, coronary heart disease, high blood pressure, and many more. Furthermore, a system of monitoring such as Type 1 diabetic patient using mobile phone for a diet management system and web-based medical diagnosis is used to predict patient's condition.

The health care organization requires scholastics and specialists to quit offering the most recent administration designs and to help the framework to contextualize its decisions over details and controllers to enhance their insight into such details, to take educated choices, and to contribute in the advancement of those territories where there is an aggregate deficit in execution administration, for example, group administrations.

Another dimension is the workflow and communication that define the procedure that involves to assure that patients' care tasks are done effectively. In addition, the dimensions could provide many features of the preceding dimensions. Another evidence development of HIT shows that the data claimed enable to look for nearly 17,000 patients over a year to discover relative risk of heart attack among patients taking both drugs and started to warn physicians through the results. One of the theory-based intervention programs is EVIDENT Program. The program is useful for giving information about healthy nutrition and exercise activities about the patients that they recorded daily. It also could monitor their nutrition and exercise activities as they will be provided a counseling session. Lastly, this mobile intervention programs helps to prevent an unhealthy behavior among the teenagers, such as smoking. The programs itself provide information that can aid to develop social strategies and useful activities for the adolescents included.

EHR is different from Electronic Medical Record or known as EMR. EMR is the source of EHRs, where it provides each patient's information such as drug allergies, drug-to-drug interactions, and past treatment of each patient in one hospital. EMR also consists of sensitive personal information such as sexually transmitted diseases, abortions, emotional problems, and physical abuses. Personal Health Record helps EMR to keep the patient information data and to help the patients by sharing the information with other departments or hospitals and thus may have better cure for the patients.

The technological advancement in terms of electronic patient records, clinical applications, and health management information system has encouraged many health care organizations to use it. In physician's case, they have difficulty in using EMR, as most physicians have practice in paper-based medical records, and now they have to change into EMR environment. In addition, EMR gives them difficulties to all the patient medical needs, and they hoped that EMR will give them a hard time to all the patient medical needs, and they expect EMR help them to work efficiently. However, by implementing this system, it can be hard to implement in both public organizations and health care organizations. A methodical examination of these systems is valuable for administrators and medical experts who endeavor to enhance care quality under health care organization. The functions in IT involve interchange with any content of information, which leads to continuation or communication of health care management between departments in the organizations and helps the end user to seek any scientific documentation into practice, treatment procedures, and health care system security.

REFERENCES

Ahlan, A.R., Ahmad, B.I., 2014. User acceptance of health information technology (HIT) in developing countries: a conceptual model. Procedia Technology 16, 1287–1296. http://dx.doi.org/10.1016/j.protcy.2014.10.145.

Blumenthal, D., 2009. Stimulating the adoption of health information technology. New England Journal of Medicine 360 (15), 1477–1479.

The Brookings Institution, 2010. Using information technology to support better health care: one infrastructure with many uses. Engelberg Center for Health Care Reform at Brookings. Retrieved from www.brookings.edu/~/media/events/2010/5/14-health-information/final-issue-brief-51310.pdf.

Buntin, M.B., Burke, M.F., Hoaglin, M.C., Blumenthal, D., 2011. A review of the recent literature shows predominantly positive results health affairs. The Benefits Of Health Information Technology 30 (3), 464–471. http://dx.doi.org/10.1377/hlthaff.2011.0178.

Chen, C.M., Kuang, M.K., Judith, W.A., 2016. A survey-based study of factors that motivate nurses to protect the privacy of electronic medical records. BMC Medical Informatics and Decision Making 16, 3. http://dx.doi.org/10.1186/s12911-016-0254-y.

Ciciriello, S., Johnston, R.V., Osborne, R.H., Wicks, I., deKroo, T., Clerehan, R., O'Neill, C., Buchbinder, R., 2013. Multimedia educational interventions for consumers about prescribed and over-the-counter medications. The Cochrane Library.

Frantzi, K., Ananiadou, S., Mima, H., 2000. Automatic recognition of multi-word terms: the c-value/nc-value method. International Journal on Digital Libraries 3 (2), 115–130.

Guo, K.L., 2008. Quality of health care in the US managed care system. International Journal of Health Care Quality Assurance 21 (3), 236–248. Retrieved from http://www.emeraldinsight.com.ezproxy.ubd.edu.bn/doi/pdfplus/10.1108/09526860810868193.

Harutyunyan, P., Moldoveanu, A., Moldoveanu, F., Asavei, V., 2015. Health-related impact, advantages and disadvantages of ICT use in education, compared to their absence in the past. eLearning and Software for Education 1, 557–563. Paper presented at the http://search.proquest.com/docview/1681255417?accountid=9765.

Hersh, W., 2009. A stimulus to define informatics and health information technology. BMC Medical Informatics and Decision Making 9 (1), 24.

Jardim, S.V.B., 2013. The electronic health record and its contribution to healthcare information systems interoperability. Procedia Technology 9, 940–948. http://dx.doi.org/10.1016/j.protcy.2013.12.105.

Jerald, D.H., Thomas, M.S., Jonatan, J., 2012. Adoption of electronic health care records: physician heuristics and hesitancy. Procedia Technology 5, 706–715. http://dx.doi.org/10.1016/j.protcy.2012.09.078.

Kauppi, J.P., Pajula, J., Tohka, J., 2014. A versatile software package for inter-subject correlation based analyses of fMRI. Frontiers in Neuroinformatics 8 (2).

Karoly, B., Bill, D., Kevin, P., 2015. Social influence on health IT adoption patterns of the elderly: an institutional theory based use behaviours approach. Procedia Computer Science 62, 516–523. http://dx.doi.org/10.1016/j.procs.2015.08.378.

Karsh, B.T., Weinger, M.B., Abbott, P.A., Wears, R.L., 2010. Health information technology: fallacies and sober realities. Journal of the American Medical Informatics Association 17 (6), 617–623.

Lega, F., Vendramini, E., 2008. Budgeting and performance management in the Italian National Health System (INHS). Journal of Health Organization and Management 22 (1), 11–22. Retrieved from http://www.emeraldinsight.com.ezproxy.ubd.edu.bn/doi/pdfplus/10.1108/14777260810862371.

Mihalcea, R., Tarau, P., 2004. TextRank: Bringing order into texts. Association for Computational Linguistics.

Millery, M., Kukafka, R., 2010. Health information technology and quality of health care: strategies for reducing disparities in underresourced settings. Medical Care Research and Review Supplement 67 (5), 268S–298S. Retrieved from http://mcr.sagepub.com.ezproxy.ubd.edu.bn/content/67/5_suppl/268S.full.pdf+html.

Motomura, S., Ohshima, M., Zhong, V., 2015. Usability study of a simplified electroencephalograph as a health-care system. Health Information and Science and Systems 3, 4. http://dx.doi.org/10.1186/s13755-015-0012-z.

Paul, R.J., Ezz, I., Kuljis, J., 2012. Healthcare information system: a patient-user perspective. Health Systems 1, 85–95. Retrieved from http://www.palgrave-journals.com/hs/journal/v1/n2/pdf/hs201217a.pdf.

Sittig, D.F., Singh, H., 2010. A new sociotechnical for studying health information technology in complex adaptive healthcare systems. BMJ Quality & Safety. Retrieved from http://qualitysafety.bmj.com/content/19/Suppl_3/i68.short.

Smeulers, M., Lucas, C., Vermeulen, H., 2014. Effectiveness of different nursing handover styles for ensuring continuity of information in hospitalised patients. The Cochrane Library.

Vedpal, Arya, Deshmukh, S.G., Bhatnagar, N., 2015. High technology health care supply chains: issues in collaboration. Science and Behavioral Sciences 189, 40–47. http://dx.doi.org/10.1016/j.sbspro.2015.03.190.

Wei, L., Bo, C.C., Rui, W., Jonathon, N., Nigel, M., 2015. A genetic algorithm enabled ensemble for unsupervised medical term extraction from clinical letters. Health Information Science and Systems 3, 5. http://dx.doi.org/10.1186/s13755-015-0013-y.

ACRONYMS AND GLOSSARY

Acronyms

ARRA	American Recovery Reinvestment Act
CAD	Computer Aided Design
CAM	Computer Aided Manufacture
CanCOR	Cancer Care Outcomes Research and Surveillance Consortium
CDSS	Clinical Decision Support System
CHIT	Consumer Health Information Technology
DSS	Decision Support System
EHR	Electronic Health Record
EMR	Electronic Medical Record
HIE	Health Information Electronic
HIM	Health Information Management
HIS	Health Information System
HIT	Health Information Technology
HMIS	Health Management Information System
ICT	Information and Communication Technology
ICTAM	Information and Communication Technology Acceptance Methodology
ICU	Intensive Care Unit
LIS	Laboratory Information System
NLM	National Library of Medicine
NLP	Natural Language Processing
PHR	Personal Health Record
R & D	Research and Development
SET	Sensor Emerging Technology
SET-ICT	Sensor Emerging Technology through Information and Communication Technology
TAM	Technology Acceptance Methodology
XML	Extensible Markup Language

Glossary of Terms

Cancer Care Outcomes Research and Surveillance Consortium project uses much information that requires IT such as demographic, contact, and medical information to investigate information on lung and colorectal cancers.

Clinical Decision Support System software to assist decision maker in clinic.

Computer Aided Design is the use of computer systems (or workstations) to aid in the creation, modification, analysis, or optimization of a design. CAD software is used to increase the productivity of the designer, improve the quality of design, improve communications through documentation, and to create a database for manufacturing. CAD output is often in the form of electronic files for print, machining, or other manufacturing operations.

Computer Aided Manufacture is the use of software to control machine tools and related ones in the manufacturing of workpieces. CAM may also refer to the use of a computer to assist in all operations of a manufacturing plant, including planning, management, transportation and storage. Its primary purpose is to create a faster production process and components and tooling with more precise dimensions and material consistency, which in some cases, uses only the required amount of raw material (thus minimizing waste), while simultaneously reducing energy consumption. CAM is a subsequent computer-aided process after computer-aided design (CAD) and sometimes computer-aided engineering (CAE), as the model generated in CAD and verified in CAE can be input into CAM software, which then controls the machine tool.

Consumer Health Information Technology an IT system to support customer health activities.

Decision Support System software applications that integrates patient data with a knowledge-base and an inference mechanism to produce patient-specific output in the form of care recommendations, assessments, alerts, and reminders to actively support practitioners in clinical decision making.

Electroencephalograph a new health technology that monitors individual brain waves from home; as a function to aid diagnosis of a disease through telemedicine.

Electronic Health Record a Patient record on the health behavior.

Electronic Medical Record a Patient record on the medical and drug history.

Extensible Markup Language a programming language that provides quick access to the information needed, which makes it more efficient and beneficial to find the definitions and synonyms for medical terms.

Health Information Electronic gained interest as health information of the patient can be exchanged.

Health Information Management managing medical electronic records.

Health Information System to ensure efficiency and security of information flows and efficiency and proximity of health system.

Health Information Technology IT system to support health activities.

Health Management Information System a health system to manage information traffic and exchange between healthcare organization and patient.

Intensive Care Unit a unit to handle emergency situation.

Laboratory Information System complicated machine that functions in information systems, integrated clinical information systems, and Electronic Laboratory Records (EMRs).

Natural Language Processing an unsupervised approach to manage a clinical document; there are three types introduced.

NC-value helps to enhance the accuracy and quality of the term data entered.

NVivo9 was introduced to improve the ability to access information and make better and accurate decisions for the patients' care within short time C-value; it helps to produce a unit-hood score depending on the length of the phrase entered.

Personal Health Record keeping the patient information data and helping the patients by sharing the information, to help other hospitals in a better cure of the patients.

Sensor Emerging Technology technology supporting health through sensors.

Technology Acceptance Methodology a methodology to find out how the technology is well accepted by the user.

FALL DETECTION AND MOTION CLASSIFICATION BY USING DECISION TREE ON MOBILE PHONE

10

Fang-Yie Leu*, Chia-Yin Ko*, Yi-Chen Lin*, Heru Susanto[†,‡], Hsin-Chun Yu*

*Tunghai University, Taiwan †Department of Information Management, Tunghai University, Taiwan ‡The Indonesian Institute of Science, Indonesia

10.1 INTRODUCTION

In recent decades, aging population is a popular phenomenon in many countries. Between 2015 and 2030, the number of people in the world aged 60 years or over is projected to grow by 56 percent, from 901 million to 1.4 billion (United Nations, 2015), and by 2050, the global population of older persons will be more than double the size in 2015, reaching nearly 2.1 billion. With the increase of age, an elderly will gradually lose his/her ability of self-care. How to properly take care of the elderly has been a hot research topic nowadays. To successfully do this, we must look into its unavoidable problems. Elderlies above the age of 65 risks a one-third chance of falling each year (CDC, 2015; The Falls Management Program, 2012). Research indicates that the risk of falls increases proportionally with aging. But only less than one-half of the cases were reported to their doctors (CDC, 2015). Once a fall occurs, the chance of fall again doubles in the coming one year. In fact, one-fifth of falls causes serious injury, including hip fractures and traumatic brain injuries. Such injuries are the main cause of hospitalization of elderlies. Hip fractures very often are resulted from falling sideways (The Falls Management Program, 2012).

The consequences of falls tend to be huge, especially for older people, very usually causing reduced quality of life, serious injuries, increased risk of death, increased fear of falling and restriction of activities (Ko et al., 2015). The injuries are not only physical but also psychological. Ironically, the fear of falling make an elderly to have a higher risk of falling. Previous studies have shown that their proposed systems only determine whether a fall occurs, e.g., Albert et al. (2012) only discriminated four kinds of falls. In fact, falls are complex events. In addition, the use of machine learning can enhance system ability to more accurately analyze falls and classify activity of daily living (ADL). Therefore, in this chapter, we develop a fall management system, named Fall Detection System using Mobile Phones (FDSMP), which utilizes the microelectromechanical systems (MEMSs) installed in a mobile phone, including a Triaxial accelerometer and a gyroscope, and decision tree techniques to detect and classify falls and ADLs. Once a fall occurs, through the instant messaging communications and the convenience of the FDSMP, caregivers can know the event immediately and provide required assistance when necessary.

Our experimental results show that the FDSMP demonstrates a success of body movement classification, which in turn enhances the accuracy for determining fallings, and correctly identifies the fall types, such as Falling forward, Falling backward, Falling right, Falling left, Body turning right while falling backward and Body turning left while falling backward, which are the six types of falls we choose in this study. The accuracy of all types of detections is 98.46% and that of fall classification is 96.57%.

The rest of this chapter is organized as follows. Section 10.2 introduces background and literature review of this chapter. Section 10.3 describes the materials and methods used in this study. Section 10.4 demonstrates our proposed approach and its implementation. Evaluation on this approach is presented in Section 10.5. Sections 10.6 and 10.7 conclude this chapter and outlines our future studies, respectively.

10.2 BACKGROUND

A lot of fall-detection approaches have been proposed (Bai et al., 2012; He et al., 2012; Zhang et al., 2006; Bourke et al., 2007; Sposaro and Tyson, 2009; Hwang et al., 2012; Doukas and Maglogiannis, 2008). Bai et al. (2012) used an Accelerometer and a GPS installed in a smart phone to design a fall monitor and position identification system. They analyzed the change of acceleration and six typical actions of human beings. The latter includes going upstairs, going downstairs, standing up, sitting down, running, and jumping. He et al. (2012), utilized calculated signal magnitude area, signal magnitude vector (SMV), and tilt angle to determine the threshold for a fall. They presented a system which classified human motions in real-time with a smartphone mounted on the user's waist. A built-in Triaxial accelerometer collects data of body motions, with which the smartphone is also able to classify the motions into five different patterns, including vertical activity, lying, sitting (or static standing), horizontal activity, and fall.

Zhang et al. (2006) employed a Triaxial accelerometer equipped in a cellphone to collect SMV values, and utilized 1-Class Support-Vector Machine algorithm for preprocessing and Kernel Fisher Discriminant, and k-NN (K-Nearest Neighbor) algorithm for precise classification. They found that this method can detect falls effectively and make less disturbance to people's daily living than the general wearable sensors do. Bourke et al. (2007) used simulated falls performed under supervised conditions, and ADLs done by elderly subjects. The ability to discriminate between falls and ADLs relies on Triaxial accelerometer sensors mounted on the trunk and thigh. Data was analyzed using the MATLAB to determine the peak accelerations recorded for eight different types of falls, including forward falls, backward falls, lateral falls left, and lateral falls right (four movements), performed on two conditions, i.e., legs straight and flexed (so a total of 8 ($= 4 * 2$)). Falls detection algorithms were devised using threshold techniques.

Sposaro and Tyson (2009) presented an alert system which used common commercially available electronic devices to detect falls, and alerted authorities on an Android-based smart phone equipped with an integrated Triaxial accelerometer. Data from the accelerometer is evaluated with several thresholds-based algorithms and position data to determine a fall. The threshold is adaptive based on parameters, such as height, weight, and level of activity, provided by users.

Some researchers added other auxiliary sensors to their detectors. Hwang et al. (2012) designed an algorithm for the detection of falls using smartphones equipped with Triaxial accelerometer and magnetometers. They also proposed a fall detection method that recognizes a fall if the magnitudes of an acceleration and angular displacement exceed given thresholds. Experiments for detecting falls are performed in four directions: fall forward, fall backward, fall left, and fall right. Data is collected from 200 experimental falls, in each of which a smartphone is fastened on a belt worn around the waist. An overall detection rate is 95%, and those of fall forward, backward falls, leftward falls, and rightward falls are 94%, 100%, 94%, and 92%, respectively.

In Doukas and Maglogiannis (2008), sensors equipped with accelerometers and microphones are attached to the body of patients, and transmit patients' movements and sound data wirelessly to the monitoring unit. The sound and movement are classified by using Support Vector Machines. Rakhman et al. (2014) and Dai et al. (2010) added a gyroscope as an assistance device. Dai et al. (2010) first designed a fall detection algorithm based on mobile phone platforms and proposed PerFallD, which is a pervasive fall detection system implemented on mobile phones. Noury et al. (2007) proposed 20 scenarios and 10 fall types to evaluate fall sensors. Li et al. (2009) classified falls into 6 types and 12 kinds of ADLs. Other state-of-the-art fall detection systems can be found in Delahoz and Labrador (2014), El-Bendary et al. (2013), Lgual et al. (2013).

Zhao et al. (2012) divided human movements into four types, including stationary, walking, running, and fall, and analyzed relevant data with a Decision tree. The results show that the precision is 100% and the recall is 75.8%. Aguiar et al. (2014) discussed the accuracies for Decision tree, K-nearest neighbors (K-NN), and Naïve Bayes, and claimed that Decision tree outperforms the others. The sensitivity of the case in which mobile phones is placed in pockets is 94% and on belts is 90.6%. The specificity in pockets is 90.2% and on belts is 96.1%, and the accuracy in pocket is 92% and on belts is 93.7%. In the classification systems presented by Zhao et al. (2012) and Aguiar et al. (2014), the precision is 86%, and recall is 100%. After adding a gyroscope as an auxiliary device, the precision is up to 97.4%, and recall is 100%. That means a gyroscope is helpful in classification.

Albert et al. (2012) found that there was a ceiling phenomenon, which is the case when acceleration of gravity is greater than a specific value, no higher value can be measured and shown due to the limitation of the maximum (minimum) acceleration that a Triaxial accelerometer can detect. He et al. (2012) using the Lenovo Le-Phone in 2012 and Beauvais (2012) utilizing the Samsung Galaxy SII in 2014 all have the same situation. In fact, when we conduct experiments with the mobile phones produced earlier than 2015, the values obtained are not quite accurate. It is recommended to use the smartphones manufactured in 2015 or after, and suitability of chip used needs to be ascertained.

As most people like to use mobile phones, the sizes of which are bigger than four inches, it is hard to place them on the waist or leg, particularly prone to cause serious damage to the phone at the time of running, jumping or sitting down. Fang et al. (2012) showed that when mobile phones are placed on the chest, waist, and thigh, the sensitivities were 72.22%, 56.67%, and 53.33%, respectively, and specificities were 73.78%, 66.39%, and 57.22%, respectively. Accuracy of placing the mobile phones on the chest is apparently higher than those of others. Therefore, in this study, during experiments, the used mobile phone is placed in testers' chests.

10.3 PROBLEM DEFINITION
10.3.1 RESEARCH METHODS

At present, studies that utilize mobile phones to detect falls mainly rely on Triaxial accelerometers. However, like the cases in Rakhman et al. (2014) and Bai et al. (2012), a gyroscope as an auxiliary micro-electronic device is employed by the FDSMP to detect a user's body rotation angle. A Decision tree is also utilized to classify a movement in a systematic way. The FDSMP further uses an industrial technology which integrates both mechanical engineering and microelectronics technology as a MEMS. An MEMS usually contains a microprocessor and multiple micro-sensors to obtain external information Waldner (2008). The size of an MEMS is determined by the sizes of its components, which are often between 1 and 100 microns (0.001 to 0.1 mm). The sizes of MEMS devices typically range between 20 microns and one millimeter. With the advancement of sensor techniques, the complexity and efficiency of MEMSs continue to improve. Within a mobile phone, the Triaxial accelerometer and gyroscope sense the phone's Triaxial accelerations and directional changes, respectively. Experiments are conducted with a user standing right behind the mobile phone, i.e., facing the same direction with the phone.

(1) Triaxial Accelerometer

The acceleration detected by a Triaxial accelerometer as shown in Fig. 10.1 is divided into x, y, and z three sub-accelerations. The unit of the triaxial accelerometer is g (9.8 m/s^2). The direction of triaxial accelerometer is as follows. The left side of the person, who stands behind the mobile phone, is the positive direction of x-axis.

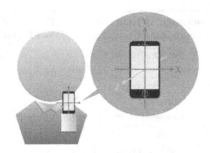

FIGURE 10.1

The three axes of a Triaxial accelerometer equipped in a mobile phone.

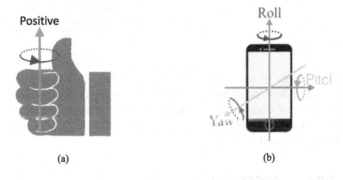

FIGURE 10.2

Pitch, Roll, and Yaw of a gyroscope equipped in a mobile phone. (a) Direction of a rotation; (b) Directions of a gyroscope.

The right side, of course, is negative. The upside (down side) of the y-axis is positive (negative). The forward (backward) of the z-axis is positive (negative).

(2) Gyroscope

A gyroscope, as a device designed to measure and maintain a sense of rotational direction based on the theory of conservation of angular (Penrose, 2007; Oxford Dictionaries, 2015), mainly consists of a rotatable rotor. When the gyroscope starts to rotate, the rotor angular momentum tends to resist the change of direction.

Gyroscopes are popularly utilized by navigation and positioning systems. The unit of the gyroscope is degree (°). The directions of its three axes are shown in Fig. 10.2b. Pitch is the rotation along the x-axis; the forward is positive. Of course, the backward is negative. Roll is the rotating along the y-axis. Rotating to the left (right) is positive (negative). Yaw is the rotating along the z-axis; rotation to the right (left) is positive (negative).

In our everyday lives, many ADLs can be found. However, in this study, we only choose standing up, sitting down, walking, and running since they are our most frequent daily activities. Falls according to Noury et al. (2007), as listed in the first

Table 10.1 The Classification in Noury et al. (2007) Are Listed in the First Two Columns. The Third Column is the Classification of the FDSMP

Fall Direction		Ending State	FDSMP
Backward		lying	Falling backward
		In lateral position	
		In the knees	
Forward	Directly	With forward	Falling forward
		Lying flat	
	With rotation	In lateral right position	
		In lateral left position	
Lateral	Right	Lying flat	Falling right
	Left	Lying flat	Falling left
–		–	Body turning right while falling backward
–		–	Body turning left while falling backward

two columns of Table 10.1, can be classified into Backward fall ending lying, Backward fall ending in later position.... In this study, as mentioned above, we classify the movements of falls into six types, including Falling forward, Falling backward, Falling right, Falling left, Body turning right while falling backward and Body turning left while falling backward.

10.3.2 SIGNAL PROCESSING

After receiving signals from a smart phone, we use SMV and angular variation, denoted by A_t, to calculate the gravity strength and the angle of a fall, where the SMV and A_t are calculated as follows.

$$\text{SMV} = \sqrt{X^2 + Y^2 + Z^2} \tag{10.1}$$

in which X, Y, and Z, respectively, represents the acceleration values of gravity in the Triaxial (i.e., X, Y, and Z) directions, where A_{t-1} is the last angular variation, ω is the measured angular velocity by gyroscope.

$$A_t = A_{t-1} + \omega \cdot \Delta t. \tag{10.2}$$

In the signals received, the maximum SMV value of each human action, and maximum and minimum values of the variation of rotation angles are identified. Note that, the Pitch that turns forward, as an example, is positive, the angle changing the most in the forward direction is the maximum, and the angle changing the most in the backward direction is the minimum. The Roll's and Yaw's are similar.

Table 10.2	Features of the Six Fallings					
	Classification					
Axis	Falling Forward	Falling Backward	Falling Right	Falling Left	Body Turning Right While Falling Backward	Body Turning Left While Falling Backward
Pitch	Maximum	Minimum	–	–	Minimum	Minimum
Roll	–	–	–	–	Minimum	Maximum
Yaw	–	–	Maximum	Minimum	–	–

10.4 PROPOSAL APPROACH

10.4.1 FEATURES OF SIX FALLING MOVEMENTS

The maximum and minimum of Pitch, Roll, and Yaw of the 6 falling movements we choose are shown in Table 10.2, in which Falling forward (Falling backward) has the maximum (minimum) values of Pitch. Falling right (Falling left) mainly is dominated on the angle of Yaw, so it has a maximum (minimum) value. Body turning right while falling backward and Body turning left while falling backward trigger both the Pitch and then Roll (see Table 10.2). Since both the two fallings are reclined (falling backward), there is a minimum in the negative direction of Pitch. The body rolling in the right spin (turning right) produces a minimum in the negative direction of Roll. Rotating/turning to the left during falling results in a maximum in the positive direction of Roll.

10.4.2 CLASSIFIERS

The Weka (Smith and Frank, 2016), as a data mining tool, provides different classification mechanisms, called classifiers, including decision tree, Rule-PART, Bayes Theorem, Rule-DTNB, Meta-Bagging, and so on. In this study, we choose some of them to analyze and classify the data collected by using the two utilized MEMSs. Their classification accuracies are listed in Table 10.3.

(1) Decision Tree

Decision tree is a popular decision support tool often used to decide the outputs for input data (e.g., determining the possible results of random events), the outcomes on given resources, and the cost effectiveness of a constructed structure (e.g., a tree or model). Decision Trees are commonly used in operation research, especially to determine the most that a strategic objective can achieve. The decision rules of a decision tree can be linearized (Quinlan, 1987), wherein the result is the content of a leaf node, e.g., node N, and decision making is the process determining the path that will finally reach N as the final result based on the values of input parameters. In fact, the path can be expressed as rules with the format: if $AND_{i=1}^{n}$ Condition i, then *Result*, meaning there are n conditions, $n \geq 1$, i.e., there are n links from the root

Table 10.3 Classification Accuracies of Classifiers Tested in This Study

Classifier	Tree			Rules		Bayes		Meta	
	J48	NBTree	BFTree	PART	DTNB	BayesNet	Attribute Selected Classifier	Bagging	Filtered Classifier
Accuracy	98.46%	98.46%	96.92%	98.46%	96.92%	98.4615%	98.46%	96.92%	96.92%
TP	1	1	0.973	1	0.973	1	0.973	1	1
FP	0.036	0.036	0	0.036	0.036	0.036	0	0.071	0.036
Precision	0.974	0.974	1	0.974	0.973	0.974	1	0.949	0.974
Recall	1	1	0.973	1	0.973	1	0.973	1	1
F-Measure	0.987	0.987	0.986	0.987	0.973	0.987	0.986	0.974	0.987

to N, and *Result* is the content of the leaf node. In other words, the path has $n + 1$ nodes, including the start node, i.e., root node, and N.

J48 is an open source of the C4.5 decision tree algorithm implemented with Java (Kaur and Chhabra, 2014) by Quinlan (Hssina et al., 2014). This algorithm can be used to classify input data. So it is often called statistical classifier (Quinlan, 1993), and it is a remarkable mining algorithm Ranked # 1 announced by the Springer LNCS (Wu et al., 2008).

(2) Rule – PART

PART, as a scheme used to generate accurate rule sets for a decision tree, is proposed by Frank and Witten (1998). The PART has two dominant schemes for rule-learning, i.e., C4.5 and RIPPER, and shows how good rule sets can be learned without any need for global optimization. They present an algorithm for inferring rules by repeatedly generating partial decision trees, consequently combining the two major paradigms, i.e., creating rules from decision trees and the separate-and-conquer rule-learning technique, for rule generation.

(3) Bayes Theorem

Bayes' Theorem is developed based on a probability theory. There is a random variable with the probability of the conditions and the edge probability distribution. The conditioned probability of random events A and B is defined as follows (Kendall et al., 1994).

$$P(A \mid B) = \frac{P(B \mid A)P(A)}{P(B)} \tag{10.3}$$

where $P(A|B)$ is a possibility in the case of when B occurs, A occurs.

(4) Rule – DTNB

DTNB (Hall and Frank, 2008) investigates a simple semi-naive Bayesian ranking method that combines naive Bayes with an induction of decision tables. Naive Bayes and decision tables can both be trained efficiently, and the same holds true for the combined semi-naive model.

(5) Meta – Bagging

Bagging (Breiman, 1996) predictors is a mechanism for generating multiple versions of a predictor, with which to construct an aggregated prediction scheme. The aggregation averages over the versions when predicting a numerical outcome, and triggers a plurality vote when predicting a class. The multiple versions are created by bootstrapping duplications on the learning set, and then using these versions as new learning sets. Tests performed by the author of Breiman (1996) on real and simulated data sets utilized classification and regression trees. Its subset selection with a linear regression shows that bagging can substantially gain in accuracy.

Although the Decision tree, rules, and Bayes shown in Table 10.3 obtain good accuracies, decision tree is simpler, and has less computation. Therefore, we choose it as the classification mechanism.

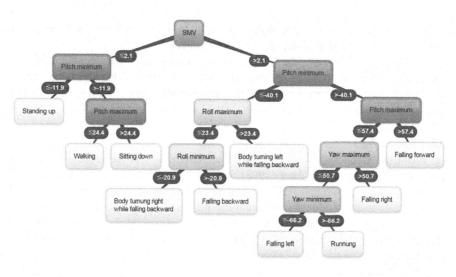

FIGURE 10.3

The decision tree obtained by inputting our experimental results to the Weka.

10.4.3 OUR CLASSIFICATION TOOL

According to Table 10.2, we put the experimental result data, collected by using the Triaxial accelerometer and gyroscope into weka and obtain the Decision Tree shown in Fig. 10.3.

Firstly, a threshold value 2.1 g on SMV is chosen. If the SMV value is less than the threshold, we consider that the corresponding body movement is an ADL, which may be standing up (when the minimum of Pitch is smaller than or equal to −11.9 degrees), walking (when the minimum of Pitch is larger than −11.9 degrees and the maximum of Pitch is smaller than or equal to 24.4 degrees) or sitting down (when the Pitch is larger than −11.9 degrees and the maximum of Pitch is larger than 24.4 degrees). Otherwise, it is a falling or running.

When the SMV is larger than 2.1 g and the minimum of Pitch is smaller than or equal to −40.1 degrees, it is a body turning left when falling backward (when the maximum of Roll is larger than 23.4 degrees) or a body turning right while falling backward (when the minimum of Roll is smaller than or equal to −20.9 degrees). When the minimum of Roll is larger than −20.9 degrees, the movement is a falling backward. When SMV is larger than 2.1 g, the minimum of Pitch is larger than −40.1 degrees and the maximum of Pitch is larger than 57.4 degrees, it is a falling forward. Otherwise, it is a falling right (when minimum of Pitch is larger than −40.1 degree, the maximum of Pitch is smaller than or equal to 57.4 degrees and the maximum of Yaw is larger than 50.7 degrees) or Running (when the maximum of Yaw is smaller than or equal to −50.7 degrees and minimum of Yaw is larger than −66.2 degrees).

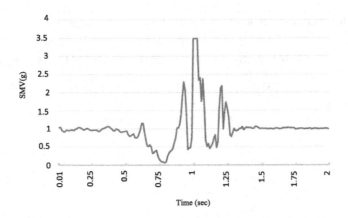

FIGURE 10.4

Ceiling phenomenon on iPhone 4.

If the minimum of Yaw is smaller than or equal to −66.2 degrees, there will be a falling left.

10.5 EVALUATION

In this section, we use the data collected by using Triaxial accelerometer and gyroscope to analyze the accuracy of the Decision Tree generated by the Weka. First, we classify a movement into a falling or one of the four types of mentioned ADL. In the first experiment, we discriminate whether it is a fall or an ADL. If it is an ADL, the movement is then further classified it into sitting down, standing up, walking or running, identified as Fall and ADL-classes (F-ADL-class) classification. In the second experiment, a movement is classified into one of the six types of mentioned falling or one of the four types of chosen ADL, denoted by movement (M-class) classification.

During the experiments, we place a smart phone on the user's chest. In order to eliminate additional confounding variables, and bring the experiments closer to their ideal states, we strap the phone parallel to the chest so that the measurement of acceleration of gravity can be closer to the body's value. This may relatively be free from interference of other external factors.

10.5.1 CEILING EFFECT

In our previous study, when the acceleration of gravity of iPhone 4 is greater than 3.5 g, the ceiling phenomenon as shown in Fig. 10.4 appears, even the Apple claims that iPhone 4 can measure more than 4 g of gravity (STMicroelectronics, 2010).

Product	iPhone 4	iPhone 5	iPhone 6	
Brand	STMicroelectronics (2010)	Bosch (2011)	Bosch (2014)	InvenSense (2013)
Model	LIS331DLH	BMA220	BMA280 0	MPU-6500
Measurement Range	±2.0	±2.0	±2.0	±2.0
	±4.0	±4.0	±4.0	±4.0
	±8.0	±8.0	±8.0	±8.0
		±16.0	±16.0	±16.0

Table 10.4 The Generations of iPhone of Triaxial Accelerometer Chip

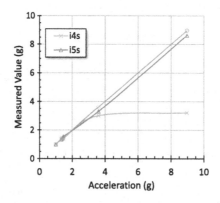

FIGURE 10.5

Measured values of Triaxial accelerometer for three iPhone models, including iPhone 4–iPhone 6.

We first test the Apple products, including iPhone 4–iPhone 6 and compare their performance.

The Triaxial accelerometer chips for the three iPhones models are shown in Table 10.4 (STMicroelectronics, 2010; Bosch, 2011, 2014; InvenSense, 2013).

The experimental results are shown in Fig. 10.5, in which the three models measure the similar values when the gravity is smaller than 3 g. When the acceleration of gravity is higher than 3 g, the ceiling phenomenon of iPhone 4 appears. But this time its acceleration cannot be larger than 3.2 g. We can see iPhone 5 and iPhone 6 can measure up to 9 g.

10.5.2 EVALUATION CRITERIA

As shown in Table 10.5, the precision of F-ADL-class classification was 98.5%, and its recall was 98.5%. The precision of M-class classification was 97.2%, and recall is 96.5%. Here,

$$\text{Precision} = \frac{TP}{TP + FP} \tag{10.4}$$

Table 10.5 Accuracies of F-ADL-Class Classification and M-Class Classification (%)							
Classification	Accuracy	TP	FP	FN	Precision	Recall	F-measure
F-ADL-class classification	98.5	98.5	2	2	98.5	98.5	98.4
M-class classification	96.5	96.5	0.4	3.5	97.2	96.5	96.8

Table 10.6 The Confusion Matrix of F-ADL-Class Classification					
	The FDSMP Classification				
Actual Classification	Walking	Running	Sitting Down	Standing Up	Falling
Walking	50	0	0	0	0
Running	0	49	0	0	1
Sitting down	0	0	50	0	0
Standing up	0	0	0	50	0
Falling	0	0	0	0	150

and

$$\text{Recall} = \frac{TP}{TP + FN} \tag{10.5}$$

where TP standing for true positives is defined as the case in which a fall occurs and the detection system accurately discovers it. FP standing for false positives is defined as that when the detection system announces that there is a fall, but actually, it is not a fall. TN standing for true negatives is defined as the case in which there is a non-fall movement (i.e., an ADL), and the system also classifies the movement as an ADL. Moreover, FN standing for false negatives is defined as that a fall occurs, but the system classifies it as an ADL.

$F_{measure}$ as a metrics taking into account both precision and recall is applied to assess the effectiveness of information retrieval (Rijsbergen, 1979), and is defined as:

$$F_{measure} = \frac{2}{(\frac{1}{Precision} + \frac{1}{Recall})} = \frac{2 \times Precision \times Recall}{(Precision + Recall)}$$
$$= \frac{2 \times TP}{(2 \times TP + FP + FN)}. \tag{10.6}$$

Accuracy of F-ADL-class classification in determining whether there is a fall or it is an ordinary ADL movement is 98.46%. However, after further classifying fallings, the accuracy of M-class classification is reduced to 96.57%. Table 10.6 is the confusion matrix of F-ADL-class classification, in which only one running was misclassified to falling. The reason is that running is a discontinuous jump with a greater gravity, and is then prone to false positives

Table 10.7 is the confusion matrix of M-class classification, in which the miscarriage of justice appears at the judgment of falling right and falling left movements, mainly because the two falls are not simple ones. They could be a falling left (right) and tumbling down to the left (right) at the same time. They are prone to the false positives. From F-ADL-class classification and M-class classification, we found that the

Table 10.7 The Confusion Matrix of M-Class Classification

Actual Classification	The FDSMP Classification									
	Walking	Running	Sitting Down	Standing Up	Falling Forward	Falling Backward	Falling Right	Falling Left	Body Turning Right While Falling Backward	Body Turning Left While Falling Backward
Walking	50	0	0	0	0	0	0	0	0	0
Running	0	49	0	0	1	0	0	0	0	0
Sitting down	0	0	50	0	0	0	0	0	0	0
Standing up	0	0	0	50	0	0	0	0	0	0
Falling Forward	0	0	0	0	25	0	0	0	0	0
Falling Backward	0	0	0	0	0	25	0	0	0	0
Falling Right	0	0	0	0	0	0	24	0	1	0
Falling Left	0	0	0	0	0	0	0	24	1	0
Body turning right while falling backward	0	0	0	0	0	1	0	0	24	0
Body nulling left while falling backward	0	0	0	0	0	0	0	0	0	25

accuracy of determining whether or not there is a fall without further falling classification or the movement is an ordinary ADL (with further ADL classification) is high. However, when detecting the movement of falls, the situation in reality is not that simple like our classification, e.g., it is possible that falling right and turning to the left are performed at the same time, causing a large gravity in the SMV, consequently leading to high difficulties in correct falling judgment and ADL classification.

10.5.3 SIGNAL ANALYSIS

The signals including Triaxial accelerations and gyroscope angles collected for different movements are comprehensively compared and analyzed. Beauvais (2012) also analyzed signals for falls. The actions of a fall can be divided into three phases, i.e., free falling, the body stroking the floor, and the body staying lying. We also record the actions of a fall with a video, besides the data collected by using gyroscope/Triaxial accelerometer, so that we can identify the correspondences between an action and its recorded signals.

The four types of our chosen ADLs are mainly distinguished by Pitch (see the left subtree of Fig. 10.3). In fact, the Roll and Yaw are the body's natural swing, produced by the control of body center of gravity. They do not constitute an important basis in ADL movement analysis. The time interval for the FDSMP to access the acceleration values and the angle variation is 0.01 second.

The six types of falling need to distinguish by the three-axes values of the gyroscope. As mentioned above, falling is usually accompanied by other fall patterns, and therefore, cannot be simply classified. The compound falling activities appear mainly because of the body's self-protection actions and instability control of the center of gravity. The purpose is to reduce the physical injury caused by the falling. In these actions, body is often curled or tumbled even simultaneously. In the following, due to high density of data, all figures are hard to use circles, triangles, and crosses to differentiate different lines representing pitch, roll, and yaw (x-axis, y-axis, and z-axis). We will show the lines in the captions.

(1) Standing Up

The signals we record for standing up, including its SMVs, Triaxial accelerations, and gyroscope angles, are shown in Fig. 10.6. Before standing up, we need to bend our upper bodies to move the center of gravity forward. Fig. 10.6b shows that the experimental Pitch is about 23 degrees of ups and downs, and that the Roll and Yaw have only a slight rotation around the body. But due to different weight transfers and personal habits, different people may present different results. However, the difference in results is often small. Comparing Figs. 10.6a and 10.6b, it can be seen when the body leaned forward the maximum, the SMV goes up and the peak is about 1.7 g. In Fig. 10.6c, y-axis starts at -1 g, because the gravity of earth is in the negative direction of y-axis. When his/her buttocks leave the chairs, there is a smooth peak and an acceleration goes upward, causing the Triaxial acceleration on y-axis to be about -1.6 g. While the user bends over, the gravity of earth shifts a portion to the positive direction of z-axis. That is why acceleration on z-axis increases.

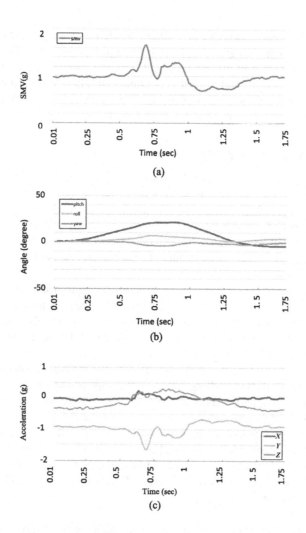

FIGURE 10.6

The SMVs, accelerations, and Gyroscope angles on a standing-up movement. (a) SMVs of standing up; (b) Gyroscope angles including Pitch, Roll, and Yaw of standing up (at time = 0.75, the upper line is pitch, the middle is roll, and the lower is yaw); (c) Accelerations of standing up (at time = 0.25, the upper line is x-axis, the middle line is z-axis, and the lower line is y-axis).

(2) Sitting Down

Before sitting down, we also bend over to move the center of body gravity forward to balance our body. The SMVs, Triaxial accelerations, and gyroscope angles, are plotted in Fig. 10.7. In Fig. 10.7b, the maximum of Pitch is about 45 degrees, meaning the upper body bends over 45 degrees. The angles of Roll and Yaw are relatively small. In Fig. 10.7c, we can also see that gravity of y-axis is negative due to gravity of

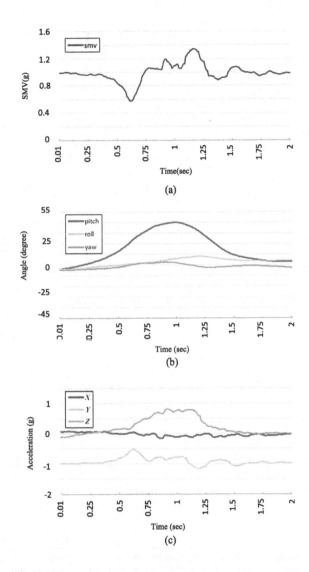

FIGURE 10.7

The SMVs, accelerations, and gyroscope angles on a sitting down movement. (a) SMVs of sitting down; (b) Gyroscope angles including Pitch, Roll, and Yaw of sitting down (at time = 1.25, the upper line is pitch, the middle is roll, and the lower is yaw); (c) Accelerations of sitting down (at time = 1, the upper line is z-axis, the middle is x-axis, and the lower is y-axis).

earth. When his/her body starts moving downward, y-axis has a smooth peak, which is about -0.5 g. When the user bends over, like that on standing up, the gravity of earth shifts a part to the positive direction of z-axis. That is why the acceleration in z-axis increases.

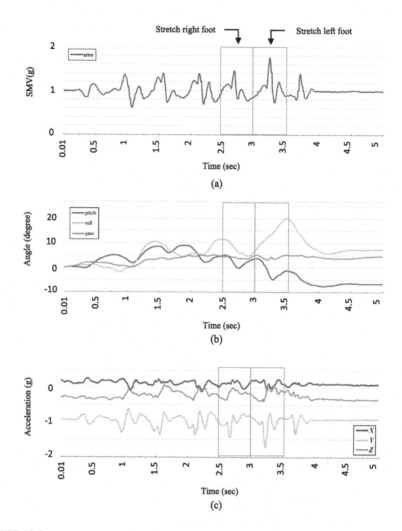

FIGURE 10.8

The SMVs, accelerations, and gyroscope angles on a walking movement. (a) SMVs of walking; (b) Gyroscope angles including Pitch, Roll, and Yaw of walking (at time = 3.5, the upper line is roll, the middle is yaw, and the lower is pitch); (c) Accelerations of walking (at time = 2, the upper line is x-axis, the middle is z-axis, and the lower is y-axis).

(3) Walking

Walking and running are movements with cyclical characteristics. So we can observe all steps of them. The recorded SMVs, Triaxial accelerations, and gyroscope angles, are shown in Fig. 10.8, which illustrates that the user walked a total of 6 steps, and from which we can see the ups and downs of Pitch on each step. In Fig. 10.8, each rectangle contains the signals of a step. Although we do not obviously feel the

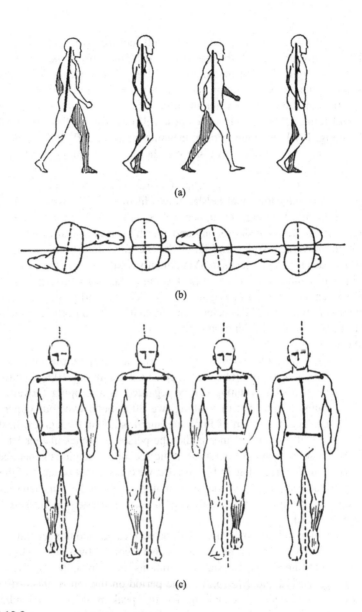

FIGURE 10.9

Schematic of Walking (Chang, 2016). (a) Side look of Walking; (b) Overlook of Walking; (c) Front look of Walking.

change of Pitch of the body while walking, we can see the second and third portraits shown in Fig. 10.9a from the left. When our legs stretch forward, body will slightly be tilted forward to balance the center of body gravity.

In Fig. 10.9b, when the user moves his/her left foot forward, as that in the second rectangle, the corresponding Roll is positive, which comes from the fact that his/her buttocks turn to right (see the first portrait in Fig. 10.9b), but his/her chest turns left a little to balance the center of body gravity, resulting in a small positive for Roll. Fig. 10.9b also shows that when the user steps out the right foot, similarly the buttocks will rotate to the left. But the chest will turn right a little. So in the first rectangle in Fig. 10.8b, Roll goes down, meaning chest turns right. Of course, when his/her buttocks turn left, meaning he/she stretches his/her right leg, Roll has a small rotation in its negative direction.

Now it is easy to recognize which leg the user stretches in the second rectangle, Roll goes up, indicating that he/she moves his/her left leg forward. Fig. 10.9c illustrates that while walking, our upper bodies will slightly vibrate along z-axis, generating Yaw angles (see the second and fourth portraits). But different persons have different movement styles. In Fig. 10.8b, the Yaw almost remains unchanged. In Fig. 10.8a, each step has 2 peaks in SMV, mainly resulting from the shake of accelerations in y-axis and z-axis. The first peak occurs when a foot touches the ground, which causes an acceleration in negative y-axis. The second peak is resulted from the body moving forward. The acceleration of gravity shifts a portion to the positive direction of z-axis as the acceleration in that axis.

(4) Running

Fig. 10.10 illustrates the recorded SMVs, Triaxial accelerations, and gyroscope angles for the person who ran a total of 10 steps. Like that in Fig. 10.8, each rectangle contains signals of a step. Running can be regarded as a jumping forward. At the beginning of running, like that of walking, we often bend over our upper bodies for the first step. The maximum of Pitch measured is about 19 degrees. In the first rectangle, since Roll goes up, it means that the person stretches his/her left leg for running. Of course, the signals contained in the second rectangle is generated when he/she stretches his/her right leg. In Fig. 10.10c, we can see that because of the swing of the body, the acceleration of x-axis has a small peak about 2 g. On each step, when a foot touches the ground, an acceleration about -2.5 g in y-axis is measured.

(5) Falling Forward

In the following, falling experiments will be performed, and the generated signals will be analyzed. The entire falling process is also recorded by a video recorder.

Fig. 10.11 plots the SMVs, Triaxial accelerations, and gyroscope angles for falling forward. In Fig. 10.11a and 10.11b, the time period on the left of the vertical line, rather than the dotted lines, is free falling. The first peak of SMV, as the indicated 1, is the impact when knees hit the ground. There is an acceleration toward to the ground (see y-axis in Fig. 10.11c). There is also a peak in z-axis, indicating that there is a z-axis acceleration, i.e., the body is still forwarding. The second peak, as the indicated 2, is generated when the palms touch the ground. In Fig. 10.11c, there is a small peak in x-axis, which means the user's right hand touches the ground before his/her left hand does, causing an acceleration on negative direction of x-axis. Meanwhile, z-axis acceleration still exists since his/her body keeps forwarding, resulting in another peak in z-axis.

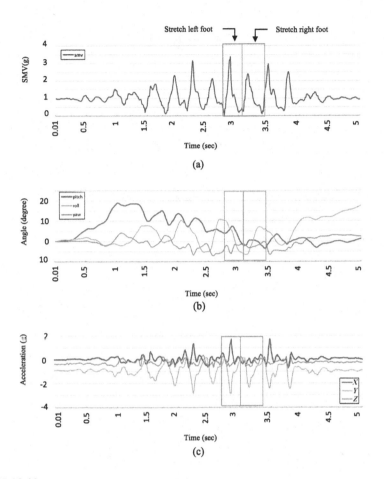

FIGURE 10.10

The SMVs, accelerations, and gyroscope angles on a running movement. (a) SMVs of running; (b) Gyroscope angles including Pitch, Roll, and Yaw of running (at time = 1.5, the upper line is pitch, the middle is roll, and the lower is yaw); (c) Accelerations of running (at time = 1, the upper line is x-axis, the middle is z-axis, and the lower is y-axis).

The last peak, as the indicated 3, is formed when the body collides with the ground. There is a peak in y-axis which is almost overlapped with the peak in z-axis. Because when touching the ground, his/her body continues going forward in the y-axis. The two small peaks between 1 and 2 and between 2 and 3 are created by the rebound of the phone in the user's pocket. When the body hits the ground (number 3), because the device is under pressure of the body, the strength of the impact is the largest with no rebound of the phone. Fig. 10.11c clearly shows that the three peaks are most dominated by the z-axis.

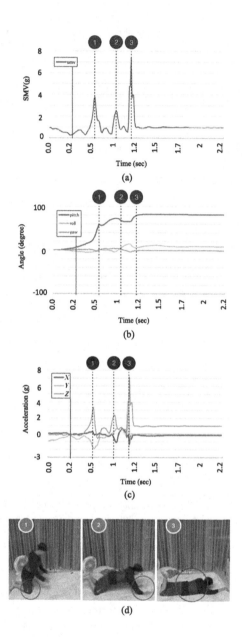

FIGURE 10.11

The SMVs, accelerations, and gyroscope angles on a falling forward movement. (a) SMV of falling forward; (b) Gyroscope angles including Pitch, Roll, and Yaw of falling forward (at time = 1.2, the upper line is pitch, the middle is roll, and the lower is yaw); (c) Accelerations of falling forward (at time = 0.5, the upper line is z-axis, the middle is x-axis, and the lower is y-axis); (d) The pictures recorded by video recorder.

(6) Falling Backward

Fig. 10.12 plots the SMVs, Triaxial accelerations, and gyroscope angles for falling backward. In Fig. 10.12a and 10.12b, like those in Fig. 10.11, the time period on the left of the vertical line, rather than the dotted lines, is free falling. The first peak of SMV, as the indicated 1, occurs when the person curled up the body forward in order to slow the speed of the falling so as to reduce body injury. The Pitch is then positive. The second peak, as the indicated 2, is generated when the buttocks hit the ground, forming the largest impact. The last peak, as the indicated 3, occurs when the user tries to balance his/her body, but failed at the end. Finally, user's head and body touch the ground.

After the maximum of the peak (between 2 and 3), there are several small peaks, which are produced by the rebound of the phone in the user's pocket. In Fig. 10.12b, Pitch between 1 and 2 goes down since user's upper body gradually moves backward, and at last becomes negative. Its final angle is about −80 degrees. Roll and Yaw almost remain unchanged. In Fig. 10.12c, we can see the second peak is produced by the accelerations in y-axis and z-axis. The biggest is that in y-axis, because the body (buttocks) hits the ground vertically. We know that it probably causes injury of vertebral. In z-axis, when the person's body hits the ground, there is an acceleration backward which is negative. It is clear that at this time point, the acceleration in negative z axis is high. It is also the reason why the person's body continues moving backward. We can also see the acceleration in x-axis at 3 fluctuates. This is because when user's body hits the ground, his/her left shoulder touches the ground first (positive in x-axis) and then the right shoulder (negative in x-axis). At 3, there is also a small peak in negative z-axis, meaning that when the user's head and body touch the ground, the impact is not serious.

(7) Falling Right

Fig. 10.13 illustrates the SMVs, Triaxial accelerations, and gyroscope angles of falling right. Before the maximum of the peak shown in Fig. 10.13a, Pitch shown in Fig. 10.13b, is positive, which is about 15 degrees since the user curls up his/her body forward in order to slow the falling speed so as to reduce body injury. The maximum of the peak is produced when the user's buttocks hit the ground. Roll is about +15 degrees since the user turns his/her body left a little, and uses his/her right-hand elbow to touch the ground in order to mitigate falling injury. Yaw at last is about +120 degrees. As shown in Fig. 10.13a, the SMV peak is about 6 g, which as shown in Fig. 10.13c is created by the acceleration of the three-axes. The acceleration in x-axis is about −3 g. It is negative because the body hits the ground rightward, and rightward is the negative direction of x-axis. The acceleration in y-axis is about −3 g, which is negative because of falling down. The acceleration in z-axis is about 4 g because the user tries to touch the ground with his/her right-hand elbow, and the body forwards a little. That is why during the maximum impact, Roll goes down.

After the maximum impact, there are several moderate peaks (in Fig. 10.13a, they are circled). They are generated due to the body hitting the ground. The acceleration in x-axis is about −2 g. It is negative because the body hits the ground rightward. The acceleration in y-axis is about 0.5 g, which is positive because the user's body

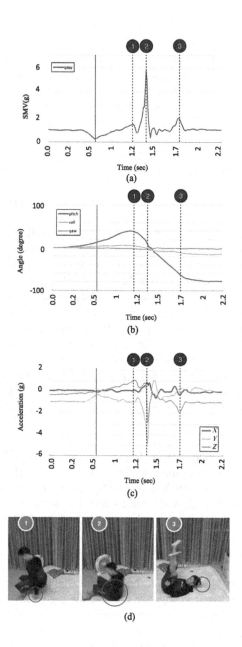

FIGURE 10.12

The SMVs, accelerations, and gyroscope angles on a falling backward movement. (a) SMVs
of falling backward; (b) Gyroscope angles including Pitch, Roll, and Yaw of falling backward
(at time = 1.2, the upper line is pitch, the middle is roll, and the lower is yaw);
(c) Accelerations of falling backward (at time = 1.2, the upper line is z-axis, the middle is
x-axis, and the lower is y-axis); (d) The pictures recorded by video recorder.

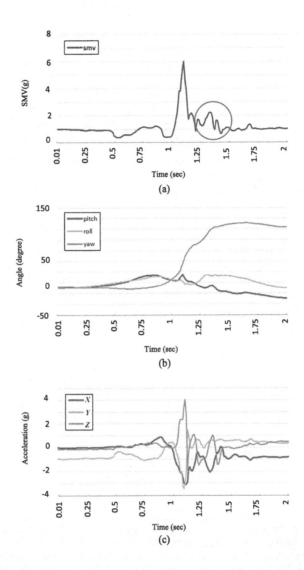

FIGURE 10.13

The SMVs, accelerations, and gyroscope angles on a falling right movement. (a) SMVs of falling right; (b) Gyroscope angles including Pitch, Roll, and Yaw of falling right (at time = 1.5, the upper line is yaw, the middle is roll, and the lower is pitch); (c) Accelerations of falling right (at time = 1.5, the upper line is y-axis, the middle is z-axis, and the lower is x-axis).

is still moving toward his/her head direction (the positive direction of y-axis). The acceleration in z-axis is about 1g because the user's right-hand elbow is still trying to relief the impact of his/her body with the ground. But at this moment, the acceleration is smaller.

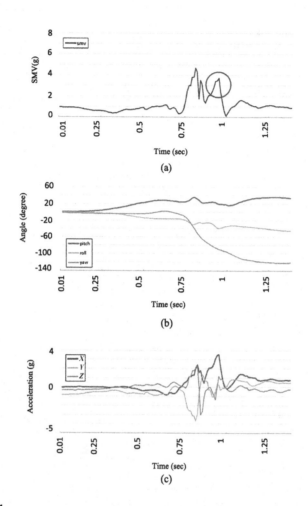

FIGURE 10.14

The SMVs, accelerations, and gyroscope angles on a falling left movement. (a) SMVs of falling left; (b) Gyroscope angles including Pitch, Roll, and Yaw of falling left (at time = 1, the upper line is pitch, the middle is roll, and the lower is yaw); (c) Accelerations of falling right (at time = 0.5, the upper line is z-axis, the middle is x-axis, and the lower is y-axis).

(8) Falling Left

Fig. 10.14 illustrates the SMVs, Triaxial accelerations, and gyroscope angles of falling right. Before the maximum of the peak shown in Fig. 10.14a, Pitch as plotted in Fig. 10.14b is positive, which is about 25 degrees since the user curls up his/her body forward in order to slow the falling speed and reduce his/her body injury. The maximum of the peak illustrated in Fig. 10.14a is produced when the user's buttocks hit the ground. The SMV is about 4.8 g. Roll is about −20 degrees, meaning that the user turns the body a little rightward and uses his/her left-hand elbow to touch the ground in order to reduce falling injury. Yaw is about −40 degrees, which is negative since this is a falling leftward. As shown in Fig. 10.14c, the maximum of the peak is created by the acceleration in three axes. The acceleration in x-axis is about 2.5 g, which is positive because the body hits the ground leftward. The acceleration in y-axis is about −3.9 g, which is negative because of falling down. The acceleration in z-axis is about 2 g, which is positive because when the body touches the ground, the user turns his/her body a little forward. After the maximum impact, there is a circled peak shown in Fig. 10.14a, which is created when the body hits the ground. The acceleration in x-axis is about 3.5 g, which is positive because the user's body hits the ground leftward. The acceleration in y-axis is about 0.5 g, which is positive because the user's body continues going toward his/her head direction (the positive direction of y-axis). The acceleration in z-axis is about 1 g. It is because the body forwards a little. The acceleration of gravity shifts a portion to positive direction of z-axis.

(9) Body Turning Right while Falling Backward

Fig. 10.15 plots the SMVs, Triaxial accelerations, and gyroscope angles for turning right while falling backward. Before the maximum of the peak shown in Fig. 10.15b, Pitch is positive, which is about 10 degrees since the user curls up his/her body forward in order to slow the falling speed and reduce body injury. On the maximum of the peak, Roll is about −60 degrees. It is the time when his/her buttocks hit the ground. As indicated in Fig. 10.15a, the SMV is about 7 g, which as shown in Fig. 10.15c, is formed from the accelerations in the three axes. The accelerations in x-axis is about −4 g, which is negative because the body turns rightward before it hits the ground. But the body continues falling down a little backward. That is why there is a negative acceleration in z-axis, which is −5.5 g. The acceleration in y-axis is about −5.2 g. It is negative because of falling down. Before touching the ground, since the body turns rightward, Yaw is finally turned to 80 degrees. The circled peak is produced when the body hits the ground. Because the user tries to touch the ground with his/her right-hand elbow to reduce falling injury. The acceleration in x-axis is about −2 g, which is negative because the body turns right before hitting the ground. The acceleration in y-axis is about 2 g, which is positive because when the right side of body hits the ground, the user's body continues going toward his/her head direction. The acceleration in z-axis is about −2 g. Because when touching the ground, the body is a little backward.

(10) Body Turning Left while Falling Backward

Fig. 10.16 plots the SMVs, Triaxial accelerations, and gyroscope angles of turning right while falling backward. Before the maximum of the peak shown in Fig. 10.16b, Pitch is stable and almost 0 degree. Roll is about 40 degrees, which is positive since the user feels unsafe, and tries to turn left a little. The degree of the left turn is about 40. The peak SMC is produced when the buttocks hit the ground. As shown in Fig. 10.16a, the SMV is about 7.9 g. From Fig. 10.16c, we can see the maximum of the peak is the combination of the three-axes accelerations. The acceleration in x-axis is about 4.8 g, which is positive because the body turns to left and the buttocks

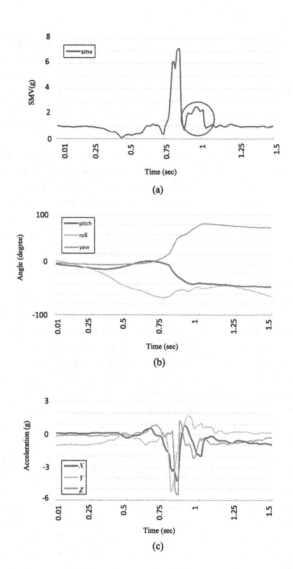

FIGURE 10.15

The SMVs, accelerations, and gyroscope angles on a body turning right while falling
backward movement. (a) SMVs of the body turning right while falling backward;
(b) Gyroscope angles including Pitch, Roll, and Yaw on the body turning right while falling
backward (at time = 1, the upper line is yaw, the middle is pitch, and the lower is roll);
(c) Acceleration of the body turning right while falling backward (at time = 1.5, the upper
line is y-axis, the middle is z-axis, and the lower is x-axis).

hit the ground. The acceleration in y-axis is about -4.8 g, which is negative because
of falling down. The acceleration in z-axis is about -4.5 g also due to falling backward. After the maximum impact, the body continues turning to left. At last, Yaw has

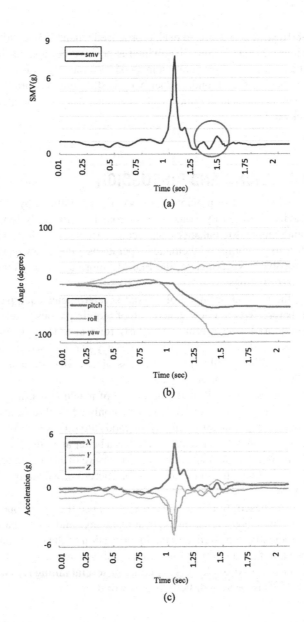

FIGURE 10.16

The SMVs, accelerations, and gyroscope angles on a body turn left while falling backward movement. (a) SMVs of the body turning left while falling backward; (b) Gyroscope angles including Pitch, Roll, and Yaw of the body turning left while falling backward (at time = 1.25, the upper line is roll, the middle is pitch, and the lower is yaw); (c) Accelerations of the body turning left while falling backward (at about time = 1.1, the upper line is x-axis, the middle is y-axis, and the lower is z-axis).

turned about-80 degrees. The circled peak is produced when the body hits the ground. The acceleration in x-axis is about 1 g, which is positive because of the body turns left and hits the ground. The acceleration in y-axis is about 1 g. It is positive because when the body touches the ground, his/her body is still going toward the positive direction of y-axis. The acceleration in z-axis is about -1 g. It is negative because of the falling backward.

10.6 CONCLUSIONS AND DISCUSSION

As mentioned above, most of previous studies classified fallings by checking accelerations of gravity collected by Triaxial accelerometers. But only few of them use machine learning methods to enhance classification accuracy, while others employed a class of matrix-like complex mathematical operations. Only a few types of fall classification are involved. Nevertheless, the FDSMP more accurately classifies ADLs and the six types of fallings.

Nowadays, the sensing accuracy of triaxial accelerometer equipped in a phone is higher than before. When a new generation of mobile phones is released, the old data may often be no longer useful, particularly for real time applications. So data need to be re-collected and re-analyzed, before they can be used to judge whether it is an ADL or a falling. Although the process of re-collection and re-analysis is complicated, the accuracy is improved a lot.

Also many studies have indicated that the use of machine learning will improve fall detection accuracies compared to those when only SMV thresholds are used. In this study, a total of nine machine learning methods are compared. We also analyze the data currently on hand. But a decision tree has a better analytical result than other machine learning methods do. Especially the F-measure of J48 is the best among the night methods compared (shown in Table 10.7). Therefore, we adopt J48 to analyze the collected data.

From the above description, we found that if we use the gyroscope and Triaxial accelerometer simultaneously, the distinguish of fallings and ADLs can be more effectively and efficiently improved, and the identification of the direction of a fall can also be enhanced. In fact, a falling movement in reality is very complex, and is not so idealistic like our study. Normally it is a mixture of several falling types, thus causing classification of falls to be very difficult in real world.

10.7 FUTURE STUDIES

Currently, our preliminary data show the superiority of our method. In the future, we will apply our experimental analysis to some complex information and further enhance our classification method. Some studies mentioned that if we can send a multimedia messaging service and GPS coordinates telling caregivers to carry out the necessary relief work when an elderly fall, it will be more helpful to give the elder required help as soon as possible. We also like to explore other relevant techniques to speed up fall detection.

Now a newly released smart phone can sense more objects and the accuracies are often higher than previous ones. Besides, many wearable devices are now more powerful than before. They can connect and transmit signals to mobile phones. In the future, we wish that a smart phone can also detect accidents, besides the elderly care, to make it really be applied to medical applications and cares.

REFERENCES

Aguiar, B., Rocha, T., Silva, J., Sousa, I., 2014. Accelerometer-based fall detection for smartphones. In: IEEE International Symposium on Medical Measurements and Applications, 2014, pp. 480–485.

Albert, M.V., Kording, K., Herrmann, M., Jayaraman, A., 2012. Fall classification by machine learning using mobile phones. PLoS ONE. http://dx.doi.org/10.1371/journal.pone.0036556.

Bai, Y.W., Wu, S.C., Tsai, C.L., 2012. Design and implementation of a fall monitor system by using a 3-axis accelerometer in a smart phone. In: IEEE Transactions on Consumer Electronics, pp. 1269–1275.

Beauvais, B.S., 2012. MyVigi: an Android application to detect fall and wandering. In: The Sixth International Conference on Mobile Ubiquitous Computing, Systems, Services and Technologies, pp. 156–160.

Bosch, 2011. BMA220 Digital, Triaxial Acceleration Sensor. Data Sheet.

Bosch, 2014. BMA280 Digital, Triaxial Acceleration Sensor. Data Sheet.

Bourke, A.K., O'Brien, J.V., Lyons, G.M., 2007. Evaluation of a threshold-based tri-axial accelerometer fall detection algorithm. In: Gait Posture, pp. 194–199.

Breiman, L., 1996. Bagging predictors. In: Machine Learning. University of California, Berkeley, pp. 123–140.

Centers for Disease Control and Prevention (CDC), 2015. Important Facts About Falls.

Chang, Y.S., 2016. http://www.eric-web.shu.edu.tw/100_2D_animation.html (in Chinese, accessed on September 1, 2016).

Dai, J., Bai, X., Yang, Z., Shen, Z., Xuan, D., 2010. PerFallD: a pervasive fall detection system using mobile phones. In: IEEE International Conference on Pervasive Computing and Communications, pp. 292–297.

Delahoz, Y.S., Labrador, M.A., 2014. Survey on fall detection and fall prevention using wearable and external sensors. Sensors 14 (10), 19806–19842.

Doukas, C., Maglogiannis, I., 2008. Advanced patient or elder fall detection based on movement and sound data, pervasive computing technologies for healthcare. In: The Second International Conference Pervasive Health, pp. 103–107.

El-Bendary, N., Tan, Q., Pivot, F.C., Lam, A., 2013. Fall detection and prevention for the elderly: a review of trends and challenges. International Journal of Smart Sensing and Intelligent Systems 6 (3), 1230–1266.

The Falls Management Program, 2012. A Quality Improvement Initiative for Nursing Facilities. Agency for Healthcare Research and Quality, December 2012.

Fang, S.H., Liang, Y.C., Chiu, K.M., 2012. Developing a mobile phone-based fall detection system on Android platform. In: Computing, Communications and Applications Conference (ComComAp), pp. 11–13.

Frank, E., Witten, H., 1998. Generating accurate rule sets without global optimization. In: Proceedings of the Fifteenth International Conference on Machine Learning, pp. 144–151.

Hall, M., Frank, E., 2008. Combining naive Bayes and decision tables. In: Proceedings of Twenty-First International Florida Artificial Intelligence Research Society Conference. AAAI Press, Coconut Grove, Florida, USA, pp. 15–17.

He, Y., Li, Y., Yin, C., 2012. Falling-incident detection and alarm by smartphone with Multimedia Messaging Service (MMS). E-Health Telecommunication Systems and Networks, 1–5.

Hssina, B., Merbouha, A., Ezzikouri, H., Erritali, M., 2014. A comparative study of decision tree ID3 and C4.5. International Journal of Advanced Computer Science and Applications, 13–19. Special Issue on Advances in Vehicular Ad Hoc Networking and Applications, 1, pp. 1–5. http://thesai.org/Downloads/SpecialIssueNo10/Paper_3-A_comparative_study_of_decision_tree_ID3_and_C4.5.pdf.

Hwang, S.Y., Ryu, M.H., Yang, Y.S., Lee, N.B., 2012. Fall detection with three-axis accelerometer and magnetometer in a smartphone. In: Computer Science and Technology, pp. 65–70.

InvenSense, 2013. MPU-6500 Register Map and Descriptions Revision 2.1. https://www.invensense.com/wp-content/uploads/2015/02/MPU-6500-Register-Map2.pdf.

Kaur, G., Chhabra, A., 2014. Improved J48 classification algorithm for the prediction of diabetes. International Journal of Computer Applications 98 (22), 0975.

Kendall, M.G., Stuart, A., Ord, J.K., 1994. Kendall's Advanced Theory of Statistics. Wiley–Blackwell.

Ko, C.Y., Leu, F.Y., Lin, I.T., 2015. Using a smartphone as a track and fall detector: an intelligent support system for people with dementia. In: Advanced Technological Solutions for e-Health and Dementia Patient Monitoring, pp. 272–295.

Lgual, R., Meduano, C., Plaza, I., 2013. Challenges, issues and trends in fall detection systems. Biomedical Engineering 12 (66). Online Journal.

Li, Q., Stankovic, J.A., Hanson, M.A., Barth, A.T., Lach, J., Zhou, G., 2009. Accurate, fast fall detection using gyroscopes and accelerometer-derived posture information. In: 2009 Sixth International Workshop on Wearable and Implantable Body Sensor Networks. Berkeley, CA, pp. 138–143.

Noury, N., Fleury, A., Rumeau, P., Bourke, A.K., ÓLaighin, G., Rialle, V., Lundy, J.E., 2007. Fall detection-principles and methods. In: Annual International Conference of the IEEE Engineering in Medicine and Biology Society, pp. 1663–1666.

Oxford Dictionaries, 2015. Gyroscope. http://www.oxfordreference.com/view/10.1093/acref/97801953-92883.001.0001/m_en_us1252970.

Penrose, R., 2007. Chap. 17.4 the principle of equivalence. In: The Road to Reality: A Complete Guide to the Laws of the Universe. Alfred A. Knopf, New York, pp. 393–394.

Quinlan, J.R., 1987. Simplifying decision trees. International Journal of Man-Machine Studies 27 (3), 221–234.

Quinlan, J.R., 1993. C4.5, Programs for Machine Learning. Morgan Kaufmann Publishers.

Rakhman, A.Z., Nugroho, L.E., Widyawan, Kurnianingsih, 2014. Fall detection system using accelerometer and gyroscope based on smartphone. In: 2014 1st International Conference on Information Technology, Computer and Electrical Engineering (ICIT ACEE), pp. 99–104.

Van Rijsbergen, C.J., 1979. Information Retrieval, 2nd ed. Butterworths, London.

Smith, T.C., Frank, E., 2016. Introducing machine learning concepts with WEKA. In: Statistical Genomics: Methods and Protocols. Springer, New York, NY, pp. 353–378.

Sposaro, F., Tyson, G., 2009. IFall: an Android application for fall monitoring and response. In: 2009 Annual International Conference of the IEEE Engineering in Medicine and Biology Society, pp. 6119–6122.

STMicroelectronics, 2010. MEMS Digital Output Motion Sensor Ultra Low-Power High Performance 3-Axes Nano Accelerometer. http://www.st.com/en/mems-and-sensors/lis3dh.html.

United Nations, 2015. World Population Ageing 2015. Report.

Waldner, J.B., 2008. Nanocomputers and Swarm Intelligence. ISTE, John Wiley & Sons, London.

Wu, X., Kumar, V., Quinlan, J.R., Fhosh, J., Yang, Q., Motoda, H., McLachlan, G.J., Ng, A., Yu, P.S., Zhou, Z.H., Steinbach, M., Hand, D.J., Steinberg, D., 2008. Umd.edu – top 10 algorithms in data mining. Knowledge and Information Systems 14, 1–37. http://www.cs.uvm.edu/~icdm/algorithms/10Algorithms-08.pdf.

Zhang, T., Wang, J., Liu, P., Hou, J., 2006. Fall detection by embedding an accelerometer in cellphone and using KFD algorithm. International Journal of Computer Science and Network Security, 277–284.

Zhao, Z., Chen, Y., Wang, S., Chen, Z., 2012. FallAlarm: smart phone based fall detecting and positioning system. Procedia Computer Science 10, 617–624.

ACRONYMS AND GLOSSARY

List of acronyms with explanation

ADL	Activity of Daily living
FDSMP	Fall Detection System using Mobile Phone
FN	False negatives
FP	False positives
F-ADL-class classification	Distinguishing the fall and the ADL motion
GPS	Global Positioning System
M-class classification	Classification of a fall motion and an ADL motion
MEMS	microelectromechanical systems
SMV	Signal Magnitude Vector
TN	Truth negatives
TP	Truth positives

Glossary of terms with explanation

Ceiling effect The ceiling effect is the level at which an independent variable no longer has an effect on a dependent variable, or the level above which variance in an independent variable is no longer measured or estimated.

Confusion matrix In the field of machine learning and specifically the problem of statistical classification, a confusion matrix, also known as an error matrix, is a specific table layout that allows visualization of the performance of an algorithm, typically a supervised learning one (in unsupervised learning it is usually called a matching matrix).

Decision Tree A decision tree is a decision support tool that uses a tree-like graph or model of decisions and their possible consequences, including chance event outcomes, resource costs, and utility. It is one way to display an algorithm.

Gyroscope A gyroscope is a device for measuring or maintaining orientation, based on the principles of angular momentum. It can be used to measure the quantity of rotation of a body, which is the product of its moment of inertia and its angular velocity.

Mobile Phone A mobile phone (or called cell phone or cellular phone) is a smartphone (or smart phone) with more advanced computing capability, features, and connectivity than basic feature phones.

Pitch One of the angular degrees of freedom, which along with roll and yaw is an aspect of aircraft flight dynamics.

Precision Common statistical usage defines precision as the reciprocal of the variance, and the precision matrix as the matrix inverse of the covariance matrix. Some particular statistical models define the term precision differently.

Recall (also called the true positive rate, the recall, or probability of detection in some fields) used to measure the proportion of positives that are correctly identified, e.g., the percentage of sick people who are correctly identified as having the condition.

Roll One of the angular degrees of freedom, which along with pitch and yaw is an aspect of aircraft flight dynamics.

Triaxial Accelerometer Accelerometer is a meter that converts acceleration into an electrical signal. A tri-axial accelerometer is a device used to sense the accelerations on X, Y, and Z axes.

Weka Waikato Environment for Knowledge Analysis (Weka) is a popular suite of machine learning software written in Java, developed at the University of Waikato, New Zealand. It is free software licensed under the GNU General Public License.

Yaw One of the angular degrees of freedom, which along with pitch and roll is an aspect of aircraft flight dynamics.

APPROACHING HARDWARE SOLUTIONS FOR MASSIVE E-HEALTH SENSOR DATA ANALYSIS

11

Mario Barbareschi[*,†], **Sara Romano**[*,†], **Antonino Mazzeo**[*,†]

DIETI – Department of Electrical Engineering and Information Technologies, University of Naples Federico II, Italy †CeRICT scrl – Centro Regionale Information Communication Technology, Italy

11.1 INTRODUCTION

In recent analysis conducted by the European Commission on economic and budgetary projections for the European Union (EU) Member States (2010–2060) (eHealth Action Plan 2012–2020 – http://ec.europa.eu/health/ehealth/docs/com_2012_736_en.pdf) it is estimated that public health expenditure is growing from 7.2% of GDP in 2010 up to 8.5% of GDP in 2060 due to the aging population and other socio-economic and cultural factors. In fact, the demographic structure in EU is radically changing due to low birth rates and increasing life expectations. In particular, the number of the elderly (65 $\geq$) and very old (80 $\geq$) EU population is expected to grow respectively from 17.4% of the total EU population in 2010 up to 30.0% in 2060 and from 4.7% in 2010 to 12.1% in 2060. At the same time, the EU population within the working age (15–64 years old) is expected to dramatically decrease from 61% to 51% of the total. These demographic changes impact on the public budgets, a decreasing number of health personnel, higher incidence of chronic diseases, and growing demands and expectations from citizens for higher quality services and social care.

The main outcome underscored by these motivations suggests that Information and Communication Technologies (ICT) applied to health and healthcare systems can increase their efficiency, improve quality of life, and unlock innovation in health markets. The e-Health (Electronic Health) is going to change the way how patients and health care providers interact. E-health has led to a growth of the health organization: multiple actors in the health care sector, with different interests, must be brought together to work towards a common goal in which the central position of patients within the care process is essential. Thus, the challenge of e-Health is to contribute to good healthcare by providing value-added services to the health care actors (patients, doctors, etc...) and, at the same time, by enhancing the effi-

Smart Sensors Networks. DOI: 10.1016/B978-0-12-809859-2.00014-0

ciency and reducing the costs of complex informative systems through the use of information and communication technologies. E-Health is the area with high growth potential and possibilities for innovation. Efficient knowledge management, organization and sharing has become a critical success factor. In this scenario, several research efforts are devoted to provide innovative and not-intrusive health monitoring systems to supervise in real-time the state and behavior of patients through the adoption of non-invasive technologies, based on wearable sensors, applications on smart tablets and sensors integrated in smart home environments (Tartarisco et al., 2012; Abo-Zahhad et al., 2014; Banos et al., 2014). Health Monitoring Systems (HMSs) provide alternatives to the traditional management of patients reducing hospitalization and the cost of formal health care, and allowing disease prevention and related lifestyle changes. In literature there has been a surge in development of so called *remote* health monitoring systems. Remote HMSs are typically combined with mobile communication systems and wearable monitoring technology and they are adopted in case distance separates healthcare professionals and patients. Those systems rely on information and communication technology in order to dispatch patients vital data provided by sensors. The remote HMS can be adopted at home as well as in hospitals. The general architecture of remote HMSs is depicted in Fig. 11.1.

Remote HMSs can be used to monitor several vital parameters within a variety of ranges. In fact, those monitoring systems rely on heterogeneous data acquisition from sensors, video, historical and simulated data, performing inferences, and data elaboration in order to provide alternatives to the traditional management of patients, e.g. allowing them to manage their health conditions at home. Depending on the functionalities to implement, the amount of data that has to be elaborated could represent the bottleneck of a monitoring system and it is critical in real-time applications. Thus in real-time applications the data increasing implies that the computational power should increase in order to reduce latencies providing actionable intelligence at the right time. To achieve an increment on computational power, hardware solutions should be adopted. There are two main classes of approaches: 1) using general purpose CPUs as multi-core processors and/or computer clusters to run the data mining software; 2) using hardware special purpose units to compute specific parts of an algorithm, reducing the computational effort. Indeed special purpose machines may not be suitable as they are not programmable and need tuning and reprogramming to achieve high accuracy. In this chapter we focus on the second solution focusing on data elaboration aspect with the respect to a massive monitoring infrastructure exploiting Field Programmable Gate Array (FPGA). Due to the low costs and the re-configuration properties, FPGAs are widely used to prototype functionalities in many industrial areas. FPGAs are integrated circuits containing programmable logic blocks and a hierarchy of reconfigurable interconnections specified by means of a hardware description language (HDL). An FPGA contains a matrix of configurable logic blocks and a hierarchy of reconfigurable interconnects that allow the blocks to be connected together. A configurable logic block contains look-up tables, multiplexers, and flip-flops which, together with interconnections, allow performance of complex combinatorial and sequential functions and implementation of

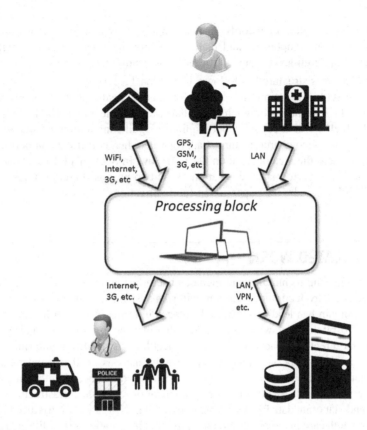

FIGURE 11.1

General architecture of a remote health monitoring system.

a wide variety of digital systems as well. Modern FPGAs also contain specialized memory, arithmetic, and communication blocks which enable more efficient implementations of digital systems. By allowing implementation of custom computational architectures, FPGAs provide opportunities for exploitation of deep parallelism in implementing algorithms.

In this chapter we present a layered architecture infrastructure for data analysis, based on two Decision Tree predictor hardware implementations. The first one is a high performance architecture, which guarantees high elaboration capacity, but requires a significant amount of area resource. The second one is a lightweight architecture which is appropriate when constraints on performance are not tight and the hardware area occupation has to be low. In particular, we propose a scalable architecture in order to make affordable to deal with a wide data volume and preserving a good hardware resources efficiency. Moreover, dealing with big data means that classification task should achieve a high accuracy rate since even a 2% of misclassification affects million of records. In this chapter we will show how we improved the classification components and feature selection in order to achieve high-level throughput and high-level accuracy for the classification task of big data.

The chapter is structured as follows. In Section 11.2 we describe related work to HMS architectures focusing on solutions for massive health related data processing. In Section 11.3 we propose two different hardware architectures. The first one guarantees high elaboration capacity, but requires a significant amount of area resource. Conversely, the second architecture is appropriate when constraints on performance are not tight and the area occupation has to be low. In Section 11.4 we show the results in terms of performance evaluation of the two proposed architectures. Finally, Section 11.5 concludes the paper giving future directions.

11.2 **RELATED WORK**

Health monitoring techniques and methods had a fast evolution during the past two decades. The introduction of Health monitoring systems and applications within the medical domain has lead to several advantages in terms of overall healthcare costs reducing hospitalization, waiting lists, consultation time. In order to reduce hospitalization that directly impacts on the overall health costs, health monitoring systems should be able to supervise patients performing analysis on their vital parameters and generate alerts for remote care giver (doctors, ambulance, family) in case any anomaly is detected. Health monitoring applications rely on infrastructures which collect and elaborate data from sensors. In particular, HMSs were introduced in order to assist health care professionals to take care of elderly patients (Cardile et al., 2010; Taleb et al., 2009; Shih et al., 2010) affected by chronic diseases (e.g. Parkinson, Alzheimer) (Lin et al., 2006; Taub et al., 2011; Chen et al., 2011; Cheng and Zhuang, 2010; Patel et al., 2009). During the past decade there has been a surge in development of so called *remote* health monitoring systems (Ren et al., 2010; Safavi et al., 2010; Pollonini et al., 2012; Yoshizawa et al., 2010). Remote HMSs are combined with mobile communication systems and wearable monitoring technology and they are adopted in case distance separates healthcare professionals and patients. Those systems rely on information and communication technology in order to dispatch patients vital data provided by sensors. A detailed overview of current state of the art in modeling and design of HMS is given in Baig and Gholamhosseini (2013).

Always more frequently there is a growing need in sharing health data between healthcare teams (doctors, nurses, family members) in order avoid unnecessary tests every time a health professional is consulted. Thus, the inclination in developing a HMS is to exploit, in combination with mobile technologies, the Cloud and its benefits in terms of on-demand access anywhere anytime, low costs and high elasticity (Thilakanathan et al., 2014; Kaur and Chana, 2014; Melillo et al., 2015; Fortino et al., 2012). Moreover, cloud computing provides a powerful storage and the massive data processing infrastructure to perform analysis and mining of the heterogeneous sensor data streams (Song et al., 2014; He et al., 2013; Moscato et al., 2014; Amato et al., 2012).

Depending on the functionalities to implement, the amount of data that has to be elaborated could represent the bottleneck of a monitoring system and it is critical in real-time applications. Thus in real-time applications, as the HMSs one, the data increasing implies that the computational power should increase in order to reduce latencies providing actionable intelligence at the right time. To achieve an increment on computational power hardware solutions should be adopted. Recently, many hardware implementations of data elaboration algorithm have been proposed in the literature. As monitoring systems exploit machine learning algorithms to accomplish data manipulation and elaboration, we pose our focus on prediction algorithm implemented in hardware. Authors of Li et al. (2013) proposed a packet classification on FPGA using the BTPCF algorithm. The architecture uses a *tree node memory* to store the binary tree structure and a *tree searching controller* to execute the binary tree searching. In Monemi et al. (2013) a traffic packets classifier is proposed, which exploits C4.5 machine learning algorithm. The main architectural characteristic is the programmability by the software, without loss of service, using memories that store the data model. Conversely, in Barbareschi et al. (2015c) authors adopted the same approach, but the architecture is updated by means of *dynamic partial reconfiguration* feature which is provided by the FPGA. In Lim et al. (2010) the authors show the power of C4.5 for classifying the Internet application traffic, due to the discretization of input features done by the algorithm during classification operations. Despite of previous implementation, the one proposed in Barbareschi et al. (2015b) exploits *multi-classification* approaches to enhance the performance of the prediction.

11.3 ARCHITECTURAL OVERVIEW

Mobile health monitoring applications rely on infrastructures which collect and elaborate data from sensors. In Fig. 11.2 a generic hierarchical tiered scheme for a monitoring infrastructure is depicted. Data are inherently in a hierarchical dependence, since they are generated by a massive monitoring activity, aggregate and elaborated by higher levels. Starting from bottom, the infrastructure needs for a crowd of sensors which collect and gather samples of physical monitored phenomena. They are managed by sensor nodes, which configure them and accomplish a first aggregation and elaboration process. Data from different nodes need for integration, since they could be in different formats and needs for further processing before any elaboration. The highest layer is responsible for elaborating information and extract knowledge from them, accomplishing the application logic.

Inherently, a monitoring application has to deal with 3 main design issues:

1. **Performance requirement**: data elaboration is a task which characterizes the whole infrastructure, but with different constraints. In particular, at bottom lay-

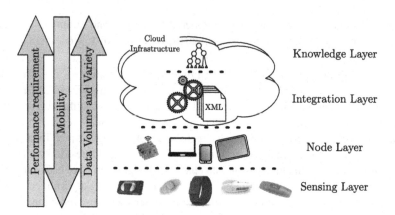

FIGURE 11.2

A generic tiered monitoring infrastructure perspective.

ers the complexity of algorithms and data amount are lower than the top, due to hierarchical aggregation of sensed samples. Indeed, data to elaborate grow each time information are aggregate and, hence, they turn out to be more difficult to elaborate;

2. **Mobility**: in m-health application, sensing equipments that monitor patients are designed to be mobile, such that patients can be monitored by devices everywhere and in every daily activity. The mobility implies an adequate design of sensing nodes: the equipment has to be non-invasive and small, and involved electronic components and software tasks have to be a low-power profile as devices are battery-powered;

3. **Data Volume and Variety**: the massive monitoring infrastructure is one of the most discussed application in Big Data literature. As regards the five 'V's, the illustrated tiered architecture is characterized by a growing data volume and variety, caused directly by aggregation of data of several types and from thousand devices.

In this chapter, our goal is to underline the data elaboration aspect by means of hardware implementation, with the respect to a massive mobile monitoring infrastructure. Thus, in this section we propose two hardware architecture implementation for data analysis suitable within the infrastructure of HMS. The first one is a high performance architecture, which guarantees high elaboration capacity, but requires a significant amount of area resource. The second one is a lightweight architecture which is appropriate when constraints on performance are not tight and the hardware area occupation has to be low. In a decision support system as a mobile HMS, data elaboration is generally accomplished by machine learning approaches (Tan, 2005; Milovic, 2012). Data modeling, data classification, and data processing can be successfully executed with the adoption of a machine learning framework in many aspects of the e-health domain: diagnosis support system, data fusion and integration, knowledge extraction. With respect to the data classification, the Decision Tree (DT) is a model

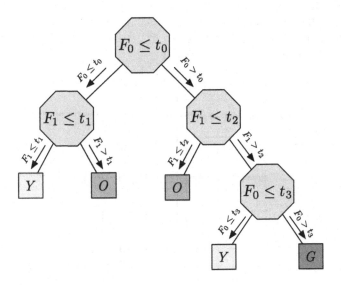

FIGURE 11.3

An example of Decision Tree with 4 internal nodes, 2 features, and 3 classes.

which represents knowledge with a tree structure. The tree scheme is clear and can be easy understand, compared to other techniques, such as neural networks. Moreover, training and prediction algorithms are significantly faster than other machine learning algorithms (Quinlan, 1993). In particular, the prediction algorithm is a tree visiting procedure: starting from the tree root, the algorithm evaluates an expression over a feature. The outcome of such operation establishes which child has to be evaluated at the next step, until a leaf is reached. For the sake of ease, in this work we consider only boolean DT, in which a node has only two children, considering that any tree can be arranged to appear in boolean form. Fig. 11.3 illustrates an example of DT: the tree is composed of 4 internal nodes, characterized by an expression over features values (F_0 and F_1), and leaves, which are labeled with the classification result. The tree classifies input data in 3 different classes, labeled as *orange*, *yellow*, and *green*.

In the next subsections, we show a high performance implementation, which intensively parallelize the algorithm using a spatial speculation, and a custom processor, which execute sequentially the tree visiting but, exploiting the proximity of samples in a temporal sequences, adopting a temporal speculation. The first architecture is suitable for high layers of the monitoring infrastructure, as it guarantees high elaboration capacity, but requires a significant amount of area resource. Conversely, the second architecture is appropriate when constraints on performance are not tight and the area occupation has to be low. We have implemented the hardware solutions for the classification task exploiting Field Programmable Gate Array (FPGA). Due to the low costs and the re-configuration properties, FPGAs are widely used to prototype functionalities in many industrial areas.

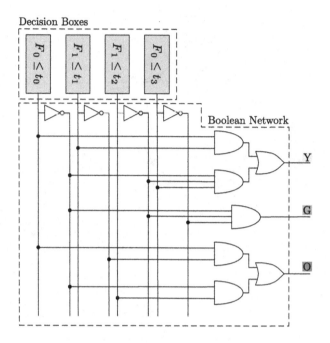

FIGURE 11.4

The high performance DT prediction implemented to execute the scheme reported in Fig. 11.3.

11.3.1 HIGH PERFORMANCE DECISION TREE PREDICTION

The sequentiality of the decision tree visiting algorithm stands in nested dependency of the cycle from the previous ones, hence to go further the algorithm needs for completion of the current evaluation. To deal with such an inherently dependence, authors in Amato et al. (2014) proposed a spatial speculation in which the tree structure is flatten on a single level and all the nodes conditions are evaluated at the same time. Once all nodes outcomes are available, a boolean net decides which leaf has been reached by the provided input.

In Fig. 11.4 is depicted the hardware implementation scheme of the prediction algorithm over the DT in Fig. 11.3. As one can notice, the speculation consist of node expressions evaluation which do not belong to the path leading to the reached leaf. In oder words, the speculative implementation evaluates conditions that the visiting algorithm does not consider.

The obtained architecture turns out to be static and cannot be programmable, being realized to execute the visiting on a specific tree model. Experiments conducted in Amato et al. (2014, 2013), Barbareschi et al. (2015a) demonstrated a notable speedup compared with software implementation, but also a significant area overhead that has to be taken into account for big tree models.

11.3.2 IMPLEMENTING A FULLY PARALLEL DT PREDICTOR

The main performance related issue of the tree visiting algorithm is the inherent sequentiality caused by the selection instruction which decides which child node has to be picked for the next iteration. Hence the procedure cannot be executed in a parallel fashion. To reach high throughput in the DT visiting avoiding sequential computations, a speculative approach can be considered.

Sequential implementations of the DT prediction algorithm may take advantages from the adoption of well-known computer architectures speculation techniques, such as the branch prediction (Barbareschi et al., 2014). Even though they are effective for real applications, sequential approaches do not exploit the high parallelism offered by the hardware.

Algorithm 1 treeVisiting(features[], DT).

Require: A feature vector **features[]**; a classification **model** DT
Ensure: The predicted class

```
1: node ← model.root
2: while !node.isLeaf() do
3:    if node.ρ(features[node.f], node.k) then
4:        node ← node.leftChild
5:    else
6:        node ← node.rightChild
7:    end if
8: end while
9: return node.label
```

In Amato et al. (2014) a new fully-parallel speculative implementation was introduced. In particular, the approach proposes to evaluate which leaf is reached by computing in parallel all the nodes conditions. Each condition is processed by fast comparators, named Decision Boxes, which work in parallel. Then, a boolean net decides, evaluating the outcomes of the Decision Boxes, which leaf, hence which class, the algorithm returns. In the evaluation performed by the hardware accelerator, the speculation stands in computing some conditions of the tree that are useless for the algorithm result, such that Algorithm 1 would never evaluate them.

Formally, let $D_\rho^k(f)$ be a node of DT which compares the feature f with the constant k by the relational operator ρ, i.e. $D_\rho^k(f) = \rho(f, k)$, $\rho \in \{<, \leq, >, \geq, =, \neq\}$. Each Decision Box implements $D_\rho^k(f)$ and takes in input the actual value for the feature f, returning a boolean result.

The boolean net is composed by different boolean functions as many as the number of classes. Formally, let BF_c be the boolean function that rises logic-1 when the classification result is the class c, logic-0 otherwise. Therefore, each BF_c has to rise logic-1 when at least one of the paths that lead to a leaf labeled as c is asserted. So each BF_c can be realized through an *or*-gate with a number of inputs (namely the fan-in) which is equal to the leaves labeled with the class c. Hence, to define a BF_c for the given class c it is necessary to find the conditions which establish if one of the paths that lead to a leaf labeled as c has been reached. To this aim, let $P_c(i)$, $i \in \{0, 1, \ldots, leavesOf(c) - 1\}$ be the i-th path that leads to the i-th leaf which is labeled with the class c. Moreover, let the assertion $A_C(i)$, $i \in \{0, 1, \ldots, leavesOf(c) - 1\}$ be the logical intersection of all decision box outcomes which belong to $P_c(i)$. Each decision box outcome is a literal of the boolean function $A_C(i)$ and can be taken direct or negated, according to the path over the tree that leads to the considering leaf. In particular, the literal associated with a node is considered directed if the path continues to its left child, negated otherwise. Thus, each assertion can be defined as:

$$A_C(i) = \prod_{D_\rho^k(f) \in P_i} \begin{cases} D_\rho^k(f) & \text{when the child is left,} \\ \overline{D_\rho^k(f)} & \text{when the child is right.} \end{cases}$$

Finally, for each class C it is possible to define the boolean function as:

$$BF_C = \sum_{i=0}^{leavesOf(C)-1} A_C(i).$$

As one can notice, the boolean net is composed by boolean functions in form of *sum of product* (SoP). Such construction is functionally equivalent to Algorithm 1, as one and only one assertion $A_C(i)$, and hence only one BF_c, rises logic-1 for each input of the circuit.

Fig. 11.4 depicts the hardware accelerator unit for the example of DT prediction model in Fig. 11.3. As one can see, decision boxes receive the actual values for both the features and their outcomes are available directed or negated. The boolean net contains three functions for the three classes and they are expressed in SoP forms. As previously stated, the speculation of the approach stands the evaluation of tree nodes which are not involved to accomplish the tree visiting algorithm. For instance, in the reported example, the boolean function associated to the class C_1 never takes into account the outcome of the decision box which evaluates $f_2 \leq d$, since in the tree such class does not depend on that node.

11.3.3 CONSIDERATIONS ON AREA OCCUPANCY AND TIME

The previously described architecture requires a number of decision boxes equal to the number of internal nodes, hence the impact on the occupied area varies linearly on the number of the tree nodes. As for the boolean net, the required resources are related to the assertions' length (*and*-gates fan-in) and to the number of leaves labeled with the same class (*or*-gates fan-in).

First of all, it is worth to notice that some assertions retrieved from a DT may be redundant. In fact, considering the example given in Fig. 11.3, the paths to reach leaves labeled with the class C_1, are: $P_{C_1}(0) = \{D_\leq^a(f_1), D_\leq^b(f_1)\}$ and $P_{C_1}(1) = \{D_\leq^a(f_1), \overline{D_\leq^b(f_1)}, D_\leq^c(f_2)\}$. The $P_I(0)$ has the associated assertion:

$A_I(0) = D^a_{\leq}(f_1) \cdot D^b_{\leq}(f_1) = D^b_{\leq}(f_1)$. Indeed, if $f_1 \leq b$, surely also $f_1 \leq a$, hence the first comparison is redundant.

Thanks to this observation, the number of literals in each assertion can be reduced. Assertions which exploit this rule are defined as essential assertions. Furthermore, considering only essential assertions, it is possible to find an upper limit to the number of literals needed to evaluate them, that is an important result for the scalability of the architecture.

Theorem 1. *Essential assertion $A_c(i)$ contains at most $2 \cdot |F|$ literals, being $|F|$ the cardinality of the features set.*

Proof. Let us consider the feature set $\{F\}$, the path $P_c(i)$ and the associated essential assertion $A_c(i)$, which is built by adding every literal corresponding to $D^k_\rho(f) \in P_c(i)$. Aiming at evaluating how many literals appear in the assertion $A_c(i)$, we can distinguish three different cases.

Degeneracy Case:
$f \in \{F\}$ is not involved in any $D^k_\rho(f) \in P_c(i)$, this means that it has not be chosen by the learner as discriminating feature for the class c.

Case 1:
$f \in \{F\}$ is a continuous feature that is involved at least once in the path $P_c(i)$. Thus $\rho \in \{<, \leq, >, \geq\}$, $f \in \mathbb{R}$. $D^k_\rho(f)$ establishes an interval for f:

- right interval of k ($[k, +\infty[$ or $]k, +\infty[$), if $\rho \in \{>, \geq\}$;
- left interval of k ($]-\infty, k]$ or $]-\infty, k[$), if $\rho \in \{<, \leq\}$.

If in $P_C(i)$ the literal appears complemented, it is possible to simply choose the complemented interval.

If f appears twice or more, $D^k_\rho(f), D^j_\lambda(f) \in P_C(i) \Rightarrow D^k_\rho(f) \cap D^j_\lambda(f) \neq \emptyset$, because the learner generates the conditions by splitting operation over them, consequently $k \neq j$. In other words, the path must be consistent in order to reach the leaf. So the intersection $D^k_\rho(f) \cap D^\lambda_j(f)$ generates:
1. a right interval if both the nodes define a right interval: $[M, +\infty[$ or $]M, +\infty[$ with $M = max(k, j)$,
2. a left interval if both the nodes define a left interval: $]-\infty, m]$ or $]-\infty, m[$ with $m = min(k, j)$,
3. an interval $[m, M[$ or $]m, M[$ or $]m, M]$ or $[m, M]$, with $m = min(k, j)$ and $M = max(k, j)$.

This process can be iterated as many times as f appears in nodes along $P_C(i)$. In the first two cases $A_C(i)$ requires only one literal which involves f, in the third an intersection (*and*-gate) between two literals.

Case 2:
f is a nominal feature $\Rightarrow \rho \in \{==, !=\}$, f can assume finite values. A $D^k_\rho(f)$ appears at most once on any path in the tree. So $A_C(i)$ requires only one literal on f. □

Being DN the number of internal nodes of the DT, an implementation of a DT predictor would ideally require only DN *and*-gates and a number of *or*-gates which is maximum equal to $\frac{DN+1}{2}$. Considering the worst case for a BF_c, the *and*-gate could have fan-in equal to $2 \cdot |F|$, and an *or*-gate a fan-in equal to $|C| - 1$. The implementation technology is able to realize gates with a maximum fan-in and, with a significant values, a compositional technique is required. On the FPGA, the gates implementation area (and delay) grows with the size of the fan-in like a step function technology, because the FPGA is able to implement each function in a k-input LUT, consequently to realize gates with higher fan-in it combines more that one LUT in a tree scheme.

As for the area occupation, being N the total fan-in an integer multiple of k, and S the depth of the tree, m_i (the number of k-gates at i-th level) is: $m_0 = \lceil \frac{N}{k} \rceil$; $m_1 = \lceil \frac{m_0}{k} \rceil$; $m_i = \lceil \frac{m_{i-1}}{k} \rceil = \lceil \frac{m_{i-2}}{k^2} \rceil = \lceil \frac{N}{k^{i+1}} \rceil$. In the last stage there is only one k-gate: $m_{s-1} = 1 = \frac{N}{k^S}$. So the relation between the tree depth S, the original fan-in N and the fixed fan-in k is: $k^S = N$, so $S = \frac{lg(N)}{lg(k)}$, hence the number of k-gates is:

$$\#k\text{-}gates = \sum_{i=0}^{S-1} m_i \simeq N \cdot \sum_{i=0}^{S-1} \frac{1}{k^{i+1}} = N \cdot \left(\sum_{i=0}^{S} \frac{1}{k^i} - 1 \right)$$

$$= N \cdot \left(\frac{1 - \left(\frac{1}{k}\right)^{S+1}}{1 - \frac{1}{k}} - 1 \right) = \frac{N-1}{k-1}, \quad k > 1.$$

At the end, considering the worst case, the total number of gates N_g for all BF_c can be defined as:

$$N_g = \sum_{\forall c} \left(\frac{2 \cdot |F| - 1}{k - 1} \cdot leavesOf(c) \right)$$

$$+ \sum_{\forall c} \left(\frac{leavesOf(c) - 1}{k - 1} \right) = \frac{2 \cdot |F| \cdot (DN + 1) - |C|}{k - 1}$$

where the first summation term is related to number of k-gates to implement the *and*-gate with the worst fan-in ($N = 2 \cdot |F|$), while the second summation term is related to the or that has a fan-in equal to the number of leaves ($N = leavesOf(c)$). The number of classes $|C|$ varies between 2 (best case) and $DN+1$ (worst case). This result is an upper bound of the area occupancy when the DT model is effectively realized over a technology target. Indeed, by applying the resource sharing techniques, subparts of an assertion could be shared among assertions which are derived from the same subpath.

As for the time, the critical path in the boolean net is given by the highest *or*-gate fan-in and *and*-gate fan-in. The worst case for the delay is generated when the DT has all the leaves labeled with a class C except one leaf, which leads to have a *or*-gate

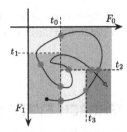

FIGURE 11.5

A 2D representation of the DT in Fig. 11.3 and a possible temporal sequence of values, in which branch mispredictions are highlighted.

with fan-in equals to DN, and one of the path of C has an assertion with $2 \cdot |F|$ literals. To tackle an excessive latency, the tree scheme of the gates can be successfully exploited to define a pipelined structure in which each gate output is registered in a dedicated flip-flop. Such technique requires registers as many as gates, but guarantees a throughput which is close to the delay of a single gate.

11.3.4 LOW AREA DECISION TREE PREDICTION

Contrary to the high performance architecture described in the previous subsection which is characterized by a significant area overhead, authors in Barbareschi et al. (2014) implemented the algorithm with an Application Specific Instruction set Processor (ASIP) approach. The processor-based architecture, called Tree Visiting Processing Unit (TVPU), is designed with the pipeline technique and execute only two instruction: one is related to the evaluation of node expressions, which cause a jump to the left or right child, the other returns the classification outcome. Since the branch strictly depends on the evaluation of condition reported in the node, it is not possible to feed the pipeline with a new instruction at each clock cycle. For this reason, the TVPU is equipped with a Branch Prediction Unit (BPU), which speculate on the next instruction to load by fetching the last branch decision. In Fig. 11.6 we reported the Register Transfer Level (RTL) of the TVPU pipeline implementation. The instructions have to be loaded into the *instruction memory*, while the actual features values have to be stored into the *feature memory* each time a prediction is required.

As for the BPU, its effectiveness is directly related to the proximity of elaborated value. Indeed, if for a condition the actual value of a feature is close to the value previously elaborated, the speculation will lead to a consistency state, saving many clock cycles. Otherwise, the processor has to flush all operations accomplished during the speculation, wasting clock cycles. To better understand, let us consider Fig. 11.5 in which we represented the 2D graph of the tree previously illustrated in Fig. 11.3: when a value assumed by the features crosses a threshold, it causes a branch misprediction. We can assume that, being physical phenomena, values smoothly vary among thresholds and, hence, the branch prediction unit can be effective in avoiding wasted clock cycles.

11.4 PRELIMINARY APPROACH EVALUATION

To demonstrate the feasibility of two architectures shown in Section 11.3 in executing the DT prediction algorithm, in this section we illustrate a real scenario based on the classification of the dataset mHealth (Banos et al., 2014). It contains records of ten volunteers about their body motion and each record is labeled with a specific physical activity. The motion was measured by accelerometers, gyroscopes, magnetometers, and electrocardiogram signals. We trained a model by employing the C4.5 algorithm integrated in the KNIME framework (Berthold et al., 2008) and it was characterized by about 2000 internal nodes and by an accuracy of 94.33%.

In order to implement both the architectures previously described on a real hardware target, we adopted a Xilinx Virtex-5 (XUPV5-LX110T).

11.4.1 HIGH PERFORMANCE ARCHITECTURE

The DT prediction unit, realized by the high performance architecture, is characterized by a throughput of about 246 Gbps, which means a classification of about 350 Msps. The unit requires, in terms of FPGA programmable resources, 17980 slices registers, and 37862 look-up tables (LUT).

We also estimated the energy consumption of the architecture by exploiting the power analysis tool integrated in Xilinx Vivado. In particular:

- We synthesized and implemented the core, verifying also the maximum clock frequency which does not violate the setup time of the registers. We also verified the presence of hold time violations.
- We defined a test bench for the core, aiming at obtain the Switching Activity Interchange format (SAIF) file, which contains the dynamic switching activity of the DT prediction unit.
- At the end, we estimated the power consumption through the employing of the power analysis tool by setting common environmental parameters (junction temperature, ambient temperature, airflow, board temperature, etc.), working frequency, and supply voltage.

The unit yield a power dissipation of about 0.745 mW.

11.4.2 LOW OVERHEAD ARCHITECTURE

In Table 11.1 we report the comparison of architectures in terms of required hardware resources (LUTs and registers), clock speed, and effective throughput. As one can

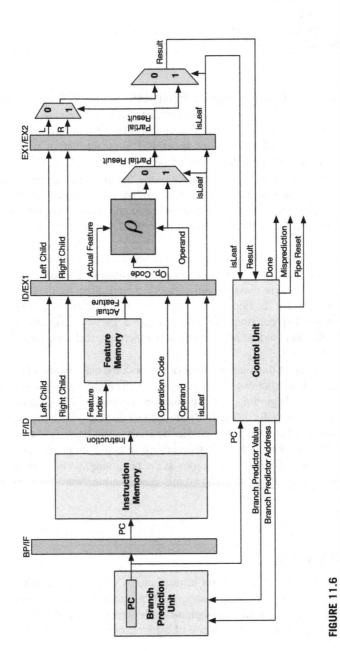

FIGURE 11.6

Detailed RTL scheme of the TVPU.

Table 11.1 Comparison Between High Performance and Low Overhead Designs of the DT Prediction Algorithm

	High Performance	Low Overhead
Slice Registers	17980	374
Slice LUTs	37862	468
Clock Speed (ns)	2.989	3.211
Throughput (Gbps)	246.24	6.34

notice, the As for the high performance architecture, we reached a throughput of about 246 Gbps, while the low overhead architecture performs the classification with an average throughput of 7.74 Gbps. Moreover, as for the branch-mispredictions, we appreciated a percentage of successes of about 82.1%, meant that on average every classification caused 3 or 4 branch-misprediction.

Results highlight the main characteristics of the two proposed architectures and their adoption in the monitoring infrastructure. As for the high performance hardware architecture, it is suitable to be adopted when requirements on performance are significantly high and when the data volume is critical, hence it can be adopted at high level of the monitoring application, within the cloud infrastructure. As for the low-overhead hardware architecture, it performs fewer classifications per time unit than the first architecture, but the occupied area is significantly lower. Therefore, it represents a good hardware classifier solution when constraints on data elaboration speed are not so tight, but the impact on computational resources has to be very low. Moreover, branch mis-predictions are able to cause a latency variation, hence the low-overhead solution could be not suitable when there are real-time constraints on data classification.

In Fig. 11.7 we report the temporal behavior of both the classification latency and the time between mis-predictions. As one can notice, the classification time turns out to be very high at the beginning of the prediction running, while at the end of the execution it is more than halved. This is due to the fact that the branch prediction unit is not yet trained such that the pipeline is able to execute speculative operations which will result in effective ones. Indeed, in terms of mis-classifications, the execution is characterized by burst of mis-classifications at the beginning, while at the end they become sporadic and the time between two mis-classification events goes up to 3500 clock cycles.

11.5 CONCLUSION

Health monitoring applications rely on infrastructures which collect and elaborate data from heterogeneous sources (sensors, medical records, clinical results, etc.). Depending on the functionalities to implement, the amount of data that has to be elaborated could represent the bottleneck of a monitoring system and it is critical

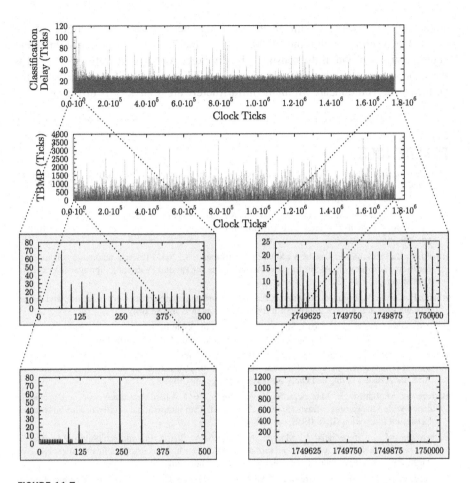

FIGURE 11.7

The temporal behavior of the classification delay and time between mis-predictions in terms of clock ticks.

in real-time applications. Thus, the data increasing implies that the computational power should increase in order to reduce latencies providing actionable intelligence at the right time. To achieve an increment on computational power, cloud computing in combination of hardware solutions should be adopted.

In this work we present a layered architecture infrastructure suitable for a generic mobile HMS and we propose two hardware architecture implementations for data analysis. The first one is a high performance architecture, which guarantees high elaboration capacity, but requires a significant amount of area resource. The second one is a lightweight architecture which is appropriate when constraints on performance are not tight and the hardware area occupation has to be low. The aim of the proposed architectures is twofold. On the one hand the high performance architecture could be adopted as a service within the Cloud. On the other hand, the low area architecture could be adopted within the sensing infrastructure. Although the algorithm used is the same, it can be used in different architectural contexts. This is possible since the performance requirements within the infrastructure are dramatically different.

REFERENCES

Abo-Zahhad, M., Ahmed, S.M., Elnahas, O., 2014. A wireless emergency telemedicine system for patients monitoring and diagnosis. International Journal of Telemedicine and Applications 2014, 4:4. http://dx.doi.org/10.1155/2014/380787.

Amato, F., Barbareschi, M., Casola, V., Mazzeo, A., 2014. An FPGA-based smart classifier for decision support systems. In: Intelligent Distributed Computing VII. Springer International Publishing, pp. 289–299.

Amato, F., Barbareschi, M., Casola, V., Mazzeo, A., Romano, S., 2013. Towards automatic generation of hardware classifiers. In: Algorithms and Architectures for Parallel Processing. Springer International Publishing, pp. 125–132.

Amato, F., Casola, V., Romano, S., Mazzeo, A., 2012. A semantic based framework to identify and protect e-health critical resources. Journal of Information Assurance and Security 7 (5).

Baig, M., Gholamhosseini, H., 2013. Smart health monitoring systems: an overview of design and modeling. Journal of Medical Systems 37 (2), 9898. http://dx.doi.org/10.1007/s10916-012-9898-z.

Banos, O., Garcia, R., Holgado-Terriza, J.A., Damas, M., Pomares, H., Rojas, I., Saez, A., Villalonga, C., 2014. mHealthDroid: a novel framework for agile development of mobile health applications. In: Ambient Assisted Living and Daily Activities. Springer, pp. 91–98.

Barbareschi, M., Battista, E., Mazzocca, N., Venkatesan, S., 2014. A hardware accelerator for data classification within the sensing infrastructure. In: 2014 IEEE 15th International Conference on Information Reuse and Integration (IRI). IEEE, pp. 400–405.

Barbareschi, M., De Benedictis, A., Mazzeo, A., Vespoli, A., 2015a. Providing mobile traffic analysis as-a-service: design of a service-based infrastructure to offer high-accuracy traffic classifiers based on hardware accelerators. Journal of Digital Information Management 13 (4), 257.

Barbareschi, M., Del Prete, S., Gargiulo, F., Mazzeo, A., Sansone, C., 2015b. Decision tree-based multiple classifier systems: an FPGA perspective. In: Multiple Classifier Systems. Springer International Publishing, pp. 194–205.

Barbareschi, M., Mazzeo, A., Vespoli, A., 2015c. Malicious traffic analysis on mobile devices: a hardware solution. International Journal of Big Data Intelligence 2 (2), 117–126.

Berthold, M.R., et al., 2008. Knime: the Konstanz Information Miner. In: Data Analysis, Machine Learning and Applications. Springer, pp. 319–326.

Cardile, F., Iannizzotto, G., Rosa, F.L., 2010. A vision-based system for elderly patients monitoring. In: 3rd International Conference on Human System Interaction, pp. 195–202.

Cheng, H.T., Zhuang, W., 2010. Bluetooth-enabled in-home patient monitoring system: early detection of Alzheimer's disease. IEEE Wireless Communications 17 (1), 74–79. http://dx.doi.org/10.1109/MWC.2010.5416353.

Chen, B. Rong, et al., 2011. A web-based system for home monitoring of patients with Parkinson's disease using wearable sensors. IEEE Transactions on Biomedical Engineering 58 (3), 831–836. http://dx.doi.org/10.1109/TBME.2010.2090044.

Fortino, G., Pathan, M., Di Fatta, G., 2012. Bodycloud: integration of cloud computing and body sensor networks. In: 2012 IEEE 4th International Conference on Cloud Computing Technology and Science (CloudCom), pp. 851–856.

He, C., Fan, X., Li, Y., 2013. Toward ubiquitous healthcare services with a novel efficient cloud platform. IEEE Transactions on Biomedical Engineering 60 (1), 230–234. http://dx.doi.org/10.1109/TBME.2012.2222404.

Kaur, P.D., Chana, I., 2014. Cloud based intelligent system for delivering health care as a service. Computer Methods and Programs in Biomedicine 113 (1), 346–359. http://dx.doi.org/10.1016/j.cmpb.2013.09.013. http://www.sciencedirect.com/science/article/pii/S0169260713003209.

Li, J., Chen, Y., Ho, C., Lu, Z., 2013. Binary-tree-based high speed packet classification system on FPGA. In: Proceedings of the International Conference on Information Networking. IEEE, pp. 517–522.

Lim, Y.-s., Kim, H.-c., Jeong, J., Kim, C.-k., Kwon, T.T., Choi, Y., 2010. Internet traffic classification demystified: on the sources of the discriminative power. In: Proceedings of the 6th International Conference on Emerging Networking EXperiments and Technologies. ACM, New York, NY, USA, pp. 9:1–9:12.

Lin, C.C., Chiu, M.J., Hsiao, C.C., Lee, R.G., Tsai, Y.S., 2006. Wireless health care service system for elderly with dementia. IEEE Transactions on Information Technology in Biomedicine 10 (4), 696–704. http://dx.doi.org/10.1109/TITB.2006.874196.

Melillo, P., Orrico, A., Scala, P., Crispino, F., Pecchia, L., 2015. Cloud-based smart health monitoring system for automatic cardiovascular and fall risk assessment in hypertensive patients. Journal of Medical Systems 39 (10), 109. http://dx.doi.org/10.1007/s10916-015-0294-3.

Milovic, B., 2012. Prediction and decision making in health care using data mining. International Journal of Public Health Science 1 (2), 69–78.

Monemi, A., Zarei, R., Marsono, M.N., 2013. Online NetFPGA decision tree statistical traffic classifier. Computer Communications 36 (12), 1329–1340.

Moscato, F., Amato, F., Amato, A., Aversa, R., 2014. Model-driven engineering of cloud components in MetaMORP(h)osy. International Journal of Grid and Utility Computing 5 (2), 107–122.

Patel, S., et al., 2009. Monitoring motor fluctuations in patients with Parkinson disease using wearable sensors. IEEE Transactions on Information Technology in Biomedicine 13 (6), 864–873. http://dx.doi.org/10.1109/TITB.2009.2033471.

Pollonini, L., Rajan, N.O., Xu, S., Madala, S., Dacso, C.C., 2012. A novel handheld device for use in remote patient monitoring of heart failure patients—design and preliminary validation on healthy subjects. Journal of Medical Systems 36 (2), 653–659. http://dx.doi.org/10.1007/s10916-010-9531-y.

Quinlan, J.R., 1993. C4.5: Programs for machine learning.

Ren, Y., Werner, R., Pazzi, N., Boukerche, A., 2010. Monitoring patients via a secure and mobile healthcare system. IEEE Wireless Communications 17 (1), 59–65. http://dx.doi.org/10.1109/MWC.2010.5416351.

Safavi, A.A., Keshavarz-Haddad, A., Khoubani, S., Mosharraf-Dehkordi, S., Dehghani-Pilehvarani, A., Tabei, F.S., 2010. A remote elderly monitoring system with localizing based on wireless sensor network. In: 2010 International Conference on Computer Design and Applications (ICCDA), vol. 2, pp. V2-553–V2-557.

Shih, D.-H., Chiang, H.-S., Lin, B., Lin, S.-B., 2010. An embedded mobile ECG reasoning system for elderly patients. IEEE Transactions on Information Technology in Biomedicine 14 (3), 854–865. http://dx.doi.org/10.1109/TITB.2009.2021065.

Song, X., Wang, C., Gao, J., 2014. An integrated framework for analysis and mining of the massive sensor data using feature preserving strategy on cloud computing. In: 2014 Seventh International Symposium on Computational Intelligence and Design (ISCID), vol. 2, pp. 337–340.

Taleb, T., Bottazzi, D., Guizani, M., Nait-Charif, H., 2009. Angelah: a framework for assisting elders at home. IEEE Journal on Selected Areas in Communications 27 (4), 480–494. http://dx.doi.org/10.1109/JSAC.2009.090511.

Tan, J., 2005. E-Health Care Information Systems: An Introduction for Students and Professionals. John Wiley & Sons.

Tartarisco, G., Baldus, G., Corda, D., Raso, R., Arnao, A., Ferro, M., Gaggioli, A., Pioggia, G., 2012. Personal health system architecture for stress monitoring and support to clinical decisions. Computer Communications 35 (11), 1296–1305. http://dx.doi.org/10.1016/j.comcom.2011.11.015. http://www.sciencedirect.com/science/article/pii/S0140366411003720.

Taub, D., Lupton, E., Hinman, R., Leeb, S., Zeisel, J., Blackler, S., 2011. The escort system: a safety monitor for people living with Alzheimer's disease. IEEE Pervasive Computing 10 (2), 68–77. http://dx. doi.org/10.1109/MPRV.2010.44.

Thilakanathan, D., Chen, S., Nepal, S., Calvo, R., Alem, L., 2014. A platform for secure monitoring and sharing of generic health data in the cloud. Future Generations Computer Systems 35, 102–113. http://dx.doi.org/10.1016/j.future.2013.09.011. http://www.sciencedirect. com/science/article/pii/S0167739X13001908. Special Section: Integration of Cloud Computing and Body Sensor Networks; Guest Editors: Giancarlo Fortino and Mukaddim Pathan.

Yoshizawa, M., et al., 2010. A mobile communications system for home-visit medical services: the electronic doctor's bag. In: 2010 Annual International Conference of the IEEE Engineering in Medicine and Biology, pp. 5496–5499.

ACRONYMS AND GLOSSARY

List of acronyms with explanation

ASIP   Application Specific Instruction set Processor
BF   Boolean Function
BPU   Branch Prediction Unit
BTPCF   Binary Tree Packet Classification
CPU   Central Processing Unit
DT   Decision Tree
EU   European Union
FPGA   Field Programmable Gate Array
GDP   Gross Domestic Product
HDL   Hardware Description Language
HMS   Health Monitoring Systems
ICT   Information and Communication Technologies
LUT   Lookup Table
RTL   Register Transfer Level
TVPU   Tree Visiting Processing Unit

Glossary

ASIP is the acronym of Application Specific Instruction set Processor and indicates a technique which exploits a custom instruction set to accomplish an algorithm.

Big Data is a term to indicate data sets which turn out to be so large (or complex) such that data processing applications are inadequate to deal with them.

Cloud computing is a paradigm of provisioning of shared computer processing resources and data as service.

CPU is the acronym of central process unit and indicates the hardware component have the task of executing instructions of programs running on a computer system.

Decision tree is a data model structure which is based on a tree structure.

Dynamic partial reconfiguration is the ability for an FPGA to change a sub-part of its functionality without interrupting the remaining part.

e-Health is part of the healthcare practice which is supported by electronic processes and communication protocols.

Feature indicates an individual measurable property of the phenomena being observed.

FPGA is the acronym of field programmable gate array, a technology of IC which allows the (re)configuration of a hardware circuit.

HDL is the acronym of hardware description language, indicating a language able to describe hardware components, their functionalities and connections.

Knowledge management is the process of creating, sharing, using, and managing the knowledge and information for a given application.

LUT is the acronym of look-up table, which is a configurable memory of the FPGA employed to realize boolean functions.

Machine learning is a research field of the computer science which studies the ability for a computer to learn without being explicitly programmed for a specific task.

Misprediction indicates the event of a wrong classification.

Test set is the part of a data set employed to test the accuracy of an extracted model.

Training set is a part of a data set used to extract a data model.

Pipeline is a hardware technique to speed-up the throughput, making the computation faster.

Real-time describes hardware and software systems subject to a time constraint which cannot be violated.

SoP is the acronym of sum-of-product and is a canonical form for describing a boolean equation.

Slice is a compound block of the FPGA which groups one or more LUTs with other low-level features, such as flip-flops as memory.

A METHOD FOR ESTIMATING STRESS AND RELAXED STATES USING A PULSE SENSOR FOR QOL VISUALIZATION

12

Sayaka Akiyama*,†, Yuka Kato*

*Tokyo Woman's Christian University, Japan †Toshiba Corporation, Japan

12.1 INTRODUCTION

The number of patients suffering from lifestyle-related diseases such as diabetes and obesity is increasing. Two third of the cause of deaths in Japan are such diseases, and the action of preventing them is very important topics of discussion because the birthrate decline and population aging continue unabated in Japan. Three kinds of factors in lifestyle-related diseases are reported in Ministry of Health and Welfare in Japan (1997), these are living environment factors, genetic factors, and lifestyle factors. Among them, lifestyle can be changed by considering the problems of the style and improved by our own determination. Therefore, until now, various studies have been made about lifestyle analysis and required lifestyle for a healthy life.

Breslow's seven lifestyle categories (Belloc and Breslow, 1972) is one of them. The categories are exercise, breakfast, eating between meals, sleep, smoking, drinking, and proper body weight. In this paper, he reported that the life expectancy is extended when the number of practical categories is large. Actually, the seven lifestyle categories are used in various places, such as a check diagram of a lifestyle on a medical examination and health care advice of a medical doctor based on the diagram. However, the check result is a just self-assessment and is obtained just nearly once a year. Therefore, it is difficult to use it for improving lifestyle actively and with awareness.

From these backgrounds, we have studied on Quality of Life (QOL) visualization system, whose target is to prevent lifestyle-related diseases by analyzing and indicating the factors which degrade QOL in a daily life. Recently, various devices, which are worn anytime and are able to collect vital information, are widely used,

Table 12.1 The Development of Sensing Technology for Vital Information

Year	Events
1896	Developed a noninvasive method of blood pressure measurement using the upper cuff, mercury sphygmomanometer, and palpation (Scipone Riva Rocci, Italy)
1903	Developed a method of electro-cardiogram measurement and recording using string galvanometer (Willem Einthoven, Netherlands)
1905	Discovered Korotkoff sounds and noninvasive method of blood pressure measurement by auscultation (Nikorai Korothov, Russia)
1928	Released the electrocardiography (ECG) using vacuum tube triode (Siemens AG, Germany)
1961	Developed a continuously recordable ECG over a long period (Holter monitor) (Norman J. Holter, USA)
1965	Released a biological information monitor (Nihon Kohden Corp., Japan)
1974	Developed the principle of pulse oximetry (Nihon Kohden Corp., Japan)
1977	Released a finger measurement type pulse oximeter (Minolta Corp., Japan)
1978	Released a home-use digital sphygmomanometer (OMRON Corp., Japan)
1979	Released a home-use automatic sphygmomanometer (Matsushita Electro., Japan)
1982	Released a portable heart rate monitor (Polar Electro., Finland)
1985	Released a home-use oscillometric type sphygmomanometer (A&D Co. Ltd., Japan)
1987	Released a wristwatch equipped with the function measuring pulse rate (Casio Computer Co. Ltd., Japan)
1988	Released a home-use finger measurement type sphygmomanometer (OMRON Corp., Japan)
1995	Released a small-sized finger measurement type pulse oximeter (Nonin Medical, Inc., USA)
1997	Released a wristwatch-type pulse oximeter (Konica Minolta, Inc., Japan)
2001	Released a home-use electrocardiograph (Parama-Tech Co., Ltd., Japan)
2014	Released a smartwatch equipped with Android Wear (LG Corp., Korea)
2015	Released Apple Watch (Apple Inc., USA)

such as Apple Watch (Apple Watch, 2015), Microsoft Band (Microsoft Band, 2015), Moto360 (Moto, 2014), Galaxy Gear of Samsung (Gear, 2014), and Flex and Zip of Fitbit (Fitbit, 2015), and the goal of the QOL visualization system is to manage an individual QOL on a daily basis by using these wearable devices. Since various sensing data, including a person's pulse, acceleration, blood sugar level, body motion, illumination can be collected from these devices, a daily health check by using data of high objectivity becomes possible in place of a subjective health check which is conducted once a year. The development of sensing technology for vital information is shown in Table 12.1. Since objective health-related data can be collected unconsciously on a long-term basis by using such wearable devices, many studies using these devices for health management have been made (Sumida et al., 2013; Imazu et al., 2013; Watanabe et al., 2006) recently. Like these, usage data for health management have been changing from subjective data to objective ones which can be obtained regularly.

In this study, we examine a module for estimating a stress state as one of the functions of the QOL visualization system and propose an estimation method using a person's pulse which is obtained from a wearable device. In addition, we propose Stress and Relaxed Value (*SRV*) as an indicator of stress status (Akiyama and Kato, 2015, 2016). In this case, we estimate the *SRV* based on LF/HF (Low Frequency/High Frequency) (Allen, 2007) which is an indicator of the psychological stress state and is derived from the fluctuation of the peak intervals of the heartbeat. Here, we represent QOL by the time series of *SRV*. Actually, there have been many studies of easy measurement methods of the heartbeat, pulse wave, and so on, and measurement devices for such vital information (Jobsus, 1977; Nakazono et al., 2009; Poh et al., 2012; Scully et al., 2011). For example, various methods estimating stress states by using brain waves (Ishihara and Yoshii, 1972), saliva (Yamaguchi et al., 2006), an electrocardiogram (Lucini et al., 2002), pulse wave (Chigira et al., 2011), and pulse rate (Shimokakimoto et al., 2014) are proposed. However, it is difficult to use them all the time in order to obtain *SRV* continuously on a long-term basis. The main target of the study is not to improve the accuracy of the measurement, and is to continuous measurement on a long-term basis by using commercial devices. In this study, we also conduct evaluation experiments using a pulse sensor and verify the effectiveness.

The contribution of this study is below:

- Defining *SRV* as a QOL indicator with considering both stress and relaxed state
- Verifying that *SRV* can be associated with lifestyle factors.

12.2 LITERATURE REVIEW

This study proposes a method estimating a stress state using vital information (i.e. objective data) obtained noninvasively from wearable devices. So far, there have been many studies on estimation methods of stress states in various research fields. In this section, we reviewed the methods using subjective data, sensing data, and vital information as related works.

12.2.1 METHOD USING SUBJECTIVE DATA

Various studies have been made about lifestyle analysis and lifestyles required for a healthy life. For example, the analysis of the relationship between the physical condition of one's health and his/her lifestyle including sleep, diet, exercise, drinking, and smoking is conducted (Belloc and Breslow, 1972). As described in Sec. 12.1, Breslow's seven lifestyle categories in this paper are widely used for a lifestyle check diagram on a medical examination and health care advice of a doctor.

However, subjective assessment bothers users and the evaluation accuracy might not be high because the check result is obtained just nearly ones a year. In addition, individual differences are not considered in the health habits mentioned in the paper.

Therefore, in this study, we aim to evaluate one's lifestyle on a daily basis over a long time while reducing user's burden. For that, we obtain the data of daily life using wearable devices. Here, we use the seven lifestyle categories to determine the lifestyle factors we focus.

12.2.2 METHOD USING SENSING DATA

As for estimation methods using objective data, there are various studies using sensing data such as acceleration data and sound data obtained from smartphones. For example, there is a research estimating a physical stress level during exercise by using acceleration data measured by a smartphone and environmental conditions (Sumida et al., 2013). In this paper, pulse rate is estimated by acceleration data obtained from the built-in sensor of the smartphone during exercise, and exercise intensity with considering load for each user is displayed. A gait improvement tool using a smartphone is also proposed (Kashihara et al., 2013). Here, gait analysis using acceleration data obtained from a smartphone is conducted. A labeling method for activity recognition is proposed as well (Murao and Terada, 2013). In this paper, the labeling is automatically done just by using an execution sequence of activities based on acceleration data and timestamps obtained from a smartphone. In addition, a method recognizing living activity using acceleration data and sound data obtained from a mobile phone is proposed (Ouchi and Doi, 2011).

These results show that lifestyles, especially human activities, can be obtained and monitored by using acceleration data, sound data etc. measured by sensors (e.g. built-in sensors of a smartphone).

12.2.3 METHOD USING VITAL INFORMATION

As for estimation methods using vital information, there are the study assessing a fatigue state of a car driver by analyzing an amylase secreted in saliva (Yamaguchi et al., 2006), the study assessing a stress state when doing some tasks, applying the mental stress, such as a calculation task and a recognition task by using the appearance of Fm-θ, which is a feature wave of electroencephalogram (EEG) (Ishihara and Yoshii, 1972), and the study estimating a stress state using heartbeat and pulse wave (Nakazono et al., 2009) to be the same as our approach. These studies can monitor a stress state continuously with high accuracy. However, there is a possibility that these methods inhibit primary behavior and conditions of users. Therefore, it is difficult to adopt them into our system because the system requires continuous data collection in everyday life. For that reason, in this study, we estimate a stress state using a person's pulse which can be easily monitored on a daily basis with commercial wearable devices though the measurement accuracy might be compromised.

As for a method of utilizing vital information obtained from devices, almost all data are used in order to detect a stress state directly such as evaluating stress during work and detecting car driver sleepiness, e.g. measuring workload during tasks (Haapalainen et al., 2010), fatigue assessment of car drivers (Yamaguchi et al.,

2006), and monitoring the effect of stress by a calculation task to EEG (Ishihara and Yoshii, 1972). On the contrary, the purpose of this study is to obtain vital information constantly in daily life and to visualize QOL based on Breslow's seven lifestyle categories. Hence, it is necessary to assess not only a transient stress state caused by some conditions but also a long-term mental condition including a relaxed state. This study solves the problem by defining a time series of *SRV* (Stress and Relaxed Value) which is an extended LF/HF.

There are studies on monitoring devices as well. For example, a mechanism measuring pulse wave at a fingertip using an area-based photo-plethysmographic sensing method is proposed (Chigira et al., 2011). In this paper, the indicator of autonomic nervous system is used to evaluate measurement accuracy of a pulse wave. Similarly, for obtaining pulse wave, a method measuring pulse wave from the blood vessel in ears by using earphones and monitoring heart rate is reported (Poh et al., 2012). In addition, there is a study detecting a physiological indicator, such as pulse rate, using a self-made wearable device (Shimokakimoto et al., 2014). The target of these studies is to develop monitoring devices of vital information. However, they calculate pulse wave intervals from pulse wave, conduct spectrum analysis and estimate a stress state using the indicator of autonomic nervous system. In this paper, we use them as the reference for calculating LF/HF.

12.3 QOL VISUALIZATION SYSTEM

First, we explain our target system, QOL visualization system. This system estimates a balance condition of the autonomous nerve (a mental stress condition) based on person's pulse obtained from wearable devices and uses the estimated value as a QOL indicator. Specifically, we use *SRV* as the indicator. The details of *SRV* will be described later. We assume that QOL is a variable resulting from Breslow's seven lifestyles mentioned above, and visualize the degree of influences. In this case, we use three indicators, which are exercise, sleep, and diet, as the lifestyle factors because this information can be obtained by wearable devices. The system visualizes these values with other factors (e.g. smoking frequency) which a user inputs manually to the system and indicates the result to the user. A GUI image of the analysis result is shown in Fig. 12.1. In the window, the current values and the target values to improve the *SRV* for each item (e.g. Sleep) are displayed.

The display content is calculated by the analyzing result of the time series data obtained. The system architecture and the analysis process are shown in Figs. 12.2 and 12.3. The QOL indicator is calculated by a pulse obtained from wearable devices as the time series of *SRV*. The factor is modeled as a function of exercise intensity estimated using an acceleration sensor, eating frequency estimated from blood sugar, sleeping time estimated by body motion and illumination, and others. The system shows this analysis result. In this study, we focus on the part of calculating *SRV* from a person's pulse.

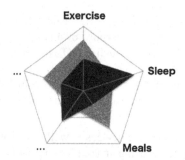

FIGURE 12.1

A GUI for the QOL visualization system. (Reprinted with permission from Akiyama and Kato (2016). © 2016 IEEE.)

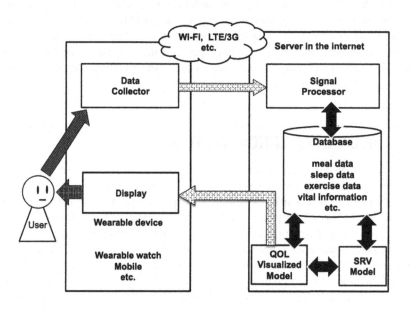

FIGURE 12.2

The system architecture of the QOL visualization system.

12.4 PROBLEM DEFINITION

The problems in this study are defined as follows.

- Problem 1: How is *SRV* formulated in?
- Problem 2: What sort of relationships are formed between *SRV* and lifestyles?

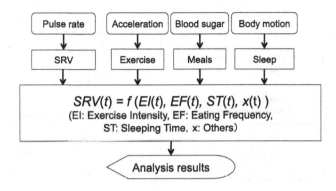

FIGURE 12.3

An overview of the QOL visualization process. (Reprinted with permission from Akiyama and Kato (2016). © 2016 IEEE.)

As for the first problem, we have to consider two issues. One is what kinds of data we obtain from wearable devices, and the other is how to calculate the indicator from the obtained data. For data types, in this study, we use the instantaneous pulse rate (*IPR*), which can be collected easily by wearable devices. This is because there is the relationship between lifestyles (mainly exercise, sleep, and diet) and the fluctuations of pulse and heart rate, and because we can calculate the value of LF/HF (LF: Low Frequency, HF: High Frequency), which is known as an indicator of stress states derived from pulse or heart beat intervals, from *IPR*. Moreover, we expect LF and HF can be used to estimate relaxed states as well as stress states because they have such features on activities of the autonomic nervous system. For calculating the indicator, we define it (i.e. the expression of stress and relaxed states) based on the indicator of the autonomic nervous system described later (Furlan et al., 1990).

As for the second problem, we calculate numerical values of long-term stress and relaxed states, grasp the trend of the time series data (the numerical data) under the condition of existence or nonexistence of the lifestyles, and associate the values with the lifestyles. Here, we conduct the experiment which continuously measures *IPR* of a test subject in a daily life by using a wearable device, and identify correlations between *SRV* and lifestyles. In this study, we focus on two lifestyle factors, which are sleep and diet, according to Breslow's seven lifestyle categories described above.

12.5 THE PROPOSED APPROACH
12.5.1 OVERVIEW

In this section, we explain an overview of the proposed method. The existing methods estimating a stress state from heartbeat and EEG conduct the estimation according to the following steps. First, they monitor an electrocardiography waveform or pulse

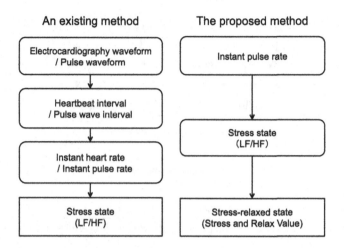

FIGURE 12.4

The procedures of the proposed method and an existing one. (Reprinted with permission from Akiyama and Kato (2016). © 2016 IEEE.)

waveform and obtain heartbeat intervals or pulse wave intervals from them. Then, they calculate LF/HF from the intervals and estimate the stress state by using the result.

The procedure of the estimation for the proposed method is the same as the existing ones. However, two steps are different. One is to monitor a person's pulse instead of an electrocardiography waveform or pulse waveform. The other is to estimate *SRV* from LF/HF. The procedures of the proposed method and an existing one are shown in Fig. 12.4. Next, we will explain each step of the procedure in detail.

12.5.2 CALCULATING LF/HF

In order to estimate *SRV*, this study conducts spectrum analysis of the fluctuation waveform of pulse wave peak intervals and uses LF/HF which is the ratio of power spectrum density in the section at the low frequency (LF) and that at high frequency (HF).

First, we explain the meanings of LF/HF. The heart rate is generally used the calculation of LF/HF. A heartbeat is the ignition cycle of the sinoatrial node that is the pacemaker cell of the heart, and the heart rate variability is fluctuating with being influenced by the brain stem events occurring periodically synchronized breathing (periodical activity in about 4-second cycle, HF), the brain stem events occurring periodically synchronized the fluctuation of a blood pressure (periodical activity occurring in about 10-second cycle, LF) and emotion (periodical activity in longer second cycle). (Heart rate variability is the fluctuation of intervals between an R wave at a point in time and the next R wave on an electrocardiographic waveform

Table 12.2 The States Exciting Sympathetic and Parasympathetic Nerves Activities and the Main Effects

Nervous System	State	The Main Effects
Sympathetic nerve	Stress state (during exercise, under tension, etc.)	Vasoconstriction, blood pressure increase, heart rate increase, sweating, bronchiectasis etc.
Parasympathetic nerve	Relaxed state (during sleep, meal, a break, etc.)	Heart rate decrease, saliva secretion, gastric juice secretion etc.

Table 12.3 The Effect of Stress States on Autonomic Nervous System

State	Active Nerve	Increasing Factor
Stress	Sympathetic nerve	The power of LF
Relaxe	Parasympathetic nerve	The power of LF and HF

(i.e. heartbeat intervals). R waves are the largest amplitude points of the waveform in Fig. 12.6, and we can use the time series data obtained from the intervals as heart rate variability.) This fluctuation is governed excitement and suppression by sympathetic activity and parasympathetic activity provided by cardiovascular and central nervous of the medulla oblongata (Lucini et al., 2002; Furlan et al., 1990; Task Force of the European Society of Cardiology and the North American Society of Pacing and Electrophysiology, 1996).

At that time, the activity of sympathetic nerve is contrary to that of parasympathetic nerve. Sympathetic nerve activity increases during stress states and parasympathetic nerver activity increases during relaxed states. Table 12.2 shows the states exciting the autonomic nerves system (sympathetic and parasympathetic nerves) activities and the main effects. Furthermore, a sympathetic nerve is transmitted by the medium of the variance component of arterial blood pressure (LF), and a parasympathetic nerve is transmitted by the medium of both of the variance components of breathing (HF) and arterial blood pressure (LF). This relationship (i.e. the effect of stress states on the autonomic nervous system) is summarized in Table 12.3. That is to say, a sympathetic nerve is an increasing factor of LF power, and a parasympathetic nerve is an increasing factor of both of LF and HF powers. Therefore, HF amplifies the power by accelerating a parasympathetic nerve, and LF amplifies the power by accelerating a sympathetic nerve and a parasympathetic nerve. Since the power of LF is amplified by both a sympathetic nerve predominant state and a parasympathetic nerve predominant state, a stress state is estimated by LF/HF.

Next, we explain the method calculating LF/HF. The procedure is shown in Fig. 12.5. First, we describe the method obtaining LF/HF from electrocardiogram waveform as follows (Montano et al., 1994).

1. Read heartbeat intervals (R-R Interval: *RRI*) from electrocardiogram waveform (see Fig. 12.6)

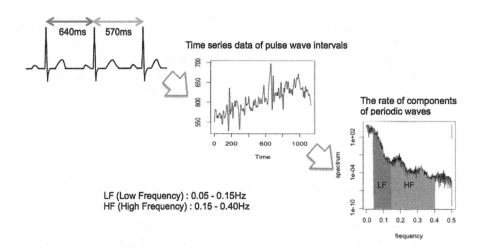

FIGURE 12.5

The procedure calculating LF/HF. (Reprinted with permission from Akiyama and Kato (2016). © 2016 IEEE.)

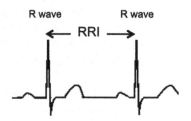

FIGURE 12.6

R-R interval. (Reprinted with permission from Akiyama and Kato, 2016. © 2016 IEEE.)

2. Calculate the instantaneous heart rate (*IHR*) by the following equation (for obtaining heart rate per 60 sec., we multiply the value by 60)

$$IHR = \frac{60}{RRI} \tag{12.1}$$

3. Approximate the time series of *IHR* by continuous function by using linear interpolation or spline interpolation etc., and execute the resampling of them at equal intervals, because *IHR* is the time series at unequal intervals

4. Conduct spectrum analysis using the time series of *IHR* after the resampling

5. Calculate the integral of the power spectrum $P(f)$ in the section at the low frequency (LF) and that at high frequency (HF). Here, f denotes the frequency

6. Calculate LF to HF ratio as LF/HF.

The main target of the study is to calculate the time series of *SRV* in a daily life by collecting vital data continuously on a long-term basis. Therefore, it is difficult to adopt the method using electrocardiogram waveform. Hence, the proposed method uses *IPR* per second which is directly measured from wearable devices and calculates LF/HF based on that. We describe the method obtained LF/HF from *IPR* as follows (Akiyama and Kato, 2015).

1. Collect the series of *IPR* from wearable devices
2. Remove outliers such as measurement errors and cover the data omission using spline interpolation
3. Conduct spectrum analysis using the time series of *IPR*
4. Calculate the integral of the power spectrum $P(f)$ in the section at the low frequency (LF) and that at high frequency (HF), where f denotes the frequency
5. Calculate LF to HF ratio as LF/HF.

In this case, we set the thresholds of outliers to 50 (minimum) and 150 (maximum), adopt auto-regressive model (AR model) for spectrum analysis, and set the range of LF and that of HF to $[0.05, 0.15]$ and $[0.15, 0.4]$ respectively. LF and HF are given by the following equations.

$$LF = \int_{0.05}^{0.15} P(f)\, df, \tag{12.2}$$

$$HF = \int_{0.15}^{0.4} P(f)\, df. \tag{12.3}$$

Spectrum analysis is a method expressing characteristics of a time series with power spectrum density function $p(f)$, which is the power of each period component. This power is given as the square of an absolute value of the Fourier coefficient and is also called spectrum. $p(f)$ is given by

$$p(f) = \sum_{k=-\infty}^{\infty} C_k e^{-2\pi i k f} \tag{12.4}$$

where k denotes the time lag, C_k denotes the autocovariance function and f denotes the frequency of $-1/2 \le f \le 1/2$ (Wiener, 1988). Here, we assume

$$\sum_{k=-\infty}^{\infty} |C_k| < \infty. \tag{12.5}$$

12.5.3 CALCULATING SRV

Next, we define Stress and Relaxed Value (*SRV*) using LF/HF. As mentioned above, LF/HF indicates a sympathetic nerve predominant state, LF indicates both sympathetic and parasympathetic nerves predominant state and HF indicates a parasympathetic nerve predominant state. In a stress state, a sympathetic nerve is active and a

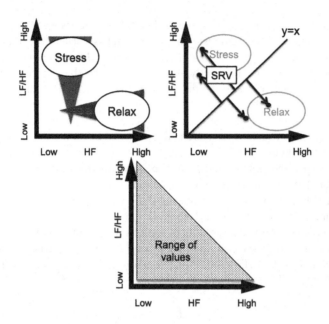

FIGURE 12.7

The relationship between stress and relaxed states, and LF/HF. (Reprinted with permission from Akiyama and Kato (2016). © 2016 IEEE.)

parasympathetic nerve is not active. On the contrary, in a relaxed state, a parasympathetic nerve is active. Therefore, the relationship between a stress state and a relaxed state, and LF/HF is shown in Fig. 12.7 where the ordinate denotes LF/HF and the abscissa denotes HF. That is to say, a stress level becomes large when LF/HF is large and HF is small because the excitement state of the sympathetic nerve is large. Moreover, a relaxed level becomes large when HF is large and LF/HF is small because the excitement state of the parasympathetic nerve is large. Therefore, we define SRV by the following equation according to the relationship.

$$SRV = \frac{1}{\sqrt{2}} \left(\frac{LF}{HF} - a \cdot HF \right). \tag{12.6}$$

In this equation, SRV is the euclidian distance between the point (x, y) and the straight line $y = x$ where the value of HF is x and the value of LF/HF is y as follows:

$$SRV = \pm \sqrt{\left(x - \frac{x+y}{2} \right)^2 + \left(y - \frac{x+y}{2} \right)^2} = \pm \sqrt{\frac{(x-y)^2}{2}} = \pm \frac{1}{\sqrt{2}} (x - y).$$
$$\tag{12.7}$$

That is to say, this equation defines the indicator as the distance from the neutral state which is neither stress nor relaxed states. This is shown in the upper right figure of

Fig. 12.7. This is a positive value in the stress state and a negative value in the relaxed state. The coefficient of HF (i.e. a) is the normalization constant, and we determine the value by the preliminary experiment in this case. Note that a range of values that (x, y) can take is the part shown in the lower figure of Fig. 12.7. This is because there is no possibility that both of LF/HF and HF will be large values at the same time from the meanings of LF and HF as described above.

From this procedure (Eq. (12.6)), we can obtain *SRV* at a certain time. This is repeated at 5-minute intervals, and the time series of $SRV(t)$ is calculated, where t denotes the time. For example, if all of the pulse wave intervals obtained are one second and the measurement time is 60 minutes, the number of the interval data is 3600 (i.e. 60×60) and we obtain 12 *SRV*s (i.e. $3600/(60 \times 5)$). This time series data are defined as the QOL indicator for the QOL visualization system.

12.6 IMPLEMENTATION

For the evaluation experiments, we implement the proposed method described in Sec. 12.5 as an experimental system. The followings are the contents.

1. Obtaining data

   The system obtains *IPR*, which is translated into the data of the most recent 60 seconds, every second from a wearable device. As the wearable device, we use a pulse oximeter of MASIMO Japan iSpO2 (iSpO2, 2015). A pulse oximeter is an instrument measuring SpO2 (arterial oxygen saturation) and a pulse rate with a probe apparatus (a sensor) attached to a finger, etc. It obtains the data by using absorption properties of a red light and an infrared light. Even though iSpO2 is a simple, portable type pulse oximeter, which is used by connecting with a smartphone, it can measure accurate values at a certain extent even under conditions of the existence of the body motion. The iSpO2 usable range of the number of measurement is 25–240 per minute and the measurement accuracy is ± 3 per minute without body motion, ± 5 per minute with body motion and ± 3 in the case of poor perfusion (blood flow).

2. Generating time series data

   First, as the preprocessing, the system detects missing data of *IPR* using timestamps obtained with the measuring data. It also detects outliers of *IPR* using the pre-defined threshold. Here, the system generates time series data by replacing the detected missing data and outliers with the mean value of *IPR*s.

3. Processing data (1)

   Next, the system carries out spectrum analysis of the time series of *IPR*. Since the time series is discrete data at 1-second interval, we adopt the Discrete Fourier Transform using AR model for the spectrum analysis. AR model constructs a data model for measurement values by estimating the most fitting order and AR coefficients. It is expressed in Eq. (12.8), where y_n denotes the current *IPR* value, y_{n-i} denotes the *IPR* value i seconds ago, a_i denotes AR coefficients, p denotes

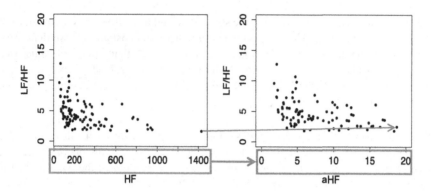

FIGURE 12.8

An image of the normalization of HF.

the order and v_n denotes the residuals following a normal distribution with mean zero and variance σ^2.

$$y_n = \sum_{i=1}^{p} a_i y_{n-i} + v_n. \qquad (12.8)$$

Here, we use the Yule–Walker method for model-fitting (AR model), and use Akaike's Information Criterion (AIC) for model comparison. The details of AR model are explained in Appendix 12.A.

4. Processing data (2)

   The system calculates the integral of the spectrum analysis result in the section at the low frequency and that at the high frequency, and obtains LF and HF values.

5. Calculating the proposed indicator

   On calculating the proposed indicator SRV, the system normalizes HF value so that there is a one-to-one correspondence between the range of X-axis and that of Y-axis on the graph of the relationship between LF/HF and HF. Fig. 12.8 shows an image of the normalization of HF (i.e. a one-to-one correspondence is illustrated in the arrows). Note that, for calculating LF/HF, the value of HF before normalizing is used. Then, the system substitutes the value obtained from the integral into the definition formula of SRV in Sec. 12.5 and calculates SRV. In this case, the normalization constant is determined as $a = 20/950$ based on the test data of HF obtained beforehand.

6. Generating time series data of the indicator

   The system repeats the procedure at 5-minute intervals and generates the time series of SRV.

12.7 EVALUATION

The proposed method calculates LF/HF by using *IPR*, and then, estimates *SRV* using the value. In order to evaluate the validation, we conduct evaluation experiments using a pulse sensor. In addition, for identifying correlations between *SRV* and lifestyles, we focus on two lifestyle factors, which are sleep and diet, and investigate the trend of the time series of *SRV* under the condition of existence or nonexistence of the factors.

12.7.1 EXPERIMENTAL CONDITIONS

12.7.1.1 Experiment 1

First, we monitor time series of pulse wave intervals which are calculated from *RRI* and *IPR* and compare the results. The test subject is a female in her twenties, and her pulse and pulse wave intervals are monitored in a relaxed condition and immediately after exercises (after several ten seconds has passed in the stable state). The measurement is conducted while the subject is sitting on the chair, and the measuring period of time is 15 minutes.

The proposed method

This experiment measures *IPR* per second by a pulse oximeter of MASIMO iSpO2 (iSpO2, 2015). This experiment obtains the data from the smartphone connected with iSpO2 and retrieves *IPR* per second as a CSV format. Then, it translates the retrieved data to *PR* by using the following Eq. (12.9). They are time series data, and *i* denotes the time. The reciprocal of the value is converted to millisecond order.

$$PR(i) = \frac{60 \times 1000}{IPR(i)}. \tag{12.9}$$

The comparison method

This experiment measures a pulse wave intervals per pulsation by a digital heartbeat sensor module of Tokyo Devices IW9PLS-MP (IW9PLS-MP, 2015). IW9PLS-MP is a sensor module measuring a heartbeat waveform (fingertip volume pulse waves), and it can retrieve a signal per pulsation as a digital pulse. This experiment obtains the pulsation timing via the serial communication channel by placing the finger on the sensor part and calculates the pulse wave intervals from the intervals between the pulsation timings.

12.7.1.2 Experiment 2

Second, we conduct an experiment to confirm that the distribution of LF/HF and HF which are calculated from *IPR* is close to the distribution between a stress state and a relaxed state shown in Fig. 12.7. This experiment measures *IPR* per second by MASIMO iSpO2 as well.

The test subject is a female in her twenties, and her pulse is monitored in two locations. One is at her home (the measurement period of time is 720 minutes), and the other is in a room of her university (the measurement period of time is 90 minutes).

12.7.1.3 Experiment 3

Third, we verify that stress and relaxed states can be estimated from *SRV*. In this experiment, we use the measurement data in experiment 2 again and record her activities (mental state) at the same time as the measurement. In the university, stress states are generated artificially by doing tasks of simple calculation problems, and relaxed states are generated by listening to music at other times. At home, relaxed states are generated much of the time.

12.7.1.4 Experiment 4

Fourth, we investigate the trend of the time series of *SRV* under the condition of existence or nonexistence of lifestyle factors in order to identify correlations between *SRV* and lifestyles. In this study, we focus on two lifestyle factors, which are sleep and diet.

As for the relationship between *SRV* and diet, the test subject is a female in her twenties, and her pulse is monitored in two cases, one is having meals and the other is not having meals. In the former case, we obtain the *IPR* data for 30 minutes just before having the meal, for 25 minutes during the meal and for 30 minutes after the meal. The total measurement time is about 95 minutes. In the latter case, we also obtain the *IPR* data for 95 minutes on an empty stomach due to skipping meals. In both cases, the measurements are conducted in a resting state sitting on a chair. In addition, in the case not having meals, drinking fluids not containing sugars (e.g. water and green tea) are allowed freely. We use iSpO2 to collect the *IPR* data. The target data for the evaluation are the data obtained just before and after the behavior.

As for the relationship between *SRV* and sleep, the test subject is also a female in her twenties, and her pulse is monitored in two cases, one is getting sleep and the other is not getting sleep. In the former case, we obtain the *IPR* data for 30 minutes before sleeping, for 185 minutes while sleeping and for 25 minutes just after awakening. The total measurement time is about 240 minutes. In the latter case, we obtain the *IPR* data for 40 minutes not getting any sleep during bedtime in a resting state sitting on a chair. We use iSpO2 to collect the *IPR* data. The target data for the evaluation are also the data obtained just before and after the behavior.

12.7.2 EXPERIMENTAL RESULTS

12.7.2.1 Experiment 1

The time series of pulse wave intervals which are calculated by the proposed method and the comparison method are plotted on graphs respectively. The experimental results are shown in Fig. 12.9. The graph on the top shows the result in a relaxed condition, and that on the bottom shows the result immediately after exercise. The

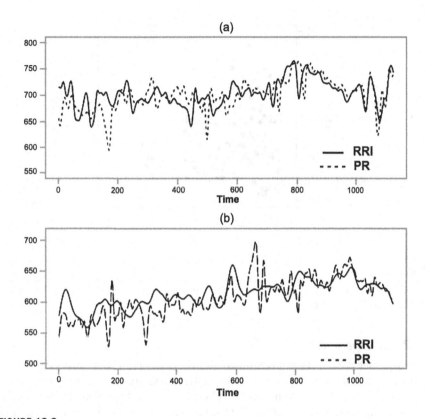

FIGURE 12.9

PR and *RRI* on relaxing (a) and after exercise (b).

shown dot line is the time series calculated by the proposed method, and the shown solid line is the data calculated by the comparison method.

The result of the comparison method (solid line) is more accurate pulse wave intervals than the proposed one. The experimental result indicates that a similar tendency is seen between them even though there is a bit of difference here. Degrees of coincidence of these two time series are high, especially after 600 seconds in a relaxed condition. On the contrary, for the data after exercises, tendencies of variation in data are similar between them, but considerable shaking exists on the proposed method and it shows that the effects of the body motion are large. In either case, we found out that it is possible to estimate a stress state by using a person's pulse.

12.7.2.2 Experiment 2

The distribution of LF/HF and HF calculated from the measured *IPR* is shown in Fig. 12.10. This result indicates that it is similar to the distribution in Fig. 12.7. Hence, we confirm that *SRV* can be calculated from Eq. (12.6).

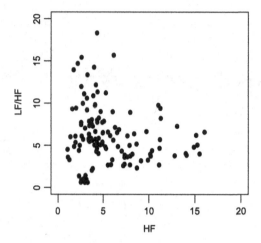

FIGURE 12.10

The distribution of the measured data (all). (Reprinted with permission from Akiyama and Kato (2016). © 2016 IEEE.)

12.7.2.3 Experiment 3

To analyze the mental states while measuring the data, among the data in Fig. 12.10, we plot them at home in Fig. 12.11, and plot them measured in a room of the university in Fig. 12.12. The circle means "in relaxed states" and the black triangle means "when doing some tasks." In Fig. 12.11 (at home), the circles are divided into two time periods, black (at night) and white (daytime). These results indicate that various states widely appear at home because the measurement time is relatively long, and relaxed states are dominant as a whole. On the contrary, in the university, many stress states exist, and relaxed states do not appear.

A part of *SRV*s calculated from the these measured data are shown in Table 12.4 (at home) and Table 12.5 (in the university). Both of positive and negative values are obtained in Table 12.4, and all of the data are positive in Table 12.5. As a result, we found out that the tendency shown in the figure is expressed as *SRV*.

In addition, the time series of *SRV*s with the states at that time are shown in Fig. 12.13 (at home) and in Fig. 12.14 (in the university). The arrows in the lower part of the figure express suspending the measurement. In Fig. 12.13, the time series data are negative as a whole, and relaxed states are dominant. *SRV*s are somewhat large while taking a meal. On the contrary, in Fig. 12.14, all of the data are positive, and *SRV* rapidly increases when starting the task.

From these experimental results, we found out that *SRV* is an indicator by which we can estimate both of stress and relaxed states. However, analyzing factors relating to QOL degradation from the tendency of long-term *SRV* fluctuation is needed when we use the model for QOL visualization systems. In the future, we will analyze long-term states as well as estimate short-term states.

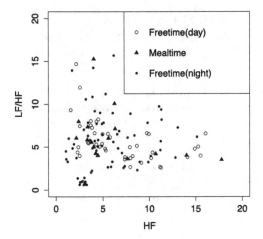

FIGURE 12.11

Data distribution (at home). (Reprinted with permission from Akiyama and Kato (2016). © 2016 IEEE.)

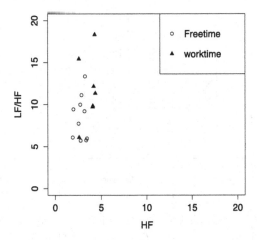

FIGURE 12.12

Data distribution (in the university). (Reprinted with permission from Akiyama and Kato (2016). © 2016 IEEE.)

12.7.2.4 Experiment 4

As for the relationship between *SRV* and diet, the experimental results are shown in Fig. 12.15 and Fig. 12.16. In Fig. 12.15, *SRV*s in the case having meals are plotted, and in Fig. 12.16, *SRV*s in the case not having meals are plotted in 5 minutes intervals.

Table 12.4 *SRV* at Home

No.	SRV	No.	SRV	No.	SRV
1	−6.11	11	0.31	21	−1.52
2	1.02	12	−4.57	22	−1.27
3	−2.42	13	0.69	23	−1.65
4	−10.06	14	−1.39	24	−1.17
5	7.95	15	−4.02	25	−0.66
6	2.70	16	1.10	26	−1.33
7	2.75	17	3.60	27	−1.55
8	2.81	18	1.06	28	−1.13
9	1.77	19	−1.19	29	0.21
10	3.94	20	−1.12	30	1.55

(Reprinted with permission from Akiyama and Kato (2016). © 2016 IEEE.)

Table 12.5 *SRV* in the University

No.	SRV	No.	SRV
1	7.14	10	3.99
2	5.82	11	1.76
3	2.46	12	1.68
4	11.37	13	5.11
5	9.90	14	5.25
6	5.64	15	2.07
7	4.92	16	3.65
8	4.05	17	4.24
9	9.07	18	2.94

(Reprinted with permission from Akiyama and Kato (2016). © 2016 IEEE.)

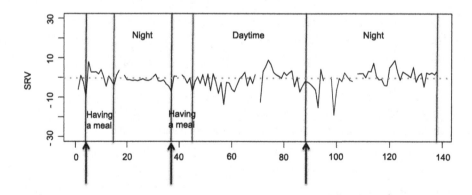

FIGURE 12.13

The time series of *SRV* estimated from the measurement data at home. (Reprinted with permission from Akiyama and Kato (2016). © 2016 IEEE.)

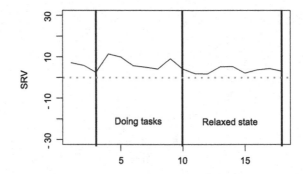

FIGURE 12.14

The time series of *SRV* estimated from the measurement data in the university. (Reprinted with permission from Akiyama and Kato (2016). © 2016 IEEE.)

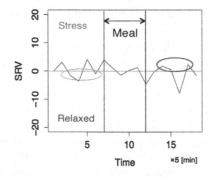

FIGURE 12.15

SRV having the meal.

Positive values plotted above the line of $SRV = 0$ are indicated stress states, and negative values plotted below the line are indicated relaxed states.

These results reveal the followings: (i) *SRV*s for 30 minutes after the meal are smaller than those for 30 minutes just before having the meal in Fig. 12.15, and relaxed states appear after the meal; (ii) Roughly 10% of *SRV*s are negative values in the case not having meals and roughly 45% in the case having the meal. The test subject in the case having the meal is more relaxed (and with fewer stress conditions) throughout the experiment period; (iii) The minimum value of *SRV*s in the case not having meals is −0.97, and the minimum value of *SRV*s for 30 minutes after the meal in the case having the meal is −7.9. The former value is larger than the latter value. This result indicates that the test subject in the case having the meal is more relaxed than in the case not having meals; (iv) The maximum value of *SRV*s in the case not having meals is 4.5, and the maximum value of *SRV*s for 30 minutes after the meal in the case having the meal is 2.3. The former value is larger than the latter value. This

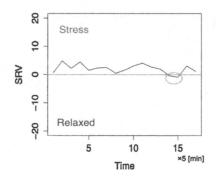

FIGURE 12.16

SRV not having meals.

result indicates that the test subject in the case not having meals is with stress states more than in the case having the meal.

In the case having the meal, relaxed states are expected to continue after the meal because stress states occurred by hunger are reduced. However stress states marked with the dark gray circle in Fig. 12.15 exist. This might be because HF values, which represent relaxed states, decrease and LF values increase temporarily just after the meal. Therefore, LF/HFs become large temporarily, and the system determines them stress states. Moreover, both in the case before having the meal and in the case not having meals, stress states are expected to continue due to skipping meals. However, relaxed states marked with the light gray circles in Fig. 12.15 and Fig. 12.16 exist. This might be because HF values, which represent relaxed states, increase due to gastrointestinal and digestive system motility by drinking fluids on an empty stomach.

As for the relationship between *SRV* and sleep, the experimental results are shown in Fig. 12.17 and Fig. 12.18. In Fig. 12.17, *SRV*s in the case getting sleep are plotted, and in Fig. 12.18, *SRV*s in the case not getting sleep are plotted in 5 minutes intervals.

These results reveal the followings: (i) *SRV*s for 30 minutes before sleeping are not very different from *SRV*s for 25 minutes just after awakening in Fig. 12.17; (ii) in Fig. 12.18, *SRV* just after the beginning of the measurement is 2.0, and *SRV* just before the end is 8.6. The latter value is larger than the former value. This result indicates that the test subject is with stress states more at the time just before the end of the measurement; (iii) *SRV*s in the case not getting sleep are larger than *SRV*s for 25 minutes just after awakening in the case getting sleep. This result indicates that the test subject is with stress states in the case not getting sleep.

In the case before sleeping and in the case not getting sleep, stress states are expected to continue, however, many relaxed states marked with the light gray circles in Fig. 12.17 and Fig. 12.18 exist. HF and LF values, which represent relaxed states, increase and LF/HF values, which represent stress states, decrease while sleeping. Here, relaxed states might appear because the similar condition with that while

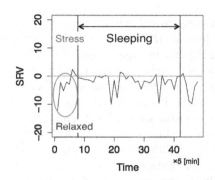

FIGURE 12.17

SRV getting sleep.

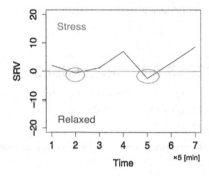

FIGURE 12.18

SRV not getting sleep.

sleeping occurs when the test subject feels drowsiness by the lack of sleep and before sleeping as well.

12.8 CONCLUSIONS

In this study, we proposed an estimation method of Stress and Relaxed Value (*SRV*) by using *IPR* obtained from wearable devices for the QOL visualization system. Moreover, we proposed to use the time series of *SRV* as a QOL indicator. Since the proposed indicator *SRV* uses vital information in calculating the value, it is possible to express individual states of physical and mental health as a numerical value. For the evaluation, we calculated *SRV* for actual vital information in a relaxed condition and in a stress state when doing some tasks. As a result, we found out that we can estimate both of stress and relaxed states by using *SRV*.

In this study, we also conducted evaluation experiments monitoring the trend of the time series of *SRV* by using *IPR* of a test subject obtained from a wearable device in order to investigate the effectiveness of *SRV*. In this experiment, both of stress states and relaxed states were generated artificially by doing tasks of simple calculation problems. As a result of this experiment, we verified that *SRV* can express both states (stress and relaxed) as an absolute scale. In this experiment, we also investigated the trend of the time series of *SRV* under the condition of existence or nonexistence of two kinds of lifestyle factors, which are sleep and diet, in order to identify correlations between *SRV* and lifestyles. As the result of this experiment, the difference of *SRV* between existence and nonexistence of the factors was observed, and we verified that *SRV* can be associated with lifestyle factors.

The contribution of this study is below:

- Defining *SRV* as a QOL indicator with considering both stress and relaxed states
- Verifying that *SRV* can be associated with lifestyle factors.

12.9 FUTURE WORKS AND CHALLENGES

In the future, we will investigate the effectiveness of the proposed *SRV* in more detail, and analyze factors relating to QOL degradation. In concrete terms, the followings are remaining as issues and challenges.

Evaluation experiments on many test subjects under various conditions

The evaluation experiments of *SRV* in this study were conducted for one test subject under limited experimental conditions. The goal of the QOL visualization system is to manage an individual QOL on a daily basis, and the verification for one test subject (an individual) is appropriate. However, more detailed investigations are needed in the future for the application scope of the proposed indicator. Specifically, we need to verify that *SRV* represents individual stress and relaxed states appropriately for multiple test subjects, and to conduct experiments which measure vital data continuously over a long time under various conditions. Furthermore, more detailed investigations are also needed for the relationship between *SRV* and lifestyles. For example, we need to examine the correlation between *SRV* and more detailed lifestyle data, such as the amount of food they eat, nutrient balance, sleeping hours, sleep quality, the intensity of exercise, exercise hours, and so on.

Trend analysis of the time series of *SRV*

The final goal of this study is that each individual can judge personally his/her lifestyle to be good or bad by using *SRV*. In order to do this, it is necessary to conduct trend analysis of the time series of *SRV* and to estimate long-term quality (good or bad) of the lifestyle. In the future, we will examine mechanisms of the time series data analysis and propose a new indicator for the purpose. Moreover, we will analyze the effect of lifestyles (diet, sleep, exercise etc.) on the time series of *SRV* in detail, and examine the results for adding to the QOL visualization system.

APPENDIX 12.A **AR MODEL**

In autoregressive (AR) model, we forecast the variable of interest in time series data using a linear combination of m past values of the variable. AR model is expressed in Eq. (12.A.1), where y_n is the time series data, a_i is AR coefficient, and v_n is normally distributed white noise with mean zero and variance σ^2.

$$y_n \; = \; \sum_{i=1}^{m} a_i y_{n-i} + v_n. \tag{12.A.1}$$

In this study, we use the model to calculate the power spectrum density function $p(f)$. The followings are the calculation procedure.

1. Calculate the order m and the parameters (AR coefficients a_i and variance σ^2) of AR model
2. Calculate the impulse response function g_i $(i = 1, 2, \ldots)$ with Eq. (12.A.2)

$$g_0 \; = \; 1,$$

$$g_i \; = \; \sum_{j=1}^{i} a_j g_{i-j}. \tag{12.A.2}$$

3. Calculate the power spectrum density function $p(f)$ by substituting the impulse response function for Eq. (12.A.3)

$$\begin{aligned}
p(f) \; &= \; \sum_{k=-\infty}^{\infty} C_k e^{-2\pi i k f} \\
&= \; \sum_{k=-\infty}^{\infty} E(y_n y_{n-k}) e^{-2\pi i k f} \\
&= \; \sigma^2 \sum_{k=-\infty}^{\infty} \sum_{j=0}^{\infty} g_j g_{j-k} e^{-2\pi i k f} \\
&= \; \sigma^2 \sum_{j=0}^{\infty} \sum_{k=-\infty}^{\infty} g_j e^{-2\pi i j f} g_{j-k} e^{-2\pi i (k-j) f} \\
&= \; \sigma^2 \sum_{j=0}^{\infty} \sum_{p=0}^{\infty} g_j e^{-2\pi i j f} g_p e^{-2\pi i p f} \\
&= \; \sigma^2 \left| \sum_{j=0}^{\infty} g_j e^{-2\pi i j f} \right|^2. \tag{12.A.3}
\end{aligned}$$

Here, we need to determine the order m and estimate the parameters (AR coefficients a_i and variance of white noise σ^2) to estimate the model.

12.A.1 DETERMINING THE ORDER

We use Akaike's Information Criteria (AIC) for determining the order of the AR model. AIC evaluates target parametric models based on the maximum log likelihood $l(\hat{\theta})$, and is defined as Eq. (12.A.4), where $\hat{\theta}$ is the maximum likelihood estimator for $\theta = (a_1, \ldots, a_m, \sigma^2)^T$ and m is the order of the model.

$$
\begin{aligned}
AIC &= -2 \cdot (\text{the maximum log likelihood}) + 2 \cdot (\text{the number of parameters}) \\
&= -2 \cdot l(\hat{\theta}) + 2 \cdot (m+1).
\end{aligned} \tag{12.A.4}
$$

The followings are the evaluation procedure using AIC for given y_n $(n = 1, \ldots, N)$.
1. Calculate the likelihood $L(\theta)$ with Eq. (12.A.5) as a product of the density function $f(y_n)$, and obtain the maximum likelihood estimator $\hat{\theta}$ by maximizing $\log L(\theta)$

$$
L(\theta) = \prod_{n=1}^{N} f(y_n). \tag{12.A.5}
$$

2. Calculate the variance estimator of white noise $\hat{\sigma}_m^2$ for AR model of order m with Eq. (12.A.6)

$$
\hat{\sigma}_m^2 = \frac{s_{m+1,m+1}^2}{N}. \tag{12.A.6}
$$

Here, $s_{m+1,m+1}$ is an element of the $(m+1) \times (m+1)$ upper triangular matrix generated by the Householder transformation for the joint matrix combining the free variables matrix with the bound variables vector of y_n. The value is constant and does not depend on $a_1, \ldots, a_m$.
3. Calculate the maximum log likelihood $l_m(\hat{\theta})$ with Eq. (12.A.7) by using the estimation value $\hat{\sigma}_m^2$

$$
l_m(\hat{\theta}) = -\frac{N}{2} \log 2\pi \hat{\sigma}_m^2 - \frac{N}{2}. \tag{12.A.7}
$$

12.A.2 ESTIMATING THE PARAMETERS

We use the Yule–Walker method for estimating the parameter of the AR model. The Yule–Walker equation for AR model of order m can be written as Eq. (12.A.8), where C_i is the autocovariance function with lag i, a_i is the AR coefficient and σ^2 is the variance of white noise of the model.

$$
\begin{aligned}
C_0 &= \sum_{i=1}^{m} a_i C_i + \sigma^2 \\
C_j &= \sum_{i=1}^{m} a_i C_{j-i}
\end{aligned} \tag{12.A.8}
$$

The followings are the estimation procedure using the Yule–Walker method.

1. Calculate the sample autocovariance function $\hat{C}_i$ of y_n.
2. Obtain the simultaneous linear equations in unknown AR coefficients as Eq. (12.A.9) by substituting the sample autocovariance function for Eq. (12.A.8)

$$
\begin{bmatrix}
\hat{C}_0 & \hat{C}_1 & \cdots & \hat{C}_{m-1} \\
\hat{C}_1 & \hat{C}_0 & \cdots & \hat{C}_{m-2} \\
\vdots & \vdots & \ddots & \vdots \\
\hat{C}_{m-1} & \hat{C}_{m-2} & \cdots & \hat{C}_0
\end{bmatrix}
\begin{bmatrix}
a_1 \\ a_2 \\ \vdots \\ a_m
\end{bmatrix}
=
\begin{bmatrix}
\hat{C}_1 \\ \hat{C}_2 \\ \vdots \\ \hat{C}_m
\end{bmatrix}.
\tag{12.A.9}
$$

3. Solve the linear equations in m unknowns, and obtain the estimated AR coefficients.
4. Calculate the estimated variance $\hat{\sigma}^2$ with Eq. (12.A.10) derived from Eq. (12.A.8)

$$
\hat{\sigma}^2 = \hat{C}_0 - \sum_{i=1}^{m} \hat{a}_i \hat{C}_i.
\tag{12.A.10}
$$

Here, we use the Levinson's algorithm by solving simultaneous linear equations in m unknown in step 3. This algorithm incrementally and efficiently solves the equations by the convolution due to the feature that the autocovariance function matrix on the left side of Eq. (12.A.9) is a symmetric diagonal-constant matrix. The followings are the calculation procedure of Yule–Walker estimates using the Levinson's algorithm. Here, a_m^i is i-th coefficients of AR model of order m.

1. For $m = 0$,
   (a) $\hat{\sigma}_0^2 = \hat{C}_0$
   (b) $AIC_0 = N(\log 2\pi \hat{\sigma}_0^2 + 1) + 2$
2. For $m = 1, \ldots, M$,
   (a) $\hat{a}_m^m = \left(\hat{C}_m - \sum_{j=1}^{m-1} \hat{a}_j^{m-1} \hat{C}_{m-j} \right) (\hat{\sigma}_{m-1}^2)^{-1}$
   (b) $\hat{a}_i^m = \hat{a}_i^{m-1} - \hat{a}_m^m \hat{a}_{m-i}^{m-1}$
   (c) $\hat{\sigma}_m^2 = \hat{\sigma}_{m-1}^2 \{1 - (\hat{a}_m^m)^2\}$
   (d) $AIC_m = N(\log 2\pi \hat{\sigma}_m^2 + 1) + 2(m + 1)$

REFERENCES

Akiyama, S., Kato, Y., 2015. A method for estimating a stress state using a pulse sensor for QOL visualization. In: Proc. of the 18th International Conference on Network-Based Information Systems (NBiS2015), pp. 232–237.

Akiyama, S., Kato, Y., 2016. An estimation model on stress and relaxed states for QOL visualization and its evaluation. In: Proc. of the 30th International Conference on Advanced Information Networking and Applications (AINA2016), pp. 232–237.

Allen, J., 2007. Photoplethysmography and its application in clinical physiological measurement. Physiological Measurement 28 (3), R1.

Apple Watch, 2015. Apple watch. http://www.apple.com/watch/.

Belloc, N.B., Breslow, L., 1972. Relationship of physical health status and health practices. Preventive Medicine 1 (3), 409–421.

Chigira, H., Maeda, A., Kobayashi, M., 2011. Area-based photo-plethysmographic sensing method for the surfaces of handheld devices. In: Proc. of the 24th Annual ACM Symposium on User Interface Software and Technology (UIST'11), pp. 499–508.

Fitbit, 2015. fitbit. https://www.fitbit.com/jp.

Furlan, R., et al., 1990. Continuous 24-hour assessment of the neural regulation of systemic arterial pressure and RR variabilities in ambulant subjects. Circulation 81 (2), 537–547.

Gear, 2014. Galaxy Gear. Samsung. http://www.samsung.com/jp/consumer/mobilephone/gear/gear/.

Haapalainen, E., Kim, S., Forlizzi, J.F., Dey, A.K., 2010. Psycho-physiological measures for assessing cognitive load. In: Proc. ACM International Conference on Ubiquitous Computing (UbiComp'10), pp. 301–310.

Imazu, S., Mizumoto, T., Sun, W., Shibata, N., Yasumoto, K., Ito, M., 2013. E-health support system for adequate caloric intake learning through simple meal survey and weight measurement. In: Proc. of IEEE Region 10 Humanitarian Technology Conference 2013, pp. 365–370.

Ishihara, T., Yoshii, N., 1972. Multivariate analytic study of EEG and mental activity in juvenile delinquents. Electroencephalography and Clinical Neurophysiology 33 (1), 71–80.

iSpO2, 2015. ISpO2. MASIMO. http://www.masimo.co.jp/ispo2/.

IW9PLS-MP, 2015. IW9PLS-MP. Tokyo devices. https://tokyodevices.jp/items/94.

Jobsus, F.F., 1977. Noninvasice, infrared monitoring of cerebral and myocardial oxygen sufficiency and circulatory parameters. Science 198 (4323), 1264–1267.

Kashihara, H., Shimizu, H., Houchi, H., Yoshimi, M., Yoshinaga, T., Irie, H., 2013. A real-time gait improvement tool using a smartphone. In: Proc. of the 4th International Conference on Augmented Human (AH'13), p. 243.

Lucini, D., Norbiato, G., Clerici, M., Pagani, M., 2002. Hemodynamic and autonomic adjustments to real life stress conditions in humans. Hypertension 39 (1), 184–188.

Microsoft Band, 2015. Microsoft band. Microsoft. https://www.microsoft.com/microsoft-Band.

Ministry of Health and Welfare in Japan, 1997. In: Annual Reports on Health and Welfare. Ministry of Health and Welfare in Japan.

Montano, N., Ruscone, T.G., Porta, A., Lombardi, F., Pagani, M., Malliani, A., 1994. Power spectrum analysis of heart rate variability to assess the changes in sympathovagal balance during graded orthostatic tilt. Circulation 90 (4), 1826–1831.

Moto, 2014. Moto360. Motorola. https://www.motorola.com/us/products/moto-360.

Murao, K., Terada, T., 2013. Labeling method for acceleration data using an execution sequence of activities. In: Proc. ACM International Conference on Ubiquitous Computing Adjunct Publication (UbiComp 2013 Adjunct), pp. 611–622.

Nakazono, Y., Ozeki, H., Mizusawa, J., 2009. Non-invasice measurement of cardiovascular stress responses by a photoplethsmography-built-in PC mouse. Journal of the Society of Biomechanisms 1 (33), 80–84.

Ouchi, K., Doi, M., 2011. A real-time living activity recognition system using off-the-shelf sensors on a mobile phone. In: Proc. of the 7th International and Interdisciplinary Conference, CONTEXT 2011, pp. 226–232.

Poh, M., Kim, K., Goessling, A., Swenson, N., Picard, R., 2012. Cardiovascular monitoring using earphones and a mobile device. IEEE Pervasive Computing 11 (4), 18–26.

Scully, C., Lee, J., Meyer, J., Gorbach, A.M., Granquist-Fraser, D., Mendelson, Y., Chon, K.H., 2011. Physiological parameter monitoring from optical recordings with a mobile phone. IEEE Transactions on Biomedical Engineering 59 (2), 303–306.

Shimokakimoto, T., Lund, H., Suzuki, K., 2014. Heart-pulse biofeedback in playful exercise using a wearable device and modular interactive tiles. Journal of Robotics, Networks and Artificial Life 1 (1), 69–73.

Sumida, M., Mizumoto, T., Yasumoto, K., 2013. Estimating heart rate variation during walking with smartphone. In: Proc. ACM International Conference on Ubiquitous Computing (UbiComp 2013), pp. 245–254.

Task Force of the European Society of Cardiology and the North American Society of Pacing and Electrophysiology, 1996. Heart rate variability standards of measurement, physiological interpretation, and clinical use. European Heart Journal 17 (3), 354–381.

Watanabe, K., Manabe, T., Yoshikawa, T., 2006. Definition of sleep indices by pulse wave and body movement and estimation of sleep stages. Transactions of the Society of Instrument and Control Engineers 42, 404–410.

Wiener, N., 1988. The Fourier Integral and Certain of Its Applications. Cambridge University Press.

Yamaguchi, M., Deguchi, M., Wakasugi, J., Ono, S., Takai, N., Higashi, T., Mizuno, Y., 2006. Hand-held monitor of sympathetic nervous system using salivary amylase activity and its validation by driver fatigue assessment. Biosensors and Bioelectronics 21 (7), 1001–1014.

ACRONYMS AND GLOSSARY

List of acronyms with explanation

AIC   akaike's information criterion
AR model autoregressive model
CSV   comma separated values
ECG   electrocardiography
EEG   electroencephalogram
Fm-θ   theta rhythm at the midline of the frontal area in the human brain
GUI   graphical user interface
HF   high frequency, the integral of the power spectrum in the section at the high frequency
IHR   instantaneous heart rate
IPR   instantaneous pulse rate
LF   low frequency, the integral of the power spectrum in the section at the low frequency
LF/HF   the value obtained by dividing LF by HF
PR   pulse rate
QOL   quality of life
RRI   R-R (between R waves) interval, heartbeat interval
R wave   the largest amplitude point of an electrocardiographic waveform
SRV   stress and relaxed value

Glossary of terms with explanation

Akaike's Information Criterion a criteria that evaluates target parametric models based on the maximum log likelihood.

Autocovariance function a function that represents the degree of similarity between a given time series and a lagged version of itself.

Auto-regressive model a mathematical model that forecasts the variable of interest in time series data using a linear combination of past values of the variable.

Breslow's seven lifestyle categories the analysis result of the relationship between the physical condition of one's health and his/her lifestyle including sleep, diet, exercise etc.

Discrete Fourier Transform discretized Fourier transform and a function from the set of complex numbers to the set of complex numbers.

Fourier coefficient coefficients of the Fourier series of a target periodic function.

Householder transformation an orthogonal transformation used for QR decomposition of a matrix.

Instantaneous pulse rate the number of pulse per unit time (the reciprocal of hearbeat intervals).

LF/HF an indicator of the psychological stress state derived from the fluctuation of the peak intervals of the hearbeat.

Likelihood a function that evaluates a target probability model (probability density) by using observation data.

Linear interpolation a form of interpolation using linear polynomials.

Parasympathetic nerve one of the autonomic nerves whose activity increases during relaxed states.

Power spectrum density the power of each period component of spectrum analysis given as the square of an absolute value of the Fourier coefficient.

Pulse oximeter an instrument measuring SpO2 (arterial oxygen saturation) and a pulse rate with a probe apparatus attached to a finger, etc.

Quality of Life the general well-being of individuals and societies including physical health, family, wealth environment etc.

Spectrum analysis a method that expresses characteristics of a time series with power spectrum density function.

Spline interpolation a form of interpolation where the interpolant is a spline curve.

SRV a QOL indicator with considering both stress and relaxed state that we propose in this study.

Sympathetic nerve one of the autonomic nerves whose activity increases during stress states.

Yule–Walker method a method that estimates the parameters of an AR model.

Time series data a series of data indexed in time order.

Wearable devices devices used by wearing on the body, such as arm and head.

LIVING LAB – EVERYDAY ACTIVITIES

CHAPTER

PROXIMITY-BASED SERVICE: AN ADVANCED WAY OF EXTENDING HUMAN PROXIMITY AWARENESS

13

Akihiro Fujihara

Department of Management Information Science, Fukui University of Technology, Fukui, Japan

13.1 INTRODUCTION

The recent widespread of mobile devices like smartphone, smartwatches, and many other mobile gadgets in the world enables to explore a possibility of novel service using them to communicate with each other by close-range wireless information-communication technologies. One of most trending topics in the recent years is *iBeacon* proposed by Apple (see Gast, 2014). iBeacon is a novel service to automatically advertise information to mobile devices like smartphone at proximity using Bluetooth Low Energy (BLE) (see Townsend et al., 2014). It enables to transmit radio waves from a small information tag called beacon module to smartphones at proximity with saving battery power. By the radio waves, the beacon module broadcasts an automatic triggering to push information to smartphone users who passes around the module with the help of an installed application for iBeacon. Typical use cases of iBeacon are automatic information transferring for in-store and out-store advertising, couponing, ordering without calling clerks. It sometimes uses as an indoor positioning system by measuring Received Signal Strength Indicator (RSSI).

Besides iBeacon, there have been many types of iBeacon like proximity-based information sharing and forwarding technologies, such as WiGig, Wi-Fi aware, Wi-SUN, and LTE direct by alliances and companies. The definition of *proximity* depends on these technologies, *i.e.*, their communication distance changes with wireless communication technologies. Moreover, in the next generation communication called 5G, the mobile phone network will become to form the similar proximity-based network. This trend might change the future direction of the Internet gradually and gradually.

Behind this background, the concept of *proximity-based service* appears to change information flows on the Internet. Tobler (1970) proposed the first law of geography:

Smart Sensors Networks. DOI: 10.1016/B978-0-12-809859-2.00017-6

"Everything is related to everything else, but near things are more related than distant things." More concretely, it is found in mobile phone communication patterns between users by analyzing Call Data Records (CDR) that there is a universal interplay between geography and communications: The probability of two users to be connected decreases with the distance r separating them, following a power law of exponent -2, which is also known as the gravity model of total communication duration between communities. It is also showed that the number of communications between two cities also obeys the gravity model (see Blondel et al., 2015). These findings indicate that the closer the distance between two users is, the more they tend to communicate with each other. Even though we cannot neglect the communication between distant users since it obeys the power laws, most frequent communications occur at close range. Therefore, if such users can be connected using peer-to-peer wireless communication technologies, most of the information flow does not necessarily come through on the Internet. This is good in the viewpoint of information security because local information flows are locally closed, which can also save needless electricity in total by avoiding global information sharing and forwarding every time. The concept of *local information production for local consumption* is an ultimate challenge of network efficiency in proximity-based service.

Furthermore, the communication distance of wireless communication technologies which have been proposed recently is extended. For example, Wi-SUN can transfer the information to 1 km distant areas at maximum, which far away when we compare with that of Bluetooth and Wi-Fi. This long-range peer-to-peer communications enable to extend our proximity awareness, which will be possible to predict what things happening in a distant area will come soon. The concept of *extended proximity* is another challenge for enabling more useful service. The potential of proximity-based service is high, but at this time we must begin to consider what we can do more in iBeacon like very limited proximity service.

As mentioned above, the proximity-based service using Bluetooth Low Energy like iBeacon is usually done by a single beacon module that corresponds to a single automatic triggering for push-type information advertising. However, if a beacon module changes an advertisement message by sensing users' context, for example where they come from, it could be useful to navigate them to their destination. This type of navigation can be applicable to guiding potential customers to a recommended shop and also to disaster evacuation guidance.

In this chapter, we propose a system for indoor route guidance using beacon modules. My image of the guidance is just like a guidance by car navigation system: When we are approaching to the next intersection soon, the system announces the next direction to turn at a right place and timing. This image of indoor route guidance can be realized when beacon modules play two types of their role: quiet and notified beacon modules. Quiet beacon module does not advertise, but it just leaves a mark in a memory space of a mobile device. Notified beacon module advertises an appropriate navigation message based on awaring users' context using information in the memory. To enable the proposed guidance system, these modules need to be deployed along passage ways in a building appropriately. We also implement a prototype of smartphone application run on Android OS. Then, we evaluate the performance of the system. As a result, the proposed system enables detecting pedestrian's moving direction and advertising a correct guidance by the history of detected beacon modules memorized on the memory space. This result opens up a possibility to aware human mobility context in a novel way.

13.2 PROXIMITY-BASED SERVICES

In this section, proximity-based service as a service for transferring information using beacon modules to mobile devices at proximity is introduced in detail. iBeacon by Apple is a representative technology of proximity-based service that provides automatic transmission of proximity specific advertisement to iOS and Android smartphones. Every iBeacon system consists of beacon module which is an advertising-specific peripheral to broadcast advertising packets to smartphones using Bluetooth Low Energy (BLE) (see Townsend et al., 2014). The protocol of iBeacon is transmit-only, *i.e.*, the beacon module is a broadcaster that transmits periodic advertising packets. The advertising packet contains three numerical identifiers: Universal Unique IDentifier (UUID), Major number, and Minor number. UUID is a 128 bit identifier to uniquely identify companies and organizations that beacon modules belong to. The major number is a 16-bit identifier used to identify main groups and the minor one is a 16-bit identifier used to identify lower level of the groups.

When receiving the advertising packet, the smartphone starts an advertisement event automatically through some installed application software. The advertisement information is shown on smartphone's display. The software sometimes interacts with a Web server on the Internet to receive additional information such as webpages, or to send the status of the received advertising packets to a Web server for location-based features. The typical conceptual image of iBeacon is illustrated in Fig. 13.1. There are many application cases proposed. Some cases are in practical use now, but some are under demonstration experiment. Typical use cases are (1) mobile advertisements and couponing, (2) orders and payments without contacting with a store clerk, (3) ticket validation, (4) indoor location positioning and navigation, (5) possession tracking. And the final one is not typical, but (6) safety confirmation in a time of disaster is also interesting enough to mention that there was an experiment using beacon modules to conduct automatic safety confirmation of disaster evacuees from tsunami at refuge towers in Japan (Aplix, 2014).

Beacon systems can also enables rough location positioning using RSSI. RSSI is the power level of the signal when it reaches the receiver and it is effective where any access to cell tower signals and GPS might not work well. However, location positioning based on RSSI depends on interference of radio waves from surrounding environments. In general, therefore, the accuracy is not so good and it takes a few seconds to measure the position which causes time lag of location positioning. This might be the cause of inaccuracy when one moves fast enough.

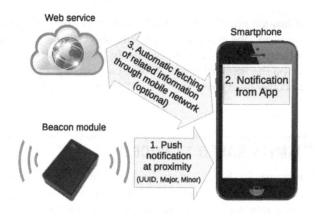

FIGURE 13.1

A typical conceptual image of iBeacon-type proximal push notification system.

Beacon modules are designed simple, small, low-cost, and low-powered to maintain easily. It is said that there are some modules that runs on one AA battery or coin cell for about one year. According to the specification of beacon modules, the interval of the advertising packets is around 100 ms. Setting the advertising interval is a balance between preserving battery life.

Recently Google has also announced a project called *the Physical Web* to standardize a mechanism to interact between things, *i.e.*, Internet of Things (IoT) using web technologies. In this project, Google's UriBeacon is used for the interaction of things.

Proximity-based service is also used in extended concept proposed by the Third Generation Partnership Project (3GPP) as *Proximity Services* (ProSe) (see Lin et al., 2014). This concept treats Device-to-Device (D2D) communication using LTE-Advanced enabling new peer-to-peer public safety networks to function when cellular networks are not available or disrupted.

13.3 PROXIMITY BEACON SYSTEM FOR INDOOR ROUTE GUIDANCE

In this section, we review our proposed system for indoor route guidance (Fujihara and Yanagizawa, 2015). We also demonstrated a system using beacon modules and smartphone application software. It can announce the next direction to turn on smartphone display before reaching the next intersection on a passage way in building at the right place and time, which is just like the car navigation system. If multiple beacon modules are properly deployed along the passage way and major and minor

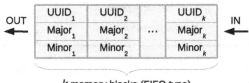

k-memory blocks (FIFO type)

FIGURE 13.2

First-in-first-out (FIFO) type memory space with size k. It can be used to remember the history of beacon detections in chronological order.

numbers are assigned to be useful for the navigation, it is possible to navigate around the intersection. We will explain them in detail below.

13.3.1 ABOUT MEMORY SPACE FOR RECORDING BEACON DETECTIONS

In a usual beacon system, one beacon module corresponds to one automatic triggering of advertising packet. Therefore, if there are n beacon modules in the system, the total number of possible combinations of advertising is n. On the other hand, consider that we are walking a long passage way where many beacon modules are allocated. In this case, smartphones can understand the direction of user movement on the way by just sensing the history of beacon detections. To do this, it is good to introduce a memory space available in smartphone applications. If the FIFO-type memory space is introduced as shown in Fig. 13.2 and it can save k sequential beacon detections, the system with n beacon modules can theoretically distinguish n^k memory states. This means that it can trigger n^k patterns of advertising with aware of the contexts of human movement.

13.3.2 A SYSTEM OF BEACON MODULES FOR INDOOR ROUTE GUIDANCE

The idea of our indoor route guidance using the memory space is shown in Fig. 13.3. Note that there are two reasons why the beacon modules are located in both sides along the passage way. (1) it increases the true-positive detection rate of human behavioral patterns and (2) the application software can detect beacon modules successfully when it is crowded with people in the passage way. Because the electromagnetic wave emitted from iBeacon can be absorbed into human bodies, which causes the decrease of context detection rate. Therefore, we need to allocate the modules as many as possible to create walls that divide the focusing area into some regions.

As shown in Fig. 13.3, we allocate four beacon modules on the left hand side of the figure with $minor = 0$ and the rest of two beacon modules on the right hand side, which is close to the intersection, with $minor = 1$. When a person walks from left

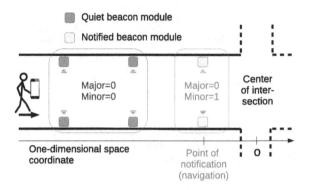

FIGURE 13.3

An image of the indoor route guidance using multiple beacon modules.

to come to the intersection, the application software first detects the beacon modules with *minor* = 0, then it also does ones with *minor* = 1. When a person walks from right, it first detects ones with *minor* = 1, then it does ones with *minor* = 0 inversely. Therefore, by recording the sequence of beacon detections with *minor* numbers, it can aware the context that the person walks from right or left. If the system provides advertisement around the detection region where there are beacon modules with *minor* = 1 only when the person walks from left, the route guidance becomes possible.

When it detects the beacon modules with *minor* = 0, it doesn't advertise, but it only saves the detection history on the memory space. Therefore, there are two types of beacon modules: (1) quiet beacon module or foot print beacon module with *minor* = 0 and (2) notified beacon module or guidance beacon module with *minor* = 1. The existence of quiet beacon module is the essential point of this system. If these patterns of two types of beacon modules are allocated in every passage way connected to the intersection as shown in Fig. 13.4, the indoor route guidance works at T-junctions and crossroads.

13.3.3 AN ALGORITHM FOR ROUTE GUIDANCE

To make the proposed system work, we also need to implement the application software to memorize the sequence of beacon detections. Therefore, it needs an algorithm for the application software. We assume that the memory space with size k has a FIFO type queue and it saves the history of beacon modules in chronological order. The information to save are *uuid*, *major*, and *minor* numbers. When user's smartphone detects a notified beacon module where its RSSI becomes equal to or more than a threshold strength S_c, the application automatically calculates the number of beacon modules with the same minor number. Then, if the number of detected beacon modules exceeds a given threshold value θ, the advertisement for indoor route

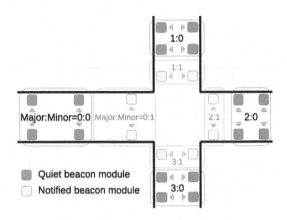

FIGURE 13.4

Major and minor number settings in the system.

guidance is provided. After the guidance, the saved information of beacon modules with the same major number is removed from the queue, which can avoid unexpected duplicate or false guidance. The detailed description of the route guidance is shown in Algorithm 1. A typical situation of the guidance and the history of its queue are also shown in Fig. 13.5.

Algorithm 1 Notification of route guidance.

Require: A queue of k-memory blocks $Q = [(uuid_1, major_1, minor_1), \cdots, (uuid_k, major_k, minor_k)]$ and thresholds S_c and θ

Ensure: Guidance $g \in \{go\_straight, go\_back, go\_right, go\_left\}$

1: **while** the application user does not reach the destination **do**
2:     **while** a beacon module with advertisement $a = (uuid_x, major_x, minor_x)$ is detected (the observed RSSI is more than S_c) **do**
3:         Insert a into the queue Q
4:         **if** $minor_x = 1$ **then**
5:             Calculate the number of blocks having the same $major_x$ in Q to save it into m
6:             **if** $m \geq \theta$ **then**
7:                 Show a correct guidance g along the route
8:                 Remove all entries having $major_x$ in Q
9:             **end if**
10:        **end if**
11:     **end while**
12: **end while**

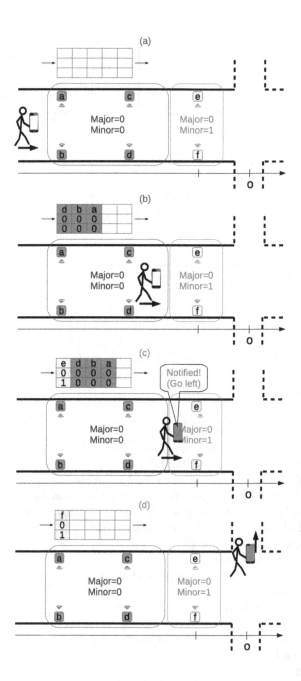

FIGURE 13.5

A typical situation of route guidance and the history of queue in the memory space.
(a) Save the history of detected beacons in the memory space; (b) Some beacons might not
be detected; (c) Get notified when the device hits a beacon with minor = 1; (d) Erase
blocks with major = 0 immediately after notified.

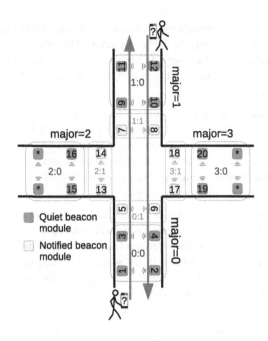

FIGURE 13.6

Experiment setting and its environment.

13.4 EVALUATION

In addition to the previous results (see Fujihara and Yanagizawa, 2015), we performed another experiment to evaluate the performance of our proposed system to work correctly. In the experiment, we use Aplix beacon (MyBeacon Pro MB004 Ac) by Aplix corporation (Aplix, 2015) as the beacon module. We allocate twenty beacon modules around the passages and their major and minor numbers are configured as shown in Fig. 13.6. We fixed the distance between the center of intersection and the position of notified beacon module being five meters. In addition to this, we try the following three experimental settings for the distance between beacon modules: (i) The distance between notified and the nearest quiet beacon modules is fixed to ten meters and that between two quiet beacon modules is also fixed to ten meters. (ii) The distance between notified and the nearest quiet beacon modules is fixed to five meters and that between two quiet beacon modules is fixed to ten meters. (iii) The distance between notified and the nearest quiet beacon modules is fixed to five meters and that between two quiet beacon modules is also fixed to five meters.

We configure the power of beacon modules maximum to make the detection range to be more than ten meters. The threshold value of the number of detected quiet beacon modules for advertising the navigation is set to $S_c = -75$ dB and $\theta = 2$. This means that $k \geq 2$ is necessary to navigate because it needs one quiet and one notified modules at least. We also created an application software for the route guidance by modifying BLE sample programs using Android SDK. We used Google Nexus 4 as the smartphone in the experiment. When we approach around the intersection in Fig. 13.6, the guidance is shown if the system successfully works. If not, the system fails to navigate a user.

We also considered three movement patterns: normal walk, fast walk, and run. We repeat these walking and running along the passage way having the smartphone that the developed application is installed to investigate the true-positive detection rate of the guidance. We also measure positions that the guidance is notified to draw a histogram of the notified position. We collected 30 samples of each movement patterns and inter-beacon-distance settings.

We introduce a true positive rate of successful detection which indicates the notification of next direction to go is guided somewhere along the passage way. We find that the true positive rate is 100% when the user moves with normal and fast walks. But, it decreases when the user runs. In the worst case, the rate goes down to 20%. These results indicate that fast movement like running tends to fail the detection, but it is successful when the movement speed is slower.

We show the histogram of sampled notified positions for the three inter-beacon-distance settings (i), (ii), and (iii) in Fig. 13.7, Fig. 13.8, and Fig. 13.9. The zero point in the horizontal axis means the intersection position. The point of notification must be located around −5 meters (five meters behind the next center of intersection). We can observe that most notified positions are between the position of the notified and the nearest quiet beacon modules. But, it sometimes notify too fast around the further quiet beacon module or too late around the center of intersection. It seems that the second distance setting (ii) is the best performance of all as shown in Fig. 13.8. This is because the following reasons. (1) The successful detection rate for walking is 100%, (2) the rate for running is still 90%, and (3) all the sampled notified positions are also stably distributed around the point of notification and the variance of notified positions is relatively small.

13.5 APPLICATION

The proposed indoor route guidance system can be applicable not only for navigating potential customers walking on a shopping mall to a shop that they are interested in, but also disaster evacuation guidance to the exit at a time of disaster. This is what we call opportunity-based service or proximity-based service. It is interesting to study how mobile and wearable devices can help us to navigate using proximity sensors from the viewpoint of Delay Tolerant Network (DTN) (Vasilakos et al., 2012) and Mobile Opportunistic Network (MON) (Denko, 2011; Woungang et al., 2013). To this end, it leads to provide truly practical support for disaster evacuation guidance (see Fujihara and Miwa, 2014).

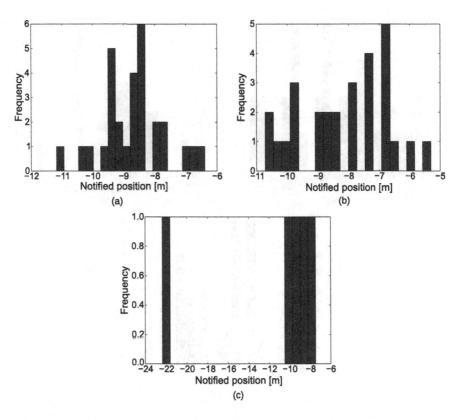

FIGURE 13.7

Histogram of sampled notified positions when the user walks slowly (a), when the user walks fast (b), and when the user runs (c) in the first inter-beacon-distance setting (i).

13.6 CONCLUSION

As an example of proximity-based service, a system for indoor route guidance using beacon modules was investigated to expand human proximity awareness and the system behaves like the car navigation system. To do this, we extend the concept of beacon module systems to advertise multiple notifications from a single beacon module using the combinations of the sequence of detected beacon modules saved on FIFO-type memory space. Furthermore, the system needs two types of beacon module: quiet and notified beacon modules on the passage way near the intersection with configuring the major and minor numbers appropriately.

We performed a simple experiment using the developed application software for the route guidance to make it working on an Android OS smartphone. We evaluated the performance by focusing on the true positive detection rate and the frequency of notified positions in some cases. Consequently, we found that the proposed system

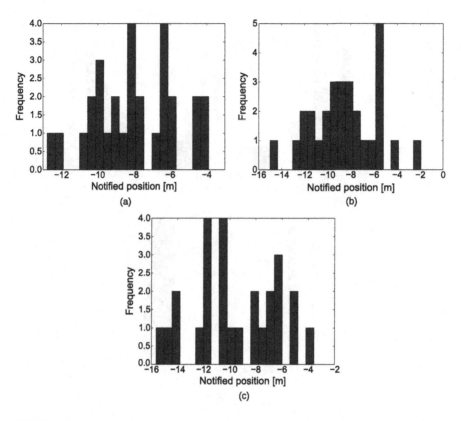

FIGURE 13.8

Histogram of sampled notified positions when the user walks slowly (a), when the user walks fast (b), and when the user runs (c) in the second inter-beacon-distance setting (ii).

can advertise navigation messages properly when we deploy beacon modules in a proper interval with understanding human movement contexts.

For future work, it needs to investigate the effectiveness of the indoor route guidance for users going through multiple intersections sequentially. It still has not been checked that radio wave interference between beacon modules of different intersections and how to control S_c to avoid the interference. It might also be interesting to challenge outdoor route guidance on the street or shopping avenues where there are possibly many people, while there was only a single person along the passage way in our experiment.

Proximity-based service can change the human sense of proximity in an extended way. The future of the Internet must evolve to become more energy efficient and also more protection of privacy. Proximity-based service is a good way to coexist them and it might realize more smart IoT with controlling connections of everything via proximity.

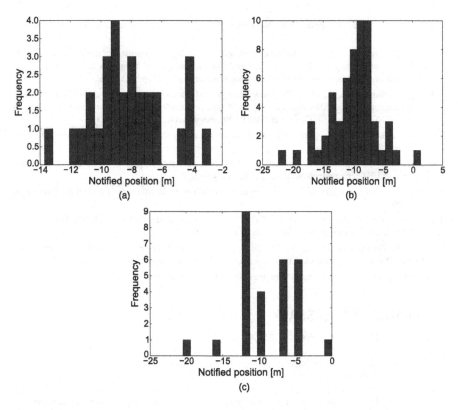

FIGURE 13.9

Histogram of sampled notified positions when the user walks slowly (a), when the user walks fast (b), and when the user runs (c) in the third inter-beacon-distance setting (iii).

ACKNOWLEDGMENTS

The author thanks Mr. Takuma Yanagizawa for supporting the application development and performing experiments. This work was partially supported by the Japan Society for the Promotion of Science (JSPS) through KAKENHI (Grants-in-Aid for Scientific Research) Grant Numbers 25870958, 26330122, and 15K21473.

REFERENCES

Aplix, 2014. Safety confirmation experiment in the time of tsunami disaster. http://www.aplix-ip.com/wp-content/uploads/09262014ZrqukDj9_PR.pdf (in Japanese).

Aplix, 2015. MyBeacon pro mb004 ac, aplix corporation. http://www.aplix.co.jp/?page_id=9463 (in Japanese).

Blondel, V.D., Decuyper, A., Krings, G., 2015. A survey of results on mobile phone datasets analysis. EPJ Data Science 4 (10).

Denko, M.K., 2011. Mobile Opportunistic Networks: Architectures, Protocols and Applications. Auerbach Publications.

Fujihara, A., Miwa, H., 2014. Disaster evacuation guidance using opportunistic communication: the potential for opportunity-based service. In: Big Data and Internet of Things: a Roadmap for Smart Environments. In: Studies in Computational Intelligence, vol. 546, pp. 425–446.

Fujihara, A., Yanagizawa, T., 2015. Proposing an extended iBeacon system for indoor route guidance. In: Proceedings of the 2015 International Conference on Intelligent Networking and Collaborative Systems. INCOS '15. IEEE Computer Society, Washington, DC, USA, pp. 31–37.

Gast, M.S., 2014. Building Applications with iBeacon: Proximity and Location Services with Bluetooth Low Energy. O'Reilly Media.

Lin, X., Andrews, J.G., Gosh, A., Ratasuk, R., 2014. An overview of 3gpp device-to-device proximity services. IEEE Communications Magazine 52 (4), 40–48.

Tobler, W., 1970. A computer movie simulating urban growth in the Detroit region. Economic Geography, 234–240.

Townsend, K., Cufi, C., Akiba, R., Davidson, R., 2014. Getting Started with Bluetooth Low Energy: Tools and Techniques for Low-Power Networking. O'Reilly Media.

Vasilakos, A., Zhang, Y., Spyropoulos, T.V., 2012. Delay Tolerant Networks: Protocols and Applications. Wireless Networks and Mobile Communications Series. CRC Press.

Woungang, I., et al., 2013. Routing in Opportunistic Networks. Springer.

ACRONYMS AND GLOSSARY

List of acronyms with explanation

BLE	Bluetooth Low Energy
PBS or ProSe	Proximity-Based Service (cf. LBS, Location-Based Service)
ProSe	Proximity Service
RSSI	Received Signal Strength Indicator
5G	The 5th Generation (of information telecommunication technology for mobile phone)
CDR	Call Data Records
3GPP	the Third Generation Partnership Project as Proximity Services (see Lin et al., 2014).
D2D	Device-to-Device (communication)
LTE	Long Term Evolution
IoT	Internet of Things
UUID	Universal Unique IDentifier
FIFO	First-In-First-Out
SDK	Software Development Kit
DTN	Delay Tolerant Network
MON	Mobile Opportunistic Network

Glossary of terms with explanation

Bluetooth Low Energy A close-range wireless communication technology designed for sensing proximity to run with low power.

Proximity-based service or Proximity Service A service provided at an opportunity of proximal encounter with beacon module (cf. Location-Based Service).

The first law of geography Waldo Tobler, a geographer, says "Everything is related to everything else, but near things are more related than distant things".

Gravity model (of migration) Like Newton's law of gravity, it explains the force between the degree of interaction between two places, for example, populations in two cities, which is proportional to the product of two populations and inversely proportional to the square of the distance between them.

Local information production for local consumption A concept for separating local information from global Internet traffics for security or privacy reasons.

Extended proximity Sensing a proximal environment using wireless communication technologies.

Beacon module A device emitting radio waves to nearby human-carried mobile devices for sending push-type advertisement packet.

iBeacon A service for sending push-type advertisement packet to mobile phones by Apple.

Physical Web A service for sending push-type advertisement packet to mobile phones by Google.

Received Signal Strength Indicator A measurement of the power of a received radio wave signal.

WiGig A wireless communication technology capable of sending a gigabit per second speed using 60 GHz frequency band.

Wi-Fi aware An extended Wi-Fi technology capable of power-efficient and interactive communication between nearby mobile devices.

Wi-SUN an IEEE 802.15.4g based wireless communications specification for peer-to-peer networking.

LTE Direct Device-to-device technology enabling the discovery of thousands of devices in the proximity of approximately 500 meters proposed by Qualcomm.

Call Data Records or Call Detailed Records A data record produced by information telecommunication systems, which contains time, duration, source and destination numbers, and others.

Device-to-Device Communication Communications without base station that devices create peer-to-peer networking.

Internet of Things The concept of networking all physical devices enabling to collect and exchange data through the Internet.

First-In-First-Out A method for organizing and manipulating a data buffer where the oldest entry is processed first.

Software Development Kit A set of software development tools that allows the creation of application for a certain software.

Delay Tolerant Network A computer network architecture capable of forming network in extreme or challenged environments with lack of continuous network connectivity because of delay, disruption or disconnect.

Mobile Opportunistic Network A computer network architecture capable of forming networks using human mobility patterns and social interactions in daily life.

WiFi TRACKING OF PEDESTRIAN BEHAVIOR

14

Andreea-Cristina Petre*, Cristian Chilipirea*, Mitra Baratchi†, Ciprian Dobre*, Maarten van Steen†

*University Politehnica of Bucharest, Romania †University of Twente, Netherlands

14.1 INTRODUCTION

Pedestrian dynamics continues to receive much attention from scientists, architects, event organizers, game designers, and many other groups. The need for understanding those dynamics is undisputed in the face of questions related to safety, experience enhancement, and planning in general. To this end, many models have been developed, not only to capture observed behavior, but also to predict what will happen in given situations. In a recent survey, Wijermans et al. (2016) observe that by far most of these models lack proper validation for the simple reason that actual data sets were not readily available.

This lack of data on pedestrian behavior has changed dramatically in recent years. Notably with the massive introduction of smartphones, sensing what is happening in a crowd has become within reach. Not only has it become easier from a technological point of view, at least as important is that with no, or only minimal intrusion, has it become possible to sense behavior at large scales. In many cases, being able to actually measure behavior of *large* groups of people is a prerequisite for model validation.

Yet, it is not only the need for validating models that drives research into sensing pedestrian behavior. The fact alone that it can be done relative easily has triggered curiosity to further understand that behavior. In other words, even without interest in models, event organizers, urban planners, and so many others, simply want to know "what is going on," in order to take better informed decisions.

In this chapter we explore how WiFi technology can be deployed for measuring pedestrian behavior. The basic idea is simple: WiFi-enabled devices regularly send messages containing a unique identifier for that device. A WiFi scanner receives those messages, which can then be processed further. Having a unique identifier enables us to (1) distinguish devices from each other, and (2) determine whether two scanners have observed the same device. In other words, we are capable of *detecting devices*. These two properties form the foundations for counting devices as well as tracking their locations when scanners are located at different places. If we know the ratio between devices and people, we can then draw conclusions on the movement of people.

Smart Sensors Networks. DOI: 10.1016/B978-0-12-809859-2.00018-8

Although other technologies exist as well, WiFi-based systems have become popular for two simple reasons. First, many people carry WiFi-enabled devices making it an ideal instrument for tracking people. Second, and very important, is that the owner of a device need not do anything to allow tracking except enabling the WiFi capabilities of her smartphone. In practice, smartphones virtually always have their WiFi enabled. Of course, this unintrusiveness with regard to tracking people imposes serious privacy issues for which often no easy solution is available.

Despite that the basic idea of WiFi-based tracking can be easily explained, accurately measuring pedestrian behavior this way is a nontrivial exercise. To make this clear, we start with presenting the technical principles of a WiFi-tracking system and pinpoint the sources of potential errors, and how easily different failures in detections can be masked or corrected. Note that many researchers and practitioners tend to ignore the technology bottlenecks one has to solve, only to be surprised by the low quality of the resulting data set of detected devices. We illustrate some of the difficulties and inaccuracies of a WiFi-tracking process with our field findings obtained in a real-world tracking experiment.

No matter how good the equipment, WiFi-based detections have inherent quality problems caused by difficulties of using a radio-based medium. The resulting data set therefore always needs to be cleaned up. We present a set of filters that remove or correct detections leading to a higher quality data set that retains the relevant properties related to pedestrian behavior of the original set. In turn, this cleaning leads to better and faster analyzes. Some of these filters are particular to WiFi but some of them can be applied to any other type of detection system.

Once we have a reasonable set of detected devices, there are many options. We illustrate some of these options in Section 14.4 by considering two different data sets. One comes from monitoring a three-day festival using some 25 scanners, which we use to discover likely paths taken by pedestrians. Another is based on scanning several locations for a long time with the purpose to *fingerprint* each location.

14.2 PRINCIPLES OF WiFi TRACKING

Tracking WiFi-enabled devices is based on the procedure which is meant for network discovery by mobile devices (Curran et al., 2011) by a WiFi access point (which we refer to as WiFi scanner in the rest of this paper). There are two ways in which a connection can be established. First, a WiFi scanner can advertise its presence by broadcasting beacon messages, to which a device can subsequently respond. However, instead of waiting for a scanner to announce itself, for mobile devices such as smartphones is generally more efficient to actively seek for scanners. To this end, a mobile device periodically broadcasts a **probe request**. Such requests are sent regardless if a connection has been established: if a better scanner is detected, the mobile device will want to connect to that access point instead of its current one.

A probe request contains a lot of information, but most importantly, it contains the MAC address of the sending device and potentially a list of network identifiers known as SSIDs, through which the device has previously connected. Usually, the perceived signal strength of the probe request is available for the receiving device (i.e., scanner). A MAC address is uniquely associated with a network interface, and can thus act as an identifier for a device. If a device has multiple network interfaces, it will thus, in principle, have multiple associated identifiers. We will denote the MAC address associated with the WiFi network interface of a mobile device as MID.

WiFi tracking works as follows. A scanner device with the identifier SID WiFi receive a probe request from a mobile device with the identifier MID at time T. This can yield into triplets $\langle SID, MID, T \rangle$. By chronologically ordering these triplets for a specific mobile device, we obtain a series of scanners that have subsequently detected that device. Knowing the location of the scanners should therefore give us information on a rough estimation about the whereabouts of the device, over a certain time span. As we explain in this chapter, matters are not that simple in practice.

14.2.1 WiFi SCANNERS: TECHNICAL BACKGROUND

WiFi tracking is based on the fact that smartphones and tablets are now ubiquitous. These **devices** have a WiFi module and are commonly used to access the internet. According to the Internet Society (Global internet report, 2015) mobile broadband already accounts for more than 50% of traffic in developing countries. WiFi devices send frames which can be received and recorded by **scanners** (more details in Section 14.3.1). The scanners can be as simple as a household WiFi router with modified software that enables the recording of any WiFi frames the router antenna receives. The same thing can be achieved with a laptop, however WiFi routers have the lowest price. Using a WiFi frame the scanner can build what we call a **detection**, a $\langle scannerid; deviceid; timestamp \rangle$ triplet. These triplets are sent to a central **server**. The process is represented in Fig. 14.1.

A WiFi scanner is a device capable of receiving WiFi frames. Virtually any device with a WiFi interface (e.g., wireless routers or smartphones) can operate as WiFi scanner, provided it supports the so-called **monitor mode** as defined in the IEEE 802.11 standard (IEEE 802.11, 2012). In monitor mode, a device can capture any (correct) WiFi frames from any other device without the need for the two to have been first explicitly associated (and able to exchange further traffic) with each other. In practice, WiFi scanning is done by WiFi routers or dedicated devices (scanners).

A practical issue is handling the captured frames is transferring them to a processing server. One way or the other, probe request are meant for allowing exchange of traffic to a network through the scanner. To this end, a scanner is generally connected to a backbone network through a separate network interface. There are many variations and combinations used in practice. For indoor scanners it is often convenient to use a separate WiFi network, or an available (wired) Ethernet network. Outdoor scanners are often equipped with additional 3G and 4G interfaces which allow sending the data over mobile networks. When no network is available to transfer the data,

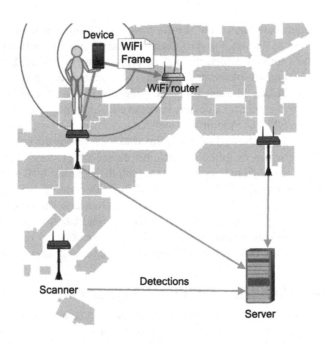

FIGURE 14.1

WiFi tracking.

it may also suffice to store captured frames locally on the scanner and to postpone data analysis.

14.2.2 COMMON ISSUES

A perfect mobility data set would be one in which the location of a mobile device is accurately known at defined intervals. This means that there is no interval when the location of the mobile device is not registered. This does not mean that a device should not trigger detections at two or more scanners simultaneously, given an intersection in their coverage range. Simultaneous detections are acceptable as long as the RSSI values can be used to calculate a realistic positioning of the device.

Unfortunately, gathering a perfect data sets through tracking WiFi-enabled devices is a challenging problem. There are many sources of errors. Some the challenges are:

Faulty Scanner

Some errors are caused by the scanners and these are often the simplest to detect and correct. For example, any interval in which a scanner is shut down or cannot receive frames will generate a clear irregularity in the density of detections over time for that scanner. In our example data set, scanners automatically reboot once every 24 hours, leading to a noticeable glitch in the detections.

Limitations of Radio-Based Detections

WiFi uses a data transmission medium which is inherently unreliable (Salyers et al., 2008). For example, most WiFi devices claim a 100-meter transmission range in ideal conditions. In reality, such specifications cannot be relied upon due to practical sources of impairment in wireless transmission, such as attenuation distortion, free space loss, noise, atmospheric absorption, multipath, and refraction (Beard and Stallings, 2016). As an effect of such impairments, in practice it is observed that tunnels extend the transmission range, while buildings and people are known to hinder transmissions. This also means that the shape or size of the area where WiFi frames can be correctly received can be highly irregular. Due to such issues, during our experiments we have come across the detection of a single device by five or more scanners at the same timestamp. However, such detections could not be explained considering the placement of scanners and a uniform detection range.

Limitations of RSSI

Using trilateration based on the received signal strength indicator (RSSI), we should, in principle, be able to pinpoint the location of a device (Liu et al., 2007). There are multiple problems to be addressed. First, RSSI measurements as taken by the scanners, are not standardized and can differ in value or strength across different types of scanners. Second, the signal strength itself can dramatically differ across multiple device manufacturers and even different devices of the same model. Solutions for the RSSI problems have been proposed (Kim et al., 2012), but only for when the mobile device is the one taking the measurements. These solutions do not directly apply to the reverse scenario. An experimental evaluation of RSSI-based localization methods is presented in Zanca et al. (2008), illustrating its inherent difficulties.

Timing Errors

Scanners timestamp detections. Consequently, their clocks may introduce many inaccurate detections if not properly synchronized between different scanners. If the clocks are not synchronized one device moving from scanner A towards scanner B, may unexpectedly be associated with records showing that the detection at scanner B *followed* by a detection at scanner A. Even when the scanners are completely synchronized it may be difficult to determine the exact time a detection belongs to. There is no way to determine if two frames received at two different scanners are actually the same, as a probe request does not include a sequence number that would permit differentiating between two separate such frames from the same device.

MAC Address Issues

There used to be a time when a MAC address could be more or less used as a stable and unique identifier for a device. This is no longer a justifiable assumption. Some devices change their MAC address seemingly at random, as also reported by Musa and Eriksson (2012a). This is known in particular in the case of some Apple devices (Stites and Skinner, 2014). Perhaps even worse when using a MAC address for device identification, is that we have noticed cases where different devices use the *same* MAC address.

Correct Inference About the Population

For various types of applications with societal relevant impacts, it is necessary the correct value about the number of people is estimated. Estimation of such numbers is yet another challenge to overcome due to the existence of randomize MAC addresses, people with multiple wifi enables devices, or people with no devices. These challenges are also alleviated when the number of people with more than one device becomes specific to spaces or occasions. For instance, when students of a course all need to bring a laptop for only one session.

Lack of Coordination

Because there is no coordination between devices and scanners, no ideal probe transmission rate can be determined or let alone set. We have witnessed a huge variation in transmission rates, caused by seemingly random behavior when a device switches its WiFi module on or off. This behavior is also dependent on the device, as reported in Cunche (2014) where a comparison between Apple and Samsung devices is presented. As a result, the effect, in combination with the unreliability of the wireless medium, is a data set with detections that can make the movement of a device seem mostly erratic. To illustrate what this may lead to, consider Fig. 14.2, which is taken from one of our measurements. In this case we have a known, non-mobile device appearing as a device that moves in a loop, with random frequency and random speed. An actual mobile device could exhibit an even more chaotic behavior. Instead of moving in what would be a straight line, the detections would show it moving in small irregular circles while eventually getting closer to its destination.

By-and-large, there are many sources that introduce *noise* into a set of detections. Table 14.1 lists the main sources of errors that introduce seemingly chaotic behaviors in datasets collected through WiFi scanning. The last column notes the levels of difficulty in dealing with these problems as experienced by us. The ones marked as hard could even be impossible to fix.

14.2.3 ALTERNATIVES

Besides using WiFi, there has also been considerable research in using other techniques. In the following, we briefly mention the most prevalent ones.

CCTV

Classically, data from crowds was obtained using visual systems. These systems are known for being able to infer information on crowds such as congestions, overcrowding, and blocking (Siebel and Maybank, 2004). However, systems that use visual data

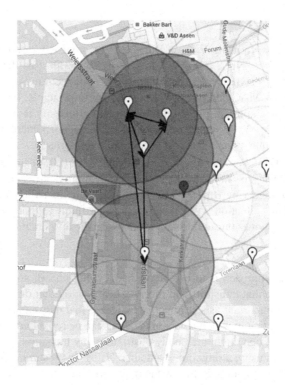

FIGURE 14.2

Movement path of static device (the circles are 100 m visual guides, they **do not** represent the cover radius).

Table 14.1 Sources of Noise in Set of Detections

Description	Scanner	Device	Method	Correctable?
	Source			
Faulty scanner	Yes	No	No	Easy
Dynamic, irregular ranges	Yes	Yes	No	Hard
RSSI issues	Yes	No	Yes	Hard
Timing errors	Yes	Yes	Yes	Hard
Multiple addr. per device	No	Yes	No	Easy
Multiple devices per addr.	No	Yes	No	Easy
Uncoordinated probes	No	Yes	No	Medium
Lost frames	No	No	Yes	Medium

are known not to scale well: First of all there is a high cost in deploying a large number of video cameras; secondly the data itself is difficult to extract and process. These systems have an important advantage of being simple to check, errors in data extraction can be corrected by manually checking video logs, but this is a timely and costly procedure.

Bluetooth

Another alternative to using WiFi is Bluetooth (or other RF based communication technologies such as NFC, RFID, ZigBee, etc). Bluetooth uses the same protocol (similar to WiFi probe requests), for establishing a connection. Such messages contain the Bluetooth MAC address, which again can, in principle, be used for device identification. An important potential advantage of Bluetooth is its more restricted transmission range. As a consequence, provided enough scanners have been positioned, we can obtain a more accurate account of trajectories.

Unfortunately, and as also reported by Schauer et al. (2014) and others, coverage by Bluetooth devices is very low. In practice, many more devices have WiFi enabled than Bluetooth, rendering Bluetooth-based tracking ineffective for many situations. It is unclear if this situation will change in the near future, although wearables that connect to smartphones through Bluetooth may change this situation.

Active Badges

A very different approach is to use active badges. In general, an active badge is a proprietary device that transmits and receives beacons to other devices and possibly also badges. An important aspect of active badges is that they can be worn in such a way that the *direction* in which a person is facing can be reliably determined. This is caused by the fact that the human body obstructs many transmissions so that we get the same effect as using a directional antenna. Using this feature, studies have been recently carried out to not only follow the trajectories of visitors at an exhibition, but to also determine at which exhibit they were looking (Martella et al., 2016). Using active badges is not efficient in the case of events with large number of participants. Another drawback of these devices is their dependence on extra equipment being carried by people.

GPS-Based Systems

The most common method to identify trajectories of individuals is given by the Global Positioning System (GPS). However the GPS receiver has an error of more than a few meters. There are numerous sources for these errors and they are analyzed in Evans et al. (2002). The accuracy of these systems is also dependent on the speed of the device as shown in Aughey and Falloon (2010). They are only meant for outdoor purposes and even when direct line of sight to the positioning satellites is hindered location data cannot be provided (e.g. existence of a forest canopy) (DeCesare et al., 2005). These problems are not local to GPS but extend to other specialized positioning systems such as the Argos satellite system (Jonsen et al., 2005; Baratchi et al., 2013) used to track animals.

In Thiagarajan et al. (2009) the authors present a way of smoothing the path taken by an individual, as given by raw GPS data. The examples they show present a data set where multiple consecutive detections move back and forth, circling the street the individual is on. This behavior is similar to the behavior we present in this article. Yet the technologies and methods used are different. To extract a clear path the authors use outlier removal with interpolation followed by Viterbi matching. Similarly Yan et al. (2010) use outlier removal and Gaussian kernel regression to smooth the paths shown by GPS data. Their methods are not directly applicable to our scenario. Data collected using GPS has a finer location accuracy than the data from WiFi scanners. In contrast, we have low number of detections and a rough approximation (in the order of hundreds of meters) of the actual position. While precise location of a device is a positive attribute of GPS systems, it is invading the privacy of the users carrying GPS equipped devices. In many civil applications based on GPS such accuracy is not required. Thereby, many GPS based data analysis is done by discritizing GPS data to a courser grained accuracy level such that the privacy of the users is not invaded (Baratchi et al., 2014).

Systems Based on Regular Cell-Phone Traffic

Finally, it is has become common to use traditional traffic generated by cell phones to locate and track people. Straightforward is to use the identifier of the cell to which the phone is connected as the base for location. However, considering that cells are relatively large, sometimes having a diameter of even a few kilometers in rural areas, these cell-identification sensing techniques are far from accurate. As explained by Ficek et al. (2013), various additional methods and techniques are necessary to improve accuracy before being able to come to any sensible tracking information.

Over all these methods WiFi does have a few clear advantages. It takes advantage of the fact that smartphones, equipped with WiFi modules are ubiquitous. This means that the cost of deploying a WiFi crowd tracking system is small and that there is no need for the cooperation of the individuals in the crowds being tracked. Furthermore, the data outputted by the system is easy to process and contains no false positives. It also has a high accuracy compared to cell-phone communication.

14.3 HANDLING RAW SENSORY INPUT

The basic concept of scanning for WiFi frames and being able to track crowds is simple. The required hardware for scanning, a WiFi enabled device, is easily accessible and cheap. This hardware does require special software in order to enable the scanning functionality.

To offer a concrete example, for our research we primarily use a platform consisting of scanners (BlueMark 1000 series) and a server to store the data trace. For communication and data transfer, we use broadband communication (a 4G mobile network). The BlueMark BM1000 scanner has 32 MB RAM, 8 MB flash, and a 384 MHz CPU. It runs openWRT (https://openwrt.org/) as its operating system, together with an application, developed by us, that collects the required data. The scanner uses a directional antenna with an antenna gain of around 12 dBi. Like the case with any wireless technology, the shape of the area in which WiFi frames can be received is irregular and inconsistent in time.

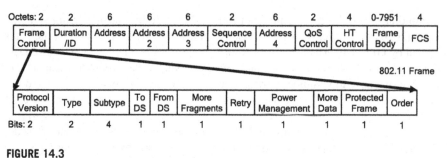

FIGURE 14.3

WiFi, 802.11 general frame format.

The scanner outputs data in SQL text format, it compresses it and periodically sends it to the server. At the server the output files are decompressed and set for long-term storage and analysis. Scanners are synchronized using NTP and reboot daily at 5AM. This last step is meant to minimize errors.

In order to build a scanner application we used the libpcap library (http://www.tcpdump.org/) to gather WiFi frames and an MD5 (Rivest, 1992) library to hash device identifiers in order to preserve the privacy of pedestrians.

14.3.1 THE 802.11 PROTOCOL FAMILY

Commonly known as WiFi, the IEEE 802.11 family of protocols (http://standards.ieee.org/getieee802/download/802.11-2012.pdf) standard defines medium access layer and physical layer specifications. At the medium access layer the transmitted data is split in frames. The standard defines 39 frame types and sub-types as well as a number of reserved ones. All the frames contain a header with information relevant to the connection itself. Even if the connection itself is encrypted, the header is sent in clear. The header may include identifiers for senders and receivers, called MAC addresses. The MAC address is meant to **uniquely** identify a wireless device. IANA (http://www.iana.org/) provides OUI (Organizationally Unique Identifier) numbers for hardware manufacturers and hardware manufacturers use these numbers to generate MAC addresses. The OUI is used as the first 24-bits of the MAC address. Because a device can be uniquely identified, detections can be correlated across multiple scanners.

Fig. 14.3 is a diagram describing the general frame format. The minimal frame format contains only the Frame control, Duration/ID, Address 1, and FCS fields. It is common that the first three MAC addresses to represent the source address (SA), destination address (DA), and the basic service set identifier (BSSID) which identifies the network. There are frame types that do not contain a source address, for instance, the Clear To Send (CTS) frame, used to signal that there are no other transmissions taking place. Out of the 39 frame sub-types only 27 have a source address. Table 14.2 lists the 27 frames.

Table 14.2 27 Frame Types/Sub-types that Contain a Source Address

Type	Sub-type		
Data	Data	Data+CF-ack	Data+CF-poll
	Data+CF-ack+CF-poll	Null	CF-ack
	CF-poll	CF-ack+CF-poll	QoS Data
	QoS Data+CF-ack	QoS Data+CF-poll	QoS Data+CF-ack+CF-poll
	QoS Null	QoS+CF-ack (no data)	QoS+CF-poll (no data)
Management	Association_Request	Reassociation_ Request	Probe_Request
	ATIM	Disassociation	Authentication
	Deauthentication	Action	
Control	Block_Ack_Request	Block_Ack	PS_Poll
	RTS		

WiFi uses a frequency band that is not common in nature. This is especially true considering that according to the standard data is encoded in a digital form, also uncommon in nature. Furthermore, all 802.11 frames contain a Cyclic Redundant Check (CRC) number in the FCS field. The CRC is used to identify transmission errors. If the CRC is missing or does not match the expected value, the frame is automatically dropped by the kernel. All these features of the WiFi protocols guarantee that a scanner can never detect a device that does not exist. In other words, there are no false positives during normal usage.

A malicious device can change its MAC address and even send frames having the MAC address of other devices as the source address. This would be an example of false positives. In the previous section we discussed the case of Apple devices which behave in this manner (Stites and Skinner, 2014). However, this type of detections can be easily identified and filtered (more details in Section 14.3.2). However, once a device is connected to a network it needs to use the same MAC address. This means that a device cannot always hide itself from the WiFi tracking. For the end user one way of preserving privacy is to keep the WiFi module offline unless in use. Most Operating Systems leave the WiFi module on by default, and so do most users. There are also convenience and usability reasons that would motivate the continuous use of WiFi.

The Frame Body field represents the payload, the useful data of a transmission, usually in the form of IP packets. This part of the frame may be encrypted if the devices are connected to a network with WEP or WPA. Scanners should completely ignore the payload even when it is sent in clear text. Scanning or recording payload data raises important privacy issues. The only relevant data for WiFi tracking is the MAC address of the transmitter, the source address.

Generally WiFi networks have one hotspot (a WiFi router) and multiple mobile devices that connect to it. The hotspot advertises its SSID (name of the network) and other information such as accepted speeds using *Beacon* frames. Mobile devices listen for *Beacon* frames and connect to the network, if it is known or after they request user input.

Table 14.3 All vs *Probe_Request* **Frames**

	High Traffic Channel	Low Traffic Channel
# All frames	2906574	18936
# Probe_Request frames	10484	12404

According to the 802.11 protocol family, mobile devices can actively scan for networks using *Probe_Request* frames. This permits a mobile device to find and connect to a hidden network, one that does not send *Beacon* frames. A *Probe_Request* is sent on every channel, once for each recently connected network. A *Probe_Request* may contain an SSID for one of the already known networks. The hotspot responds with a *Probe_Response* frame, containing data about the network capabilities. After this the mobile device and hotspot can start the process of association, which in the cases of encrypted networks is followed by authentication. If these steps are successful, then the mobile device is connected to the network and can start communicating.

Roaming represents a mobile's device capability to change the hotspot it is connected to without causing an interruption in service. This is common for mobile phones on the GSM network. A phone call is not interrupted even when moving. In order to enable roaming capabilities *Probe_Request* frames are sent even when a device is already connected to a network. This way a device can identify other hotspots with better signal and can dynamically change to them. Because they are sent at somewhat regular intervals as long as the WiFi module is on, most experiments are run filtering everything but *Probe_Request* frames.

The frequency with which *Probe_Request* frames are sent is not clearly defined and in practice it is dependent on multiple factors. Each network stack implementation has different policies and rules on setting the frequency and most commonly it is dependent on battery status. For example, a device with low battery tries to conserve it as much as possible and sets a very low frequency at which to send *Probe_Request* frames. We found it to be common for the frequency of a scan (one frame for every channel and recently connected network combination) to be set to 30 seconds.

In order to confirm the behavior of mobile devices when sending *Probe_Request* frames we conducted a small experiment. We counted the total number of frames and the number of *Probe_Request* frames detected by one of our scanners. The scanner had two antennas set on two different channels and the data was recorded for a period of one day. The channels were chosen to be the ones where we detected the most and respectively the least number of frames for a previous period of one day. The results are presented in Table 14.3. The number of *Probe_Request* frames are very similar for both channels. In contrast, the total number of frames differs with several orders of magnitude. There are multiple reasons for which the number of *Probe_Request* frames do not perfectly match, the main one being the noise in the medium. It is common for WiFi frames to be lost or dropped because of CRC errors. Because the numbers of *Probe_Request* frames on difference channels are similar, the experiment confirms that in order to scan for available networks a mobile devices sends *Probe_Request* frames on every channel.

After frames are collected by the scanners they need to be converted into detections and sent to the server. A simple approach would be to create a detection for each individual frame. But, not every frame is a good representative of a detection. For instance, it is impossible to create detection from frames that don't contain a source address as there would be no way to set the *device_id*, frames that arrive in a very short interval would generate detections that are indistinguishable or similar enough that they would be irrelevant. In order to clean this raw data set and keep only detections that are relevant we propose a number of filters, some at the scanner and some other the server level. These filters are described in detail in our previous work (Chilipirea et al., 2015a) and we present some of the basic concepts behind them in the next subsections.

14.3.2 FILTERING DATA AT THE SCANNERS

With multiple scanners and one central server the first issue when trying to achieve scalability of the system is the bandwidth usage. In order to preserve bandwidth as much data as possible needs to be removed at the scanners. Even better: unnecessary detections are discarded and not sent to the server. For instance, frames that do not have a source address need to be ignored.

Filter 1

removes frames that do not contain a source address. Only the sub-types from Table 14.2 can pass the filter. The filter also removes frames that are not generated by a mobile device. Fields $fromDS$ and $toDS$ (DS represents the Distribution System, a wired network) from the 802.11 frame, in Fig. 14.3, indicate if a frame is sent to or from a distribution system. The filter is extremely fast: it only needs to check the type of the frame and in case it is a Data frame it needs to check the $fromDS$ field.

Filter 2 and 0

are meant to work together. Filter 0 removes all *Beacon* frames but keeps a list with their MAC addresses. These are MAC addresses of WiFi access points or hotspots. Filter 2 uses these addresses to remove all frames sent by hotspots or access points as these are most likely non-mobile devices. These frames are not removed by Filter 1. In our data we encountered frames with $fromDS$ set to 0 while the transmitter address was that of a hotspot.

Filter 3

is meant to only accept frames that are far enough time wise. The WiFi protocol allows the transmission of multiple frames every second, even by the same device. However, the location of a device does not change significantly in under a second. The filter accepts only frames that have at least three seconds between them. This three second interval was chosen empirically, by looking at the data in our sources.

Table 14.4 Data Set Characteristics

	Arnhem	Assen	Student Complex
# of frames	2472380	11860349	2906574
# of scanners	5	27	1
# of days	1	3	1
Start Date	2014-07-07	2015-06-25	2014-07-04

Other values can be chosen depending on the application, for instance, in Musa and Eriksson (2012b), where a similar filter is applied, a one-second interval is used as an aggregation point. This time based filter can also be used to filter devices that are non-mobile: Devices that are detected by the same scanner for several hours in a row. In order to implement this filter there are some memory usage considerations. The filter requires a list with all devices detected as well as the first and last time when they were detected.

In Chilipirea et al. (2015a) we go into details on the performance and the results of each individual filter. To get an idea on the advantages of having these filters at the scanners we present the overall results for them. The filters can be used at the same time and all frames that pass all of them are considered to be a detection and sent to the server. We present the results on two data sets. One data set is generated using a scanner placed inside a room of a student complex of VU University Amsterdam, Netherlands. The other data set is obtained from one of the scanners we used in the city center of Arnhem. Because the two traces were run in different environments, particularities of these environments can be seen in the data. For instance, the Student Complex trace has a lot of data frames that are sent by only two devices, most likely belonging to the students closer to the scanner. Table 14.4 contains the number of frames and the dates on which the trace started.

When testing the filters we initially recorded all frames detected by the scanners and run each filter offline. We did not add any time constraint limitations. We left the software to process the data as fast as possible. This means that we analyzed a few days of data in under 3 minutes, and this has some significant effects on the effectiveness of the 3rd filter. On a real application the filters would run real-time on the scanner.

Filter 1 alone removes about 80% of the frames. The filter is also extremely fast. Filters 0 and 2 are very dependent on the environment. When there are a lot of hotspots near the scanners, there are a lot of *Beacon* frames and the filters remove a larger percentage of the recorded frames. The last filter is more aggressive in this scenario then it would be in real life. However, in real life we discovered that a lot of non-mobile devices, such as printers produce a large number of frames which would be removed by this filter. The results are presented in Fig. 14.4. Filter 0, 1, 2, and 3 are tested independently and finally we start all of them simultaneously. It is clear that the combination of all the filters is more effective than each of them used individually. This means that the filters remove different types of frames and that there is no filter which removes all the frames removed by the others.

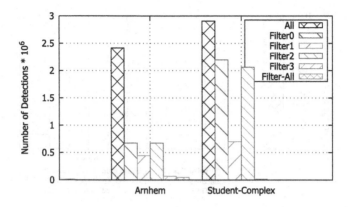

FIGURE 14.4

Scanner filters.

14.3.3 FILTERING DATA AFTER CENTRALIZATION AT SERVER

When the data arrives at the server it has the following format $(scannerId; devi-ceid, timestamp)$. Here, $scannerid$ identifies our scanner for which the physical location is known, and $deviceid$ identifies a mobile device. For privacy reasons, it should be stored as an MD5 hash. Finally, the $timestamp$ represents the date and time of the detection, when the 802.11 frame was received at the scanner. We found that this format is somewhat similar across most WiFi tracking projects.

After the data arrives at the server it can further be filtered to help with performing analysis for gaining insight on peoples movements. Some of the filters are:

Duplicates Filter

removes all detections that have the same values for all three, $scannerid$, $deviceid$, and $timestamp$. Having these duplicate detections is possible because of clock synchronization errors and buffering between the server and the scanners.

Time Filter

removes all the detections that are not part of the interest period. It is common that data is generated in the testing phase or outside of the area of interest in time.

The Malicious Device Filter

is used to remove all detections that have a randomly generated MAC address. This is especially true for Apple devices as advertised in Stites and Skinner (2014). Some Apple devices randomize their MAC address when sending $Probe\_Requests$ frames. Because the address keeps changing a device using this feature cannot be tracked over multiple sensors. To apply this filter we require that the $deviceid$ contains the OUI, alongside the hash of the entire MAC address.

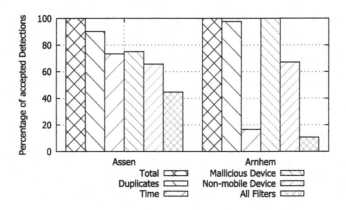

FIGURE 14.5

Server filters.

Non-Mobile Device Filter

removes all detections of devices seen at only one sensor. These devices can appear to be mobile devices such as phones or laptop, but are only used in a non-mobile fashion. An example would be a laptop that is used as a desktop workstation.

We have conducted several experiments where we gather WiFi header data and use it in order to understand pedestrian dynamics. One of our early experiments consisted of only one scanner and was performed inside of a student campus building. We later performed larger experiments which cover entire city centers and were performed during a living statues festival (http://www.worldlivingstatues.nl/) in the city of Arnhem, Netherlands and the tt music festival (http://www.ttfestival.nl/) around the MotoGP event in Assen, Netherlands. These 2 festivals each attracted about a hundred thousand visitors to these two cities. The characteristics of these data sets are presented in Table 14.4.

By applying these filters we obtained the results in Fig. 14.5. Each filter removes a number of detection from the data sets and by applying all four of them the data sets are reduced even further. After applying the four filters the Arnhem data set has been reduced to 10% of its original size and the Assen one to 44%.

14.3.4 RESULTING DATA SET

After the filters have been applied the remaining data is ready for analysis, as it contains far more accurate about the actual movements of pedestrians across those detections. We extracted one day from the Arnhem data set in order to give an example of what can be expected from a WiFi tracking data set.

One of the interesting features to observe in the data is the variation on the number of devices and detections throughout the day. In the data set a device is represented by its ID (OUI and MD5 hash of the MAC address) and by all the triplets

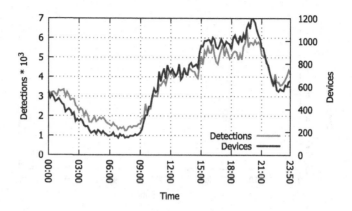

FIGURE 14.6

# of detections/devices over time.

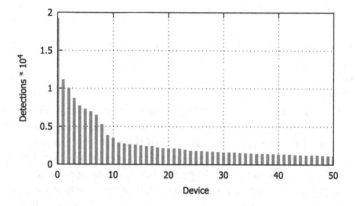

FIGURE 14.7

# of detections, first 50 devices.

$\langle scannerid; deviceid; timestamp \rangle$ that have the $deviceid = ID$. The variation in the number of devices and detections during a day can be observed in Fig. 14.6. Both the number of devices and detections increase during the day and drop during the night. This creates a day/night cycle which is common in most data sets related to human activities. The two patterns created by these numbers do not perfectly match. This is because different devices do not generate the same number of detections.

The number of detections is dependent on the mobile device usage and the probability of receiving a correct frame given the noise in the environment on the WiFi transmission frequency. To show this we counted the number of detections for each individual device. In Fig. 14.7 we present the results. To improve visibility we ordered the devices in decreasing order of the number of detections they had and we

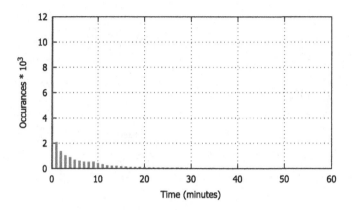

FIGURE 14.8

\# of occurrences of different time periods for which a device is in range of at least 1 scanner.

removed the ones with most detections as well as the long tail of devices with few or only one connection. The result can be approximated to a Zipfian distribution (Zipf, 1949). Many types of data sets studied in social sciences can be approximated to match a Zipfian distribution.

The Zipfian distribution is also visible in other features of a WiFi crowd tracking data set. For instance when we measure the amount of time a device is visible by any of our scanners. Fig. 14.8 represents periods of time for which a device is detected by scanners ordered decreasingly by the number of occurrences for each time value, in minutes. Again we removed the large number of occurrences that were detected only briefly (under one minute) and the long tail of devices that were detected for variously long periods on time. To measure the amount of time in which a device is visible we consider the difference between the first and last detections of a continuous set of detections that have no more than five minutes in between.

The quality of pedestrian tracking data sets is given by the frequency with which detections are registered. Having only the detections data set, if there are only few detections of a device the path taken by the device through the city could be impossible to determine. In the case of WiFi tracking this frequency cannot be controlled and is dependent on each of the mobile devices that are being tracked. We measured the frequency for our data set and counted the number of occurrences of consecutive detections of a device, at any scanner, with a fixed number of seconds between them. In Fig. 14.9 we vary the number of seconds between consecutive detections, and show the number of occurrences for each case.

The highest value in Fig. 14.9, at three seconds is most likely caused by the Filter 3 on the scanners. The filter only accepts detections if they are more than three seconds apart. Other interesting peaks are at 30, 60, and 120 seconds. These peaks are most likely caused by the frequency at which most mobile devices send *Probe_Request*

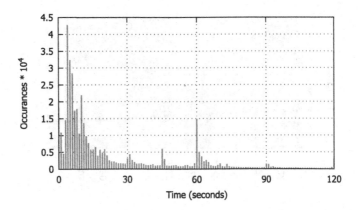

FIGURE 14.9

# of occurrences of different time periods between two consecutive detections of the same device.

frames. We also noticed this 30 second frequency, along with its multiples, on the mobile device we used to initially test the scanners with.

The feature we described above are common for all the data sets we encountered. We believe that having a WiFi tracking data set that contradicts these features could be a sign of errors.

14.4 DATA ANALYSIS

After pre-processing and cleaning the data there still remains the question of how to extract useful information from it. In this section, we give some examples of how such data can be used for this purpose. The dataset that we have collected, contains probe requests from a period where normal activities in a city is changed due to a festival. An example of the type of information that can be extracted is the change of normal usage patterns of space during such an event.

One of the approaches that can be used to acquire a general understanding of how these changes occur is performing multi-resolution analysis. Wavelet transformation is well-known for representing changes in a time series in different resolutions. For our specific dataset, we use a Haar wavelet to see how the changes in city mobility patterns are reflected in the number of people near scanners. When applied to the time series of a specific period of time, the Haar Wavelet can represent the changes between two consecutive timestamps of different duration (increase or decrease in the number of probes). Fig. 14.10 represents the wavelet coefficients over the above-mentioned period of time. Each bar in the figure represents the fluctuations in the number of visitors within a window of time to the next (with the size of a specific window size shown on the y-axis). As seen, even without going in more detail it is

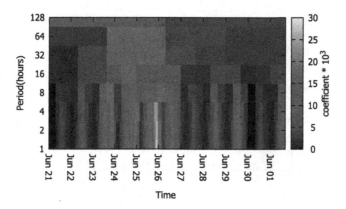

FIGURE 14.10

Multiresolution analysis of number of visitors in the city during a period of 10 days.

apparent that the coefficients appear to be brighter during the three days of the festival (24–26 June) compared to the normal "rhythm" of the city. It is also seen that outside the festival time, these fluctuations in all periods of different length have a repetitive pattern.

What is shown above is an example of a general analysis but it is also interesting to know how these changes appear at different locations. Each of the spaces covered by a scanner has general use cases during normal days. During the festival time these use cases change. For instance, empty squares turn into stages where music is played, parks turn into places where people camp, and new means of public transport are being used. Automatically understanding different semantics and extracting those semantics from probe requests is a challenging problem. Inferring such information requires extensive spatio-temporal analysis. Each of the spaces will potentially represent a different usage pattern, specific to that place. In what follows we take two of the locations covered by a scanner (a train station and a square (or a stage area), respectively) and represent how different spatio-temporal features extracted from the probe requests change over the festival and normal days.

Incoming and outgoing traffic: One of the features that changes during the festival is the pattern of incoming and outgoing traffic to a space. In other words, how people go to a place and leave it. Fig. 14.12 compares the train-station and the stage area in terms of this feature. It is seen that there is a delay between the population who come to and leave these spaces during the festival. To better understand the amount of delay between the incoming and outgoing traffic, a cross-correlation function can be used. In general, the cross-correlation graph shows how two correlated time-series follow each other. The first peak in the cross-correlation graph can represent the delay between the two time series. Fig. 14.11 shows the results of applying this function on the incoming and outgoing time series. As seen, during the festival there is a four-hour delay between people who come to and leave the city. This can be related to the

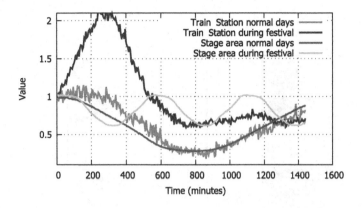

FIGURE 14.11

Cross-correlation between people coming and leaving spaces.

duration of stay of visitors in the city. On normal days, however, this delay is about 1 hour which might represent the delay between the appearance of people who leave the city and those who travel to the city for work. On normal days, there is no specific delay between people who come to a stage area and later leave. However, during the stage time there is a 50 minutes delay. This delay may represent the average time visitors stay near a stage for watching the program.

Group sizes: Another feature that is interesting to look at are group sizes. Each spot within the city attracts groups of different sizes. It is possible that more groups are formed during the festival. This could be groups of people who know each other and attend an activity together, but also people who temporarily form a group by performing the same activity (over a period of time). Groups can be identified by their synchronous appearance and disappearance near a scanner. As shown in Fig. 14.13 the number of people in groups of larger than three increases during the festival.

Duration of visits: Apart from the group size itself, it is also possible that the duration of time people spend near each scanner also changes. As a stage has a program with a specific duration running, it is possible that this also appears as a difference. Fig. 14.14 compares the duration of visits of different group sizes during the festival and normal days. As expected groups of different sizes stay considerably longer in comparison to normal days when looking at stages, but also the train station.

14.5 TOWARD LARGE-SCALE CROWD-TRACKING SYSTEMS

Crowd-tracking systems are used for analyzing or monitoring human activities, such as the flow of pedestrians through busy city centers or the density of individuals in an area, with applicability in a vast number of fields. Examples can span from infrastructure improvement and management to monitoring and security.

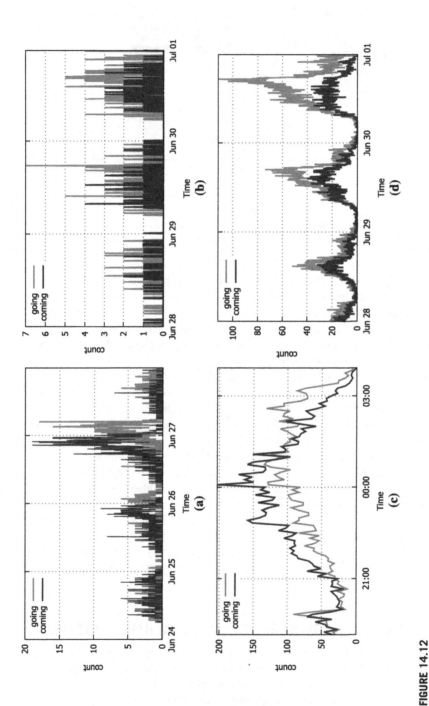

FIGURE 14.12

Comparisons of people going to and leaving a stage area and train station during festival and in normal days. (a) Train station – during festival; (b) Train station – normal days; (c) Stage area – during festival; (d) Stage area – normal days.

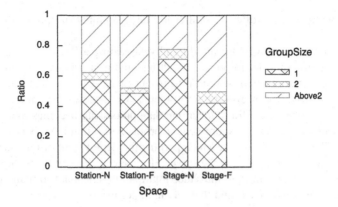

FIGURE 14.13

Stay duration distribution of different group sizes (F and N stand for Festival days and Normal days).

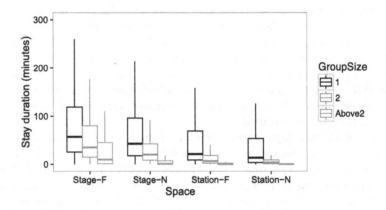

FIGURE 14.14

Ratio of visitors in different group sizes.

Having a scanning system the size of a city raises important issues. Let's compare it to a one-building scanner system. It is obvious that the area of a building is far smaller than that of a city. But, there are other differences as well. The number of people carrying WiFi devices in a city can be in the order of millions and smartphones and tablets are becoming more and more popular. Covering the area of a city requires a larger number of scanners than a building and, when high accuracy is required, the number of scanners grows even more in order to provide an appropriate granularity. Another point of difference is that moving from a room to another can be achieved by means of a small number of potential paths. However, cities, with their vast networks of streets and pathways, offer a significantly larger number of options to reach any location.

All these problems increase the difficulty in scaling a WiFi tracking system to the size of a city. In order to obtain a high resolution or accuracy on the measurements one possible solution would be to install more scanners. But increasing the number of scanners raises even more problems. The bandwidth usage at the server side increases with each extra scanner and data processing becomes more complex. Previously in this chapter we addressed the problem of minimizing bandwidth usage by having filters at the scanners, which permits the deployment of more scanners. It also minimizes the bottleneck at the server. However, with a high enough number of scanners there will be a need to introduce multiple servers and deal with inter-server communication issues. We found that, in our experiments one server was sufficient, even at peak hours.

Once you have a large-scale crowd-tracking system you need to make sense of the incoming data. This usually means flow and density analysis. Interpreting a large volume of data is it's own challenge, but when dealing with a noisy WiFi system, more difficulties appear. When thinking of a tracking system with stationary scanners one may assume that two consecutive detections of the same device at different scanners represent a movement of the pedestrian carrying the device from one scanner to the other. This is not always true. By analyzing our data sets we identified many cases where two consecutive detections of a device are at scanners which are placed at different ends of the city. This is normal if you consider the noise created by high packet loss rates and devices that can have their WiFi turned off temporarily. Even if we can assume that the device moved from one scanner to the other it is not trivial to determine the path that was used.

Even when there is high noise it is reasonable to expect that two scanners which are "close" have more consecutive detections then those that are further. The assumption is simple, when a pedestrian is detected at scanner A it is extremely likely that a following detection would be at a scanner which is placed "close" to scanner A. However, determining which scanners are "close" to each other is not trivial. By "close" we mean that it is more likely for scanners placed on adjacent streets to have consecutive detections of a device than ones placed in other locations. The problem is made even more difficult by the fact that the placement of scanners usually follows the architecture of the city. It is common for a street map to have an irregular shape.

Having the list of "close" scanners allows one to addresses scalability issues in multiple manners: it permits the deployment of smarter scanner-to-scanner aggregation algorithms; it permits faster data processing at the server level; it enables the simplification of path prediction systems.

A WiFi crowd-tracking system can be represented as an undirected graph, $G = (V, E)$, with vertices represented by scanners $V = \{v_i | v_i \text{ is a scanner}\}$ and edges $E = \{e_{ij} = (v_i, v_j) | i < j; v_i, v_j \in V\}$. G is a **Full Mesh**.

In Chilipirea et al. (2015b) we proposed several methods of creating graphs that show the "closeness" of WiFi scanners. The most direct one is created by using data obtained from the scanners themselves. We call it the **Inferred Graph** (IG)

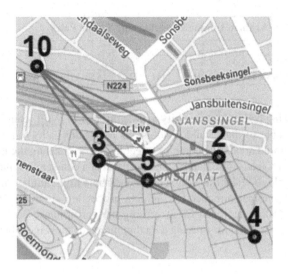

FIGURE 14.15

Arnhem scanners full mesh.

and we define it as a weighted, undirected graph with the same number of edges as the full mesh: $G_{ig} = (V, E_{ig})$ with $E_{ig} = \{e_{ij} = (v_i, v_j); w_{ij}$ weight of $e_{ij}|i < j; v_i, v_j \in V; w_{ij} =$ number of unique devices moving from v_i to v_j or viceversa$\}$. The inferred graph shows which scanners are "closer" but it does not offer a list with scanners that we consider "close" and scanners that we consider "far." In order to accomplish this we need to select a number of edges with the highest values for w. When we created Inferred Graphs for our data sets we discovered that, due to the noise and errors of a real WiFi crowd tracking data set, there is no edge in the graph with a weight w of 0. These errors can be caused by frame collision, environmental causes, or even the temporary disabling of the WiFi capability for some devices. The Inferred Graph also can be used to determine popular sequences of scanners, which can be correlated with popular paths in the city.

As an example, Fig. 14.15 represents the full mesh graph created using the scanners we used in Arnhem. The system gathered detections of devices from 5 scanners placed at reasonable distances (between 70 m and 300 m) from each other. The numbers the ID of the scanner as it was set by our system, positioned on the map at the same position at which the scanners were during the experiment. In Table 14.5 we show the number of movements from one scanner to the other, grouped by device. It is obvious, and the results show this, that the scanners that are furthest away, 4 and 10, have the fewest detection pairs. What is interesting is that the ones that are physically closer on the map, 3 and 5, do not have the highest number of devices moving from one to the other. This indicates that the map and people behavior affect the "closeness" of the scanners.

Table 14.5 Arnhem Unique Detections for Sensor Pairs

V_i	2	2	3	4	2	3	3	2	5	4
V_j	3	4	5	5	5	4	10	10	10	10
# Unique Devices	6040	3009	2331	2220	1856	1604	925	237	92	90

When tracking pedestrians in a city a large number of WiFi scanners is necessary in order to be able to draw valid conclusions about the paths followed, especially when the monitored area has a complex street layout. Increasing the number of scanners inadvertently leads to scalability issues. Even at a coarser granularity the system is still susceptible to scalability problems caused by a high number of detectable devices: poor runtime performance for the data processing algorithms and bandwidth problems. A possible approach for increasing the performance of the system under high load is using a graph for filtering data that indicates movement between scanners that are not "close."

14.6 CONCLUSIONS

Tracking pedestrians remains an open problem. There is no solution that offers perfect results in all scenarios. However, WiFi tracking offers one of the most promising alternatives. It takes advantage of how popular smartphones are and how the use of mobile internet is constantly increasing in order to provide tracking data on crowds with large numbers of individuals.

Because smartphone use keeps increasing and there are already multiple alternatives for wearable computing it is correct to assume that the accuracy and the amount of data resulting from WiFi tracking can only increase.

We showed how a WiFi tracking systems can be implemented and what are some of the difficulties of designing and managing such a system on both a small and a large scale. The resulting data can be interpreted using many techniques, a few of which we presented in this chapter.

With many use cases for tracking data, from facility management to simulations that take into account human behavior, interest in the area can only increase. In depth analysis of the resulting data may offer exciting results and opportunities.

REFERENCES

Aughey, R.J., Falloon, C., 2010. Real-time versus post-game GPS data in team sports. Journal of Science and Medicine in Sport 13 (3), 348–349.

Baratchi, M., Meratnia, N., Havinga, P.J., Skidmore, A.K., Toxopeus, B.A., 2013. Sensing solutions for collecting spatio-temporal data for wildlife monitoring applications: a review. Sensors 13 (5), 6054–6088.

Baratchi, M., Meratnia, N., Havinga, P.J.M., Skidmore, A.K., Toxopeus, B.A.K.G., 2014. A hierarchical hidden semi-Markov model for modeling mobility data. In: Proceedings of the 2014 ACM International Joint Conference on Pervasive and Ubiquitous Computing. UbiComp'14. ACM, New York, NY, USA, pp. 401–412. http://doi.acm.org/10.1145/2632048.2636068.

Beard, C., Stallings, W., 2016. Wireless Communication Networks and Systems, Global Edition. Pearson Education Limited. https://books.google.nl/books?id=LJ5VCwAAQBAJ.

Chilipirea, C., Petre, A.-C., Dobre, C., van Steen, M., 2015a. Filters for Wi-Fi generated crowd movement data. In: 2015 10th International Conference on P2P, Parallel, Grid, Cloud and Internet Computing, 3PGCIC. Institute of Electrical & Electronics Engineers (IEEE).

Chilipirea, C., Petre, A.-C., Dobre, C., van Steen, M., 2015b. Proximity graphs for crowd movement sensors. In: 2015 10th International Conference on P2P, Parallel, Grid, Cloud and Internet Computing, 3PGCIC. Institute of Electrical & Electronics Engineers (IEEE).

Cunche, M., 2014. I know your MAC address: targeted tracking of individual using Wi-Fi. Journal of Computer Virology and Hacking Techniques 10 (4), 219–227.

Curran, K., Furey, E., Lunney, T., Santos, J., Woods, D., McCaughey, A., 2011. An evaluation of indoor location determination technologies. Journal of Location Based Services 5 (2), 61–78. http://dx.doi.org/10.1080/17489725.2011.562927.

DeCesare, N.J., Squires, J.R., Kolbe, J.A., 2005. Effect of forest canopy on GPS-based movement data. Wildlife Society Bulletin 33 (3), 935–941.

Evans, A.G., et al., 2002. The global positioning system geodesy odyssey. Navigation 49 (1), 7–33. http://dx.doi.org/10.1002/j.2161-4296.2002.tb00252.x.

Ficek, M., Pop, T., Kencl, L., 2013. Active tracking in mobile networks: an in-depth view. Computer Networks 57 (9), 1936–1954. http://dx.doi.org/10.1016/j.comnet.2013.03.013. http://www.sciencedirect.com/science/article/pii/S1389128613000996.

Global internet report, 2015. Internet Society (ISOC).

IEEE 802.11, 2012. Wireless LAN medium access control (mac) and physical layer (phy) specifications. In: IEEE-SA, pp. i–22.

Jonsen, I.D., Flemming, J.M., Myers, R.A., 2005. Robust state-space modeling of animal movement data. Ecology 86 (11), 2874–2880.

Kim, Y., Shin, H., Cha, H., 2012. Smartphone-based Wi-Fi pedestrian-tracking system tolerating the rss variance problem. In: IEEE International Conference on Pervasive Computing and Communications (PerCom), pp. 11–19.

Liu, H., Darabi, H., Banerjee, P., Liu, J., 2007. Survey of wireless indoor positioning techniques and systems. IEEE Transactions on Systems, Man and Cybernetics. Part C, Applications and Reviews 37 (6), 1067–1080. http://dx.doi.org/10.1109/TSMCC.2007.905750.

Martella, C., Miraglia, A., Cattani, M., van Steen, M., 2016. Leveraging proximity sensing to mine the behavior of museum visitors. In: IEEE International Conference on Pervasive Computing and Communication. IEEE Computer Society.

Musa, A., Eriksson, J., 2012a. Tracking unmodified smartphones using Wi-Fi monitors. In: Proceedings of the 10th ACM Conference on Embedded Network Sensor Systems, pp. 281–294.

Musa, A., Eriksson, J., 2012b. Tracking unmodified smartphones using Wi-Fi monitors. In: Proceedings of the 10th ACM Conference on Embedded Network Sensor Systems. ACM, pp. 281–294.

Rivest, R., 1992. The md5 Message-Digest Algorithm.

Salyers, D.C., Striegel, A.D., Poellabauer, C., 2008. Wireless reliability: rethinking 802.11 packet loss. In: International Symposium on a World of Wireless, Mobile and Multimedia Networks, WoWMoM, pp. 1–4.

Schauer, L., Werner, M., Marcus, P., 2014. Estimating crowd densities and pedestrian flows using Wi-Fi and Bluetooth. In: Proceedings of the 11th International Conference on Mobile and Ubiquitous Systems: Computing, Networking and Services. MOBIQUITOUS '14. ICST (Institute for Computer Sciences, Social-Informatics and Telecommunications Engineering), Brussels, Belgium, pp. 171–177.

Siebel, N.T., Maybank, S., 2004. The advisor visual surveillance system. In: ECCV 2004 Workshop Applications of Computer Vision, ACV, Vol. 1. Citeseer.

Stites, D., Skinner, K., 2014. User privacy on iOS and OS X. Presented in Session 715 of Core OS WWDC14.

Thiagarajan, A., Ravindranath, L., LaCurts, K., Madden, S., Balakrishnan, H., Toledo, S., Eriksson, J., 2009. Vtrack: accurate, energy-aware road traffic delay estimation using mobile phones. In: Proceedings of the 7th ACM Conference on Embedded Networked Sensor Systems. ACM, pp. 85–98.

Wijermans, N., Conrado, C., van Steen, M., Li, J., Martella, C., 2016. A landscape of crowd-management support: an integrative approach. Safety Science 86 (7), 142–164.

Yan, Z., Parent, C., Spaccapietra, S., Chakraborty, D., 2010. A hybrid model and computing platform for spatio-semantic trajectories. In: The Semantic Web: Research and Applications. Springer, pp. 60–75.

Zanca, G., Zorzi, F., Zanella, A., Zorzi, M., 2008. Experimental comparison of RSSI-based localization algorithms for indoor wireless sensor networks. In: Proceedings of the Workshop on Real-World Wireless Sensor Networks, pp. 1–5.

Zipf, G.K., 1949. Human Behavior and the Principle of Least Effort.

ACRONYMS AND GLOSSARY

List of acronyms with explanation

WiFi   wireless fidelity
MAC   media access control
SSID   service set identifier
BSSID   basic service set identifier
SID   scanner identifier
MID   mobile device identifier
T   time
RSSI   received signal strength indicator
CCTV   closed-circuit television
RF   radio frequency
NFC   near field communication
RFID   radio-frequency identification
GPS   global positioning system
CPU   central processing unit
RAM   random-access memory
SQL   structured query language
NTP   network time protocol
MD5   message digest 5
MHz   megahertz
MB   megabyte
IEEE   institute of electrical and electronics engineers
IANA   internet assigned numbers authority
OUI   organizationally unique identifier
FCS   frame control sequence
SA   source address
DA   destination address
CTS   clear to send
QoS   quality of service
CRC   cyclic redundancy check
WEP   wired equivalent privacy
WPA   Wi-Fi protected access
GSM   global system for mobile communications
IG   Inferred Graph

Glossary of terms with explanation

Active badges wearable device that communicates with other devices of the same type, the scope is usually to gather data on how individuals meet or where they are.
Crowd-tracking gathering data on the movement of crowds or groups of people.

Detection a tuple (detection, scanner id, time stamp) that is a record of a device that has passed near to a WiFi scanner.

Filtering the process of removing noise from the data set (for instance, detections that have random MAC addresses).

Inferred graph graph obtained by using the consecutive detections of devices.

MAC address number that uniquely identifies a device, used for network level communication (usually set by the manufacturer of the device or the network module).

Proximity graph graph that has as vertices the deployed WiFi scanners and edges between the vertices. An edge exists if two WiFi scanners are reachable by following paths through the city.

Tracking gathering data on the movement of people.

WiFi-enabled device a device that has WiFi capabilities and is usually carried by an individual (usually a smartphone).

WiFi scanner scanner that can receive WiFi frames from other WiFi-enabled devices and records these frames as detections.

THE LIFE MANAGEMENT PLATFORM ACHIEVES DATA PROTECTION AND SAFE SHARING

15

Hideyuki Shimizu, Hisashi Sakamoto, Tohru Miyazaki, Masayoshi Kai

NEC Solution Innovators, Ltd., Japan

15.1 INTRODUCTION

As humans grow, they go through various stages in life, such as infancy, childhood, adolescence, adulthood, and old age. Such life environments as relationships with family members also undergo various changes — marriage, childcare, child education, child independence, and aging together. Throughout these various life stages, human beings live their lives utilizing the various support systems that surround them, such as childcare support during infancy, educational support during childhood, medical care and nursing support during old age. The support systems that each individual needs should be suitable to their age and environment.

The various support systems described above are mainly provided by local authorities, civic associations, NPOs, and private companies. Central governments give support through monetary means, such as grants for local authorities and benefits, and pensions for individuals. Support services are provided and operated based on these grants and investments for individuals.

Recently, discrepancies between available support services and the needs of citizens have begun to surface. Due to the progress of our society's maturation, sophistication, and informatization, the lifestyles of citizens are quite diverse in some aspects, such as family members, working hours, type of home, and so on. For this reason, the needs of citizens are also diverse. In contrast, service providers have regarded fairness, uniformity, and profitability as important. For local authorities, it is difficult to deal with the demand of each person thoroughly because their main aim is to serve the overall public, not individuals. Even for private companies, how to make a profit from the excessively subdivided needs of citizens becomes a problem. To solve these discrepancies, they must properly understand the needs of citizens, offer a variety of services for their diverse personalities, and maintain ease in offering various services.

In response to this recent situation, we define "life management" as the actualization of support services, which gives newly awareness of hidden needs, that are necessary for maintaining and improving QoL (quality of life), and that are individualized for various personal life environments to be used in places accessed in everyday life (at home) from birth to old age.

In this concept of life management, CPS (cyber-physical system) and big data are should be applied to bring awareness to the citizens' hidden needs. CPS is used for sensing and identifying the life behaviors of each citizen. From this data of the individuals' behaviors, big data is used to understand the citizens' overall trends (such as location or generation difference) and reveal hidden needs. With these technologies, we should take many technical things into account, such as sensing the targets, how to sense the targets, the process for identifying behaviors, and so on.

Mechanisms for data sharing are required to achieve diverse services and ease in service delivery. Generally speaking, sensor data and the analysis results for certain services are only used by the providers or the service itself and not shared with other services. This situation may cause two problems.

First is a problem with the ease of service delivery. Although some services measure the same type of data, each service uses their own respective sensor. Namely, service development costs increase because all components, from sensors to service applications, must be developed. Meanwhile, for service users, implementation and running costs increase because they must use a different sensor for each service. For example, imagine using an energy monitoring service and an elderly care service. Energy monitoring is a service that monitors and optimizes the power consumption of home appliances using power sensors. Meanwhile, elderly care service monitors power consumption and spots health risks. Elderly care service learns the user's typical energy consumption patterns, and if there is an anomaly (e.g. a pot that an elderly person uses every morning is not turned on, or a television remains on for an inexplicably long time), it alerts family members or caretakers of the possibility of a health incident (e.g. an elderly person cannot wake up due to illness, or is unconscious while the television is on). Though both services need to measure the same power consumption, they each must use their own sensor because they are isolated services. Fig. 15.1 illustrates this problem.

The second problem is the prevention of service diversity. It is thought that combining multiple services and creating new understanding is a step towards diversifying services. For example, in the area of elderly care, by combining analyzed results from individual services for early detection of such health risks as metabolic syndrome and locomotive syndrome, understanding can be expected of the overall judgment of the person's health and the mutual interaction between health risks. However, combining services is difficult given the current situation in which measured data and analysis results are not shared between services.

To solve these problems, data sharing mechanisms between services are required. Although the data can be an asset for service providers, we must pay attention to the fact that service data can include personal information. Hence, it is necessary to add the appropriate access rights and browsing rights to create a good balance between data sharing and protection. This chapter defines and suggests the IoT platform, Life Management Platform, to achieve data sharing and protection simultaneously. Life

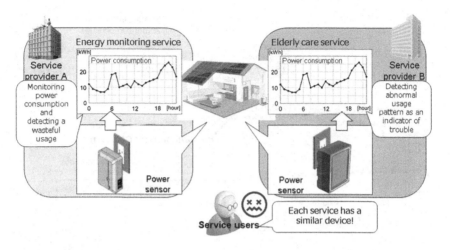

FIGURE 15.1

Problem of using multiple services.

Management Services is also defined as a service group that actualizes life management through the platform. In this chapter, we introduce our in-development model case for Life Management Services.

15.2 LIFE MANAGEMENT PLATFORM
15.2.1 OVERVIEW

In order to bring about service diversity and an ease in service delivery, the Life Management Platform must satisfy the following requirements.

(Requirement 1)   Measurement information must be protected through appropriate access rights.

(Requirement 2)   Communication protocols must not be limited (interoperability between sensors is ensured).

(Requirement 3)   Services must be able to share measurement information and analysis results to other services.

(Requirement 4)   Service providers must be easily able to provide various services.

To satisfy these requirements, the Life Management Platform encrypts sensor information, giving access rights to the people targeted for measurement or owners of the place in which sensors are set. For encryption, the platform uses information protection technology used by DRM (Digital Rights Management) for contents protection. Recently, DRM is not only used for copyright protection when digital content is distributed, such as movies or music, but also to deter the disclosure of information regarding companies' e-mails and any other documents (NEC). We are going to ap-

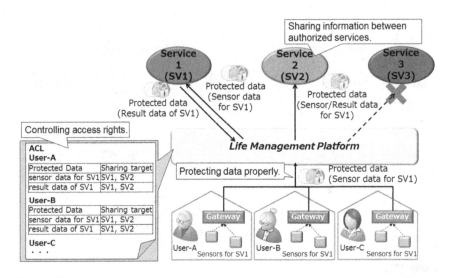

FIGURE 15.2

Effects of the Life Management Platform.

ply this technology to sensor networks and IoT platforms to properly protect personal information and sensor measurement data, even if sensor data is sent to outsiders or unintended services.

Fig. 15.2 shows the effects of the Life Management Platform. In this image, sensors for service 1 (SV1) are set up in every user's house. Life Management Platform controls the access rights for the sensor data and result data of each service. If data from SV1 is set to be shared with service 2 (SV2), SV2 can browse it. However, data sharing with service 3 (SV3) cannot occur because it is not set to share with SV3.

15.2.2 COMPARISON WITH OTHER IoT PLATFORMS

There are no IoT platforms that apply access controls to the data itself. The access management of a few existing IoT-Platforms is described below.

(1) Thread

Thread (Thread) is a platform created by the Thread Group, which was established by Nest Labs, under Google, Samsung, ARM, and Freescale in July, 2014. It mainly uses functions for the network layer and transport layer, but it has no functions for information protection. It leaves information protection to the application layer.

(2) HomeKit

Homekit (HomeKit) was developed by Apple and supports end-to-end encryption from devices to services. Due to the information not being protected in storage or in service programs, it has a risk for leakage. It also does not support access controls. In the past, Apple has promoted authentication of accessories through the MFi license (MFi license) program. Similarly for Homekit, Apple supplies their logo only to accessories that satisfy their performance criteria. Therefore, Homekit does not have enough interoperability and it is difficult to construct networks for public big data on Homekit.

(3) AllJoyn

AllJoyn (AllJoyn Framework) is a framework, developed by Qualcom and promoted by AllSeen Alliance (AllSeen Alliance), that utilizes end-to-end application level security. With its security manager, it can map applications to people. It is effective for information distribution between services because applications can be registered to security groups and certifications, and ACL policies can be managed. However, it has risks for leakage because its applications support neither information protection in storages nor access control for each person.

15.2.3 THE FUNCTION OF LIFE MANAGEMENT PLATFORM

The functions of the Life Management Platform that are necessary to achieve information protection throughout, from collecting data from sensors to providing information to service users, while satisfying the requirements described above, are listed below.

(Function 1) Data protection technology: protecting sensor data using individual access rights and retaining said rights if services share data.

(Function 2) Transparent data collection technology: transparently receiving and processing data sent from sensor networks based on various standards or specifications.

(Function 3) Database technology for horizontal data sharing: managing an appropriate range of browsing of sensor data and/or personal information that has been collected, and sharing the information and data between multiple services. Due to permissions from the owners and providers of the data, only the permitted services can share the data. Also, data traceability is ensured, allowing data owners and providers to check the state of data usage (users, amount of use, et cetera).

(Function 4) Integrated online and physical sensor data analyzing technology: executing analysis not via services, but in the platform. Without decrypting, data is used and results of the calculations are sent to applications in a re-encrypted state. With this process, the protection level of information stays the same until service delivery.

Figs. 15.3 and 15.4 show the layer structure and components of the Life Management Platform design based on functional requirements.

The function of each layer is as described below.

(1) Data Transfer Layer

First, using the Edge Computing Platform and the Data Transfer Layer, the platform achieves function 1, data protection technology. The sensors and gateways of the sensor networks, which relay measured data to the cloud side, are equivalent to the Edge Computing Platform and encrypt data measured by the sensor via individual access rights. Similar to DRM technology for contents protection, this encryption

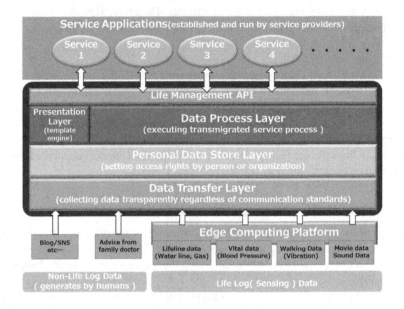

FIGURE 15.3

Layer structure of the Life Management Platform.

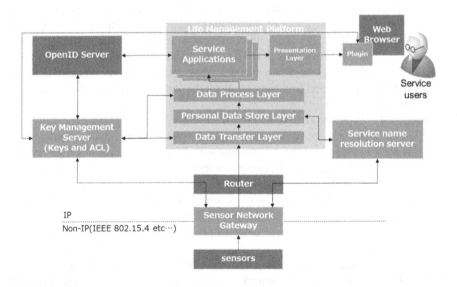

FIGURE 15.4

Components of the Life Management Platform.

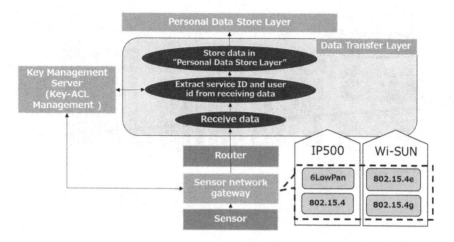

FIGURE 15.5

Data Transfer Layer.

technology defines the proper range of browsing and distribution by setting rights for data owners and viewers. The Data Transfer Layer takes over the protected data that is received and stores it in the Personal Data Store Layer non-decrypted. (See Fig. 15.5.)

Collection of measurement data by the sensors, as well as transparent data collection technology in the Edge Computing Platform and the Data Transfer Layer make function 2, can be achieved. This platform adopts IEEE 802.15.4 series as physical/MAC layer protocol, which is suitable for wireless sensor networks, and ensures interoperability between multiple standards. The platform has interoperable profiles that are especially flexible in that users can define original profiles in the upper layer, such as ZigBee (ZigBee Alliance), IP500 (IP500 Alliance), and Wi-SUN (Wi-SUN Alliance).

(2) Personal Data Store Layer

Function 3, database technology for horizontal data sharing, is implemented in the Personal Data Store Layer. All service data is stored in this layer and categorized by each service and each user. Stored data is also encrypted by different keys for each service and user. When these data are stored in storage area, Personal Data Store Layer makes indexes and these are encrypted using same key as service data. This layer is able to query data without decryption and all data remain encrypted, but indexes are decrypted during the query. When the query is end, these indexes are re-encrypted. (See Fig. 15.6.)

(3) Data Processing Layer

Function 4, integrated online and physical sensor data analyzing technology, is achieved using the Data Processing Layer and the Presentation Layer (described in the next section). Life Management Services passes its program modules for data

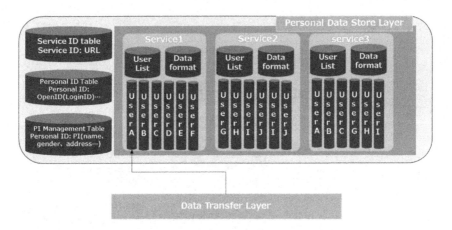

FIGURE 15.6

Personal Data Store Layer.

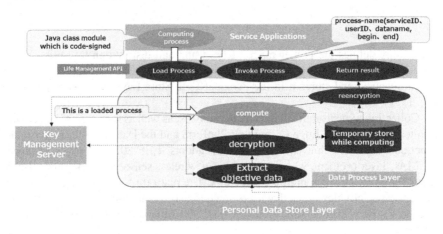

FIGURE 15.7

Data Process Layer.

analysis to the Data Processing Layer and services can analyze the encrypted data via this layer. In the Data Processing Layer, sandbox is prepared and the program modules for data analysis are executed in it. The analyzation results are re-encrypted and passed to the services. (See Fig. 15.7.)

(4) Presentation Layer

The Presentation Layer corresponds to individual measurement results that are queried from the Data Processing Layer, based on the need for the service users to browse their personal data. Data that is received is decoded and shaped into a user-friendly format, such as graphs. (See Fig. 15.8.)

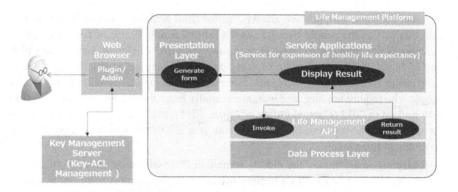

FIGURE 15.8

Presentation Layer.

The Presentation Layer uses a function that generates content for each request and provides it to the user in a protected state that conforms to EME (Encrypt Media Extensions).

According to the service user's requirements, the Presentation Layer ensures that all processed data is transmitted in a format matching the service content and in a protected state. One of modification methods is the conversion of data to images to prevent copying and pasting.

Through this layer structure, the Life Management Platform can provide both data protection and sharing under proper access rights.

15.3 LIFE MANAGEMENT SERVICE

This section introduces one of the services to expand the expectancy of a healthy life for the elderly as model case for the Life Management Services constructed on the Life Management Platform.

15.3.1 BACKGROUND OF MODEL CASE

Nowadays, maintenance and improvement of QoL, the main aim of Life management, is in demand, especially for the elderly. Due to the worldwide progression of an aging population and the increase of medical and nursing care costs for supporting the elderly, it is necessary to expand the expectancy of a healthy life for the elderly.

The ratio of people over 65 years old in 2010 was 16.6% in the United Kingdom, 20.3% in Italy, 20.8% in Germany, 16.8% in France, and 23.0% in Japan, the highest in the list. It is said that the population continuously aging. In 2050, the ratio is expected to reach 24.7% in the United Kingdom, 33.0% in Italy, 32.7% in Germany, 25.5% in France, and 38.8% in Japan (UN, 2013). For elderly people, the older they become, the more people there are who suffer from health problems. In Japan, it is said that half of the people over 65 (44.4% of males and 49.3% of females) cannot maintain their health due to diseases or injuries (Ministry of Health, Labour and Welfare of Japan, 2010). As the rate of aging grows, the number of elderly people who have health issues steadily increases. For example, in Japan, the number of elderly people who require nursing care increased rapidly from about 2.88 million in 2001 to 5.45 million in 2012 (Cabinet Office of Japan, 2015). Medical and nursing care costs for supporting these people continue to grow year after year, which puts pressure on government finances.

Therefore, making sure the elderly remain healthy within the limited budget is a social issue. The expansion of a healthy life expectancy, which is achieved by catching the subtle signs relating to changes in health in daily life early on and being aware of the risks in recovering frail conditions, is necessary to solve this issue.

There are three main factors which inhibit a healthy life expectancy. First are visceral diseases, called lifestyle-related diseases (metabolic syndrome), such as cancer, heart disease, and diabetes. The second is so-called locomotive syndrome, like falls, fractures, and lower back pain. Mental illnesses, including dementia, depression, and so on, are the third. Early detection of these factors is important for the expansion of a healthy life expectancy.

In addition, it is significant to point out that signs of these risks do appear in places at which we spend our daily lives (at home). One reason is that patients are unwilling to discuss their symptoms, although those close to them want to know. Another reason is that patients can subconsciously affect data measured in medical institutes, such as the case with masked hypertension (Dementia Care Information Network, 2009).

Thus, some IT services are supplied for home use due to the situations mentioned above. One example of such a service is for MCI (Mild Cognitive Impairment) screening (Association of Locomo Dementi prevention therapy), which is an early detection method for MCI, a possible state of dementia.

However, because the factors that induce poor health interact with each other, early detection of a certain factor does not always expand the expectancy of a healthy life. For instance, lifestyle-related diseases, such as obesity or diabetes, can be prompted by hypokinesis caused by locomotive syndrome. Recent studies also report that a sudden reduction in walking speed is not only a symptom of locomotive syndrome, but also of cognitive function (Verghese et al., 2014). As described in the previous section, because changes in health are affected by personal daily lifestyles and environments, factors that induce poor health also vary according to the individual (Tokyo Medical Association).

In this case, expansion of healthy life expectancy may be achieved not by constructing only one uniformed service, but through a method that prepares multiple basic services for each symptom that are selected and combined to cooperate with each other to help individuals. In other words, the sharing mechanisms of the Life Management Platform are essential for the expansion of healthy life expectancy.

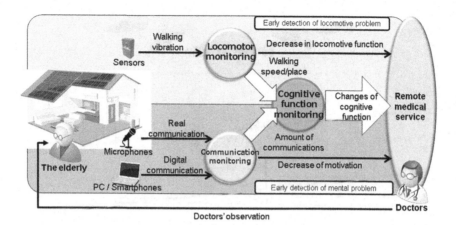

FIGURE 15.9

Conceptual diagram of Life Management Services.

Hence, we are constructing services for the expansion of healthy life expectancy as the platform's Life Management Services.

15.3.2 LIFE MANAGEMENT SERVICES FOR THE EXPANSION OF HEALTHY LIFE EXPECTANCY

The cooperation of four basic services, "Locomotor monitoring service," "Communication monitoring service," "Cognitive function monitoring service," and "Remote medical information reporting service" are described as services for the expansion of healthy life expectancy. Fig. 15.9 is a conceptual diagram of this system.

This example focuses on personal gait (e.g. walking speed and the amount of walking), which shows daily changes in locomotor and cognitive functions, and communication (e.g. the amount or tendency of communication), which shows changes in the mental state and verbal and conscious behaviors. The locomotor monitoring service collects information about walking using vibration sensors placed in the house. The communication monitoring service sets up microphones and digital devices, like PCs, to collect information related to communication. They are effective to check changes in health even as standalone services; however they can create the new value of detecting changes in cognitive function by sharing their measurement data and analysis results to cognitive function monitoring services. In addition, remote medical information reporting services can offer information about a patient's total health to doctors based on the other three services.

The following describes the content of each service in development.

(1) Locomotor monitoring service

This service observes the inhabitants' hypokinesis by measuring changes in gaits (walking speed, pace, and stride), storing data measured for a certain period, and

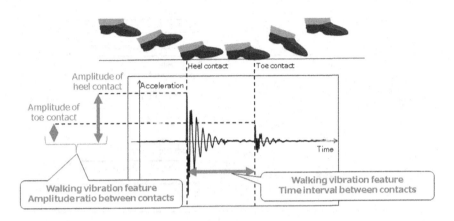

FIGURE 15.10

Example of walking vibration.

comparing these values. A feature of this service is its ability to measure and identify individual gaits using vibration sensors.

Because the privacy of inhabitants must be considered when measuring their gaits in the house, no methods that use cameras, such as using human identification sensors, were studied (NICT, 2014). The method in this example calculates walking speed from the distance and the reaction time between human detection sensors. However, because the sensors in this method only detect human presence, it is difficult to determine who caused a sensor to react if the target does not live alone. Additionally, features of personal gaits other than walking speed cannot be measured.

In this service, vibration sensors are settled in the floor and detect vibrations caused by people walking. Walking vibrations can detect personal walking features. There are some paper about identification trial using features such as walking pace estimated by walking vibrations (Ryosuke et al., 2016). The service focuses on vibrational peaks, in particular, which are caused by feet coming into contact with the floor. As shown in Fig. 15.10, there are two sequential vibrational peaks when a foot contacts the floor. One is caused by heel and the other is caused by the toe. The features between these two sequential peaks (time lag, amplitude ratio, and comparison of frequencies) may be used as individual features. In fact, it is reported that the comparison of peak frequencies can be used to distinguish males from females (Kyoko, 2006). According to this report, the frequencies for females tend to be lower for heel contact and higher for toe contact. In contrast, the frequencies for males tend to be lower for both heel and toe contact.

In addition to being able to measure walking speed using the distance and reaction time of vibration sensors using the same method as human detect sensors, walking pace can be seen in the vibration of each step. Stride can then be estimated using the simple formula: (stride) = (walking speed)/(walking pace).

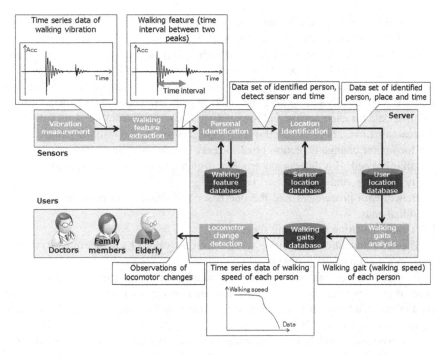

FIGURE 15.11

Block diagram of locomotor monitoring service.

Along with Fig. 15.11, this paragraph explains the process of personal identification and walking gait monitoring. In this explanation, the gait that is monitored is walking speed. We also use the time intervals between heel and toe contact as walking features for personal identification. First, when walking vibration occurs near vibration sensors, the vibration monitoring block sends time series data regarding the walking vibration to the walking feature extraction block. From the time series data for vibration, the walking feature extraction block extracts walking features and sends them to service server with the sensor identifier and measurement time. After receiving the walking features, the server identifies the individual in the personal identification block. In this procedure, this block refers the walking feature database. In the walking feature database, data sets for user ID, walking features for each user and deviation range are stored. The personal identification block checks identification criterion, whether difference of walking feature value between received data and stored data in the database is within the deviation range. If it recognizes a match and results in successfully identifying a person, the block outputs data sets for user ID, sensor ID, and measurement time. Other than just personal identification, this block also reflects newly measured and received data to the database using statistical methods, such as calculations of averages, to make this service follow the natural changes for each individual. Next, the location identification block identifies the place of the sensor in reference to the sensor location database. The sensor location database stores sensor identifiers and their setting location. By searching for the locations of sensors by sensor ID, the location identification block can put the individual, location, and time data together. This data set is stored in the user location database. Periodically (e.g. daily), the walking gaits analysis block extracts location and time data for each person and measures walking speed from the sensor distance and reaction time. This calculated walking speed is sent to the walking gait database. The locomotor change detection block detects locomotor changes from differences in walking speed. If it detects any trouble, it then notifies the users (elderly people, family members, and doctors).

Hence, with procedures like the one described above, it is possible to monitor individual gaits.

(2) Communication monitoring service

The communication monitoring service and easily detect whether inhabitants suffer from a mental health problem.

It is said that during conversation, the number of words used by patients who suffer from a mental illness, such as dementia or depression, show a downward trend (Shuko and Eiji, 2014). It is also reported that the content of the words may change (Tetsuaki et al., 2014). Therefore, the targets for measurements should be daily conversations and digital communication (such as e-mail and social networking services). Daily conversations are collected by microphone, while digital communication is measured via computer software that gathers such information. To protect privacy, the content of all collected communication is distorted to be unrecognizable by the microphone and software. For conversations, the amount of conversation is estimated by volume, speed, intonation, and frequency. For this estimation, we could use some techniques to presume speaker and utter speed from acoustic data not from contents of conversations (Yoshitomo et al., 2007; Gong et al., 2003). On digital side, frequency and reaction speed are used to measure the amount of digital communication. Fig. 15.12 illustrates the service in detail. We also develop the function for changing distorted levels of communication to make this service more effective for monitoring. Using a function to change permissions, for example, can allow for the addition of an analysis on the amount of words used.

(3) Cognitive function monitoring service

This service monitors changes in cognitive function using data collected and analyzed by the locomotor monitoring service and communication monitoring service. Without any additional special sensors, this service supplies users with the benefit of detecting changes in cognitive functions by using information from the other services.

It is said that cognitive function is closely related to both changes in locomotor behavior and changes in verbal and conscious behavior. It has been reported that a sudden reduction in walking speed, to less than 80 cm/s, may indicate a possible MCR (motoric cognitive risk) (Verghese et al., 2014). It is also reported that decreases in the amount of communication, especially changes in the amount of words used, show the progress of decline in cognitive function (Shuko and Eiji, 2014). There are also

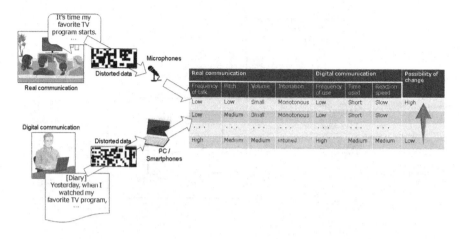

FIGURE 15.12

Image of communication monitoring service.

other signs, such as frequent and repetitive behavior and a tendency to withdrawal due to loss of motivation. Because there are numerous indicators, it is necessary to integrate a lot of information to detect changes in cognitive function.

Thus, this service integrates the following information.

(Information 1) Walking speed: analysis results from the locomotor monitoring service

(Information 2) Amount of communication: analysis results from the communication monitoring service

(Information 3) Walking route repetitiveness: index using the estimated walking route from measurement data of the locomotor monitoring service (sequence of sensors' reactions for individual walking vibration)

(Information 4) Amount of walking: index using the estimated walking route from measurement data of the locomotor monitoring service (sequence of sensors' reaction for individual walking vibration).

Fig. 15.13 shows a block diagram of this service. This service uses the following types of input information for the multiple indexes described above: walking speed, stored in the walking gaits database as the analysis results from the locomotor monitoring service; data sets for individual, location, and time in the user location database, which is a byproduct of the locomotor monitoring service; and the amount of communication in the communication amount database of the communication monitoring service. For this data, the walking speed and amount of communication are used without extra analysis. In contrast, data sets for individual, location, and time must be analyzed independently to index the amount of walking and walking repetitiveness. The blocks in Fig. 15.13 (estimation of walking route, route repetitiveness indexing, estimation of the amount of walking, and detection of change in cognitive

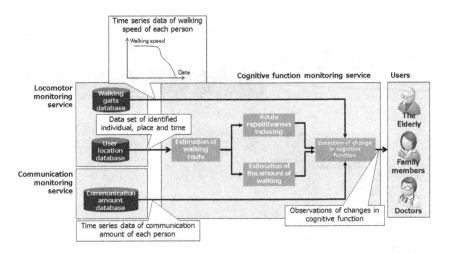

FIGURE 15.13

Block diagram of cognitive function monitoring service.

function) are necessary to do this additional analysis. Their functions are described below.

i. Block for estimating the walking route

This block estimates walking route of each person based on the data sets for individual, place, and time in the user location database of the locomotor monitoring service. Walking routes are separated, for example, by certain divisions of time or by continuous action. (See Fig. 15.14.)

ii. Block for indexing route repetitiveness

From the walking routes that the previous block estimated, this block indexes the frequency each route is repeated in a certain period and defines the overall tendencies as walking route repetitiveness. Simply put, this block derives the ratio for the number of times a user walks a certain route in a particular day to the number of times in a typical day. Fig. 15.15 shows an easy example of this ratio using two people, users A and B. The average ratio of all routes is defined as personal walking route repetitiveness. If the average value greatly exceeds 1, it may indicate a decline in cognitive function.

iii. Block for estimating the amount of walking

Using the walking routes that the block for estimating the walking route estimated, the amount of walking is defined as the sum of all routes in a given period. If the distances between sensors or rooms are clear, the amount of walking is simply calculated by summing the distances. Even if the distance information is lacking, it may be possible to use the number of movements between rooms as an index.

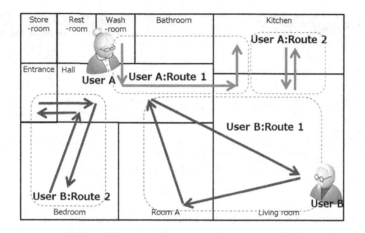

FIGURE 15.14

Estimation of walking routes.

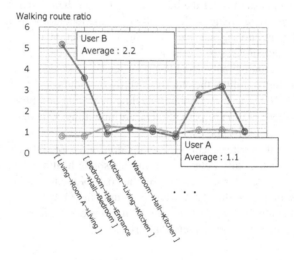

FIGURE 15.15

Walking route repetitiveness.

iv. Block for detecting changes in cognitive function

By using four types of information — walking speed from the locomotor monitoring service, amount of communication from the communication monitoring service, and the walking route repetitiveness and amount of walking from the previously described blocks — this block detects changes in total cognitive function. For example, judgments are made by a combination of changes in each index, such as in Fig. 15.16. This block sends notifications to users if a change is detected.

Walking speed	Amount of communication	Walking route repetitiveness	Amount of walking	Possibility of change in cognitive function
Slow	Low	High	Low	High
Slow	Low	High	Medium	
Slow	Medium	High	Low	
. . .	. . .	. . .	. . .	
Medium	Medium	Low	Medium	Low

FIGURE 15.16

Matrix of observations for change in cognitive function.

As described above, the cognitive function monitoring service detects cognitive changes using a combination of results from the other services and its own additional, new analysis data.

(4) Service for remotely reporting medical information

Family doctors can monitor the health conditions of their patients remotely without visiting them by referring to the results from the locomotor monitoring service, communication monitoring service, and cognitive function monitoring service. If there are any signs of risks to a patient's health, doctors can send feedback on the medical aspects to individuals using this service. These kinds of services, which enable family doctors to follow the elderly covertly by monitoring their daily lives, are effective precisely because recognizing decline of mental and physical functions is not easy, and there are many patients that are unwilling to reveal their conditions regarding cognitive functions.

As described above, the selection, combination, and cooperation of multiple basic services allow for the daily detection of various subtle health changes and promote early measures for health risks. Therefore, the costs of medical and nursing care for elderly people may be reduced. At the same time, the QoL for the elderly and the expansion of healthy life expectancy may be improved. This is an application of our model case for the Life Management Platform.

15.4 CONCLUSION

Due to the progress in our society's maturation, sophistication, and informatization, the needs of citizens are becoming more diverse and difficult to deal with individually using separate services. In this chapter, we proposed a new style of services that can respond to the diversity of citizens' needs by sharing their measured data and analysis results. To achieve such new styles, we suggested using the IoT platform, Life Management Platform. The platform adds access rights and browsing rights appropriate for services that maintain data and creates a good balance between data sharing and protection. With functions below, the platform achieves secure information sharing throughout, from collecting data from the sensors to supplying information to service users.

(Function 1)   Data protection technology: protecting sensor data using individual access rights and retaining rights if services share data.

(Function 2)   Transparent data collection technology: transparently receiving and processing data sent from sensor networks based on various standards or specifications.

(Function 3)   Database technology for horizontal data sharing: managing an appropriate range of browsing of sensor data and/or personal information that has been collected, and sharing the information and data between multiple services.

(Function 4)   Integrated online and physical sensor data analyzing technology: executing analysis not via services, but in the platform.

We also gave an example of services for the expansion of healthy life expectancy for the elderly as life management services achieved through the platform. Although maintaining the health of the elderly is an urgent issue in a rapidly aging society, it is difficult to make overall judgment with the existing individual services because health risks show various aspects and interactions. We showed a model scenario in which the selection, combination, and cooperation of multiple services on the platform allow for the daily detection of various subtle health changes and promote early measures for health risks for those with frail conditions.

15.5 FUTURE WORKS AND CHALLENGES

We plan to develop prototype of Life Management Platform based on this concept. This prototype include key exchange module, Light weight encryption module, sensor data distribution module, and access rights management module. We'll evaluate the architecture and control method of this prototype. After this evaluation, we'll step up its performance and scalability.

Along with development of the platform, we undertake feasibility verification of model case for the Life Management Services. Now, we are developing sensor prototypes and collecting experimental data for locomotor and communication monitoring services. At the next step, these measured and analyzed data will be shared on our platform and we are going to perform limited demonstration in some facilities for the elderlies to check the effectiveness of cooperation of multiple services.

Our final goal is realization of "secure IoT data distribution market." Currently, we can find some data distribution markets called "data marketplace" in the world. But, at the present situation, we cannot treat IoT data such as sensor data that represent personal conditions or secret data that limited members can access there because of security issues. Only the Life Management Platform can realize "secure IoT data distribution market" with its characteristic functions such as protecting data, setting proper access rights, and sharing protected data safely with allowed users, companies, and other organizations.

REFERENCES

AllJoyn Framework. https://allseenalliance.org/framework/.

AllSeen Alliance. https://allseenalliance.org/.

Association of Locomo Dementi prevention therapy. Methods to detect mild cognitive impairment. http://locomo.name/check_home.html.

Cabinet Office of Japan, 2015. White paper of aged society 2015. http://www8.cao.go.jp/kourei/whitepaper/w-2015/zenbun/27pdf_index.html.

Dementia Care Information Network, 2009. The trend of dementia strategy in the UK. http://www.dcnet.gr.jp/retrieve/kaigai/pdf/uk09_care_05.pdf.

Gong, X., Hiroshige, Makoto, Araki, Kenji, Tochinai, Koji, 2003. A study on a multi-dimensional acoustic parameter for speech rate estimation. IEICE Technical Report, SP2003-68(2003-08), http://ci.nii.ac.jp/naid/110003295684.

HomeKit. https://developer.apple.com/homekit/.

IP500 Alliance. http://ip500.org/.

Kyoko, S., 2006. Research of measuring method of walking based on foot pressure distribution for the purpose of material extraction. Doctoral dissertation, Tokyo Univ. http://repository.dl.itc.u-tokyo.ac.jp/dspace/handle/2261/51209.

MFi license. https://developer.apple.com/programs/mfi/.

Ministry of Health, Labour and Welfare of Japan, 2010. Estimates of national medical care expenditure of 2010. http://www.mhlw.go.jp/toukei/saikin/hw/k-iryohi/10/.

National Institute of Information and Communications Technology (NICT), 2014. Research and development for application and platform technology of social big data (2014–2015 NICT call). https://www.nict.go.jp/collabo/commission/k_178b03.html.

NEC. FileShell – NEC enterprize document protection solution. http://jpn.nec.com/infocage/fileshell/.

Ryosuke, S., Mano, Y., Kim, J., Nakajima, K., 2016. Personal identification using characteristic of floor vibration during walking. IEICE Technical Report BioX2016-1(2016-06), http://ci.nii.ac.jp/naid/40020873808/en.

Shuko, S., Eiji, A., 2014. Language processing as language ability test: an example of dementia with longterm blog writing. Collection of papers, 20th Annual Convention of the Association for Natural Language Processing. http://www.anlp.jp/proceedings/annual_meeting/2014/pdf_dir/E7-1.pdf.

Tetsuaki, N., et al., 2014. Social counselling: does SNS heal mental state of patients? DEIM forum 2014. http://mednlp.jp/PAPER/2014-DEIM-dep.pdf.

Thread. http://www.threadgroup.org/technology/ourtechnology.

Tokyo Medical Association. Guidebook for nursing care staff and regional care. https://www.tokyo.med.or.jp/kaiin/kaigo/chiiki_care_guidebook/kaigo_guide.htm.

United Nations Statistics Division (UN), 2013. Demographic yearbook system, demographic yearbook 2013. http://unstats.un.org/unsd/demographic/products/dyb/dyb2013.htm.

Verghese, Joe, et al., 2014. Motoric cognitive risk syndrome. Official Journal of the American Academy of Neurology 83 (24), 2278–2284. http://www.ncbi.nlm.nih.gov/pmc/articles/PMC4277675/.

Wi-SUN Alliance. https://www.wi-sun.org/.

Yoshitomo, et al., 2007. The relation between correct judgments ratio and frequency range in the text independent talker identification system using correlation matrix of narrow band speech envelope. IEICE Technical Report EA2007-82(2007-11), http://www.nbu.ac.jp/~fukushima/Project/ID_Project/houkokusyo/20071116_IEICE_kobashikawa/11EA_paper_Kobashikawa.pdf.

ZigBee Alliance. http://www.zigbee.org/.

ACRONYMS AND GLOSSARY

List of acronyms with explanation

ACL   Access Control List, List of information which are described data read/write/delete rights
API   Application Programming Interface
CPS   Cyber-physical system, the system which connect closely IT and real world
DRM   Digital Rights Management
EME   Encrypt Media Extensions
IEEE   Institute of Electrical and Electronics Engineers
IoT   Internet of Things
IP   Internet Protocol
MCI   Mild Cognitive Impairment, the intermediate stage between the expected cognitive decline of normal aging and dementia
MCR   Motoric Cognitive Risk, the predementia syndrome characterized by slow gait and cognitive complaints
MFi   Made for iPod/iPhone/iPad, the license which is defined by Apple corporation to certify a product conformance to iPod/iPhone/iPad
NPO   Non-profit organization
PC   Personal Computer
PI   Personal Information
QoL   Quality of life
SNS   Social Networking Service
Wi-SUN   Wireless Smart Utility Network

Glossary of terms with explanation

6LowPan IPv6 over Low-Power Wireless Personal Area Networks standardized as RFC4944.

Digital communication Communications via internet and computers such as e-mail and social networking services.

Edge Computing One of the IoT technologies that moves process from cloud-side to near user-side such as sensor network gateway or private cloud server.

Gait A manner of walking such as walking speed, pace, and stride.

IEEE 802.15.4 International standard of wireless communication developed by IEEE (Institute of Electrical and Electronics Engineers).

IP500 High-secure IoT mesh wireless communication standard developed in Germany.

Locomotive syndrome A condition of reduced mobility due to impairment of locomotive organs.

Masked hypertension A clinical condition that patients' blood pressure are normal at the doctor's office but high at home.

Metabolic syndrome A cluster of the most dangerous heart attack risk factors: diabetes, abdominal obesity, high cholesterol, and high blood pressure.

OpenID An open standard and decentralized authentication protocol.

Personal Data Store The storage which store and manage data secure and structurally.

Plugin A software component that adds a specific feature to an existing computer program.

Walking vibration Floor vibrations induced by walking.

ZigBee An IEEE 802.15.4-based specification for a suite of high-level communication protocols used to create personal area networks.

INDEX

Printed in the United States
By Bookmasters